Methods in Molecular Biology

Series Editor
John M. Walker
School of Life and Medical Sciences
University of Hertfordshire
Hatfield, Hertfordshire, AL10 9AB, UK

For further volumes:
http://www.springer.com/series/7651

Zymography

Methods and Protocols

Edited by

Jeff Wilkesman

Centre for Environmental, Biology and Chemistry Research, Facultad de Ciencias y Tecnología, Universidad de Carabobo, Valencia, Venezuela

Liliana Kurz

Centre for Environmental, Biology and Chemistry Research, Facultad de Ciencias y Tecnología, Universidad de Carabobo, Valencia, Venezuela

Humana Press

Editors
Jeff Wilkesman
Centre for Environmental, Biology and Chemistry Research
Facultad de Ciencias y Tecnología
University of Carabobo
Valencia, Venezuela

Liliana Kurz
Centre for Environmental, Biology and Chemistry Research
Facultad de Ciencias y Tecnología
University of Carabobo
Valencia, Venezuela

ISSN 1064-3745 ISSN 1940-6029 (electronic)
Methods in Molecular Biology
ISBN 978-1-4939-7109-1 ISBN 978-1-4939-7111-4 (eBook)
DOI 10.1007/978-1-4939-7111-4

Library of Congress Control Number: 2017943128

Printed on acid-free paper

This Humana Press imprint is published by Springer Nature
The registered company is Springer Science+Business Media LLC
The registered company address is: 233 Spring Street, New York, NY 10013, U.S.A.

Dedication

To our children, Jarod and Kylie
To our other children: our students

In Memoriam

To Prof. Ricardo Maldonado, former Rector-President of the University of Carabobo, who untimely passed away the night before Christmas.

Preface

Zymography, the technique where enzyme activity is detected on electrophoretic gels, has received increasing attention in recent years. This book updates the information concerning the latest techniques and protocols associated with zymography. The methods described are intended to reach a broad variety of researchers from the fields of biological and medicinal science. Isozymes and allozymes, used as gene markers, enable advances in enzymology, molecular evolution, and genetics. Many fields will benefit from a bibliographical collection regarding zymography, including clinical and diagnostic medicine, medical genetics, agricultural entomology, genetic monitoring of environmental pollution, and forensic science.

The main purpose of this volume is to bring together some of the most recent and broad zymography methods currently used. The book is of value not only to experts in the field but also to the incoming new scientists willing to learn and perform electrophoretic zymography.

Chapters have been written following the classic format employed in the *Methods and Molecular Biology* series. All chapters initiate with a brief description of the basic theory behind the method being analyzed. The Materials section lists all chemicals, reagents, buffers, and other materials necessary for the correct performance of the experiments. The Methods section includes a detailed stepwise description of the protocol. Deatails regarding problem shooting, tips, hints, and advices for key steps in experimentation are given in the Notes section, thus complementing the Method section.

Valencia, Venezuela

Jeff Wilkesman
Liliana Kurz

Contributors

ADARELYS ANDRADES • *University of Leipzig, Leipzig, Germany; Centre for Environmental, Biology and Chemistry Research, Facultad de Ciencias y Tecnología, Universidad de Carabobo, Valencia, Venezuela*

GEORGINA BARRETO • *Centre for Environmental, Biology and Chemistry Research, Facultad de Ciencias y Tecnología, Universidad de Carabobo, Valencia, Venezuela*

MONIKA BARTEKOVA • *Institute for Heart Research, Slovak Academy of Sciences, Bratislava, Slovak Republic; Institute of Physiology, Faculty of Medicine, Comenius University, Bratislava, Slovak Republic*

PÉTER BENCSIK • *Department of Biochemistry, Faculty of Medicine, University of Szeged, Szeged, Hungary; Pharmahungary Group, Szeged, Hungary*

DEBASISH BHATTACHARYYA • *Division of Structural Biology and Bioinformatics, CSIR-Indian Institute of Chemical Biology, Kolkata, India*

DIANA CASTRO • *Centre for Environmental, Biology and Chemistry Research, Facultad de Ciencias y Tecnología, Universidad de Carabobo, Valencia, Venezuela*

JUANJUAN CHEN • *Vascular Surgery Research Laboratory, Division of Vascular and Endovascular Surgery, Brigham and Women's Hospital and Harvard Medical School, Boston, MA, USA*

SHANYAN CHEN • *Department of Pathology and Anatomical Sciences, University of Missouri School of Medicine, Columbia, MO, USA; Interdisciplinary Neuroscience Program, University of Missouri, Columbia, MO, USA*

ZHENZHOU CHEN • *Department of Pathology and Anatomical Sciences, University of Missouri School of Medicine, Columbia, MO, USA*

AASTHA CHHABRA • *Peptide and Proteomics Division, Defence Institute of Physiology and Allied Sciences (DIPAS), DRDO, Timarpur, Delhi, India; Department of Biotechnology, Jaypee Institute of Information Technology, Noida, Uttar Pradesh, India*

NACK-SHICK CHOI • *Department of Biochemistry & Health Science, Changwon National University, Changwon, Changwon-si, Republic of Korea; Careside Co., Ltd., Seongnam-si, Gyunggi-do, Republic of Korea*

LELLYS M. CONTRERAS • *Centre for Environmental, Biology and Chemistry Research, Facultad de Ciencias y Tecnología, Universidad de Carabobo, Valencia, Venezuela*

JIANKUN CUI • *Department of Pathology and Anatomical Sciences, University of Missouri School of Medicine, Columbia, MO, USA; Harry S. Truman Memorial Veterans' Hospital, Columbia, MO, USA*

PÉTER FERDINANDY • *Pharmahungary Group, Szeged, Hungary; Department of Pharmacology and Pharmacotherapy, Semmelweis University, Budapest, Hungary*

MARLENY FERNÁNDEZ • *Centre for Environmental, Biology and Chemistry Research, Facultad de Ciencias y Tecnología, Universidad de Carabobo, Valencia, Venezuela*

ESTHER GAMERO-SANDEMETRIO • *Departamento de Biotecnología, Instituto de Agroquímica y Tecnología de Alimentos, CSIC, Valencia, Spain*

ROCÍO GÓMEZ-PASTOR • *Departamento de Biotecnología, Instituto de Agroquímica y Tecnología de Alimentos, CSIC, Valencia, Spain; Departament de Bioquímica i Biologia Molecular, Universitat de València, València, Spain*

OSCAR GONZÁLEZ • *Centre for Environmental, Biology and Chemistry Research, Facultad de Ciencias y Tecnología, Universidad de Carabobo, Valencia, Venezuela*
ANIKÓ GÖRBE • *Department of Biochemistry, Faculty of Medicine, University of Szeged, Szeged, Hungary; Pharmahungary Group, Szeged, Hungary*
ZEZONG GU • *Department of Pathology and Anatomical Sciences, University of Missouri School of Medicine, Columbia, MO, USA; Harry S. Truman Memorial Veterans' Hospital, Columbia, MO, USA*
HAILUN HE • *School of Life Sciences, State Key Laboratory of Medical Genetics, Central South University, Changsha, China*
ZULLY HERNÁNDEZ • *Centre for Environmental, Biology and Chemistry Research, Facultad de Ciencias y Tecnología, Universidad de Carabobo, Valencia, Venezuela*
DUYEN HOANG • *Department of Soil Science, Vietnam Nation University of Forestry, Hanoi, Vietnam*
DAE-OOK KANG • *Department of Biochemistry & Health Science, Changwon National University, Changwon, Republic of Korea*
GAVIN KERNAGHAN • *Department of Biology, Mount Saint Vincent University, Halifax, NS, Canada*
RAOUF A. KHALIL • *Vascular Surgery Research Laboratory, Division of Vascular and Endovascular Surgery, Brigham and Women's Hospital and Harvard Medical School, Boston, MA, USA*
KRISZTINA KISS • *Department of Biochemistry, Faculty of Medicine, University of Szeged, Szeged, Hungary*
LILIANA KURZ • *Centre for Environmental, Biology and Chemistry Research, Facultad de Ciencias y Tecnología, University of Carabobo, Valencia, Venezuela*
YAKOV KUZYAKOV • *Department of Agricultural Soil Science, University of Göttingen, Göttingen, Germany; Department of Soil Science of Temperate Ecosystems, University of Göttingen, Göttingen, Germany*
HAO LI • *College of Life Science and Technology, Beijing University of Chemical Technology, Beijing, China*
DAN LIU • *School of Life Sciences, State Key Laboratory of Medical Genetics, Central South University, Changsha, China*
MADATHIPARAMBIL G. MADANAN • *Regional Medical Research Centre, Indian Council of Medical Research, Port BlairAndaman & Nicobar Islands, India*
EMILIA MATALLANA • *Departamento de Biotecnología, Instituto de Agroquímica y Tecnología de Alimentos, CSIC, Valencia, Spain; Departamento de Biotecnología, Instituto de Agroquímica yTecnología de Alimentos, CSIC, Valencia, Spain; Departament de Bioquímica i Biologia Molecular, Universitat de València, Valencia, Spain*
MICHAEL MAYERHOFER • *Department of Biology, Mount Saint Vincent University, Halifax, NS, Canada*
AMBILI MECHOOR • *Department of Biotechnology Engineering, Sahrdaya College of Engineering and Technology, Thrissur, Kerala, India*
FANJUN MENG • *Department of Pathology and Anatomical Sciences, University of Missouri School of Medicine, Columbia, MO, USA*
JÉSSYKA FERNANDA SANTIAGO MONTE • *Programa de Pós-graduação em Bioquímica, Centro de Biociências, Universidade Federal do Rio Grande do Norte, Natal, Brazil*
JOANA PATRÍCIA MOLATO FIGUEIREDO LOPES MONTEIRO • *Global Health and Tropical Medicine, Instituto de Higiene e Medicina Tropical, Universidade Nova de Lisboa, Lisbon, Portugal*

CLÁUDIA JASSICA GONÇALVES MORENO • *Programa de Pós-graduação em Bioquímica, Centro de Biociências, Universidade Federal do Rio Grande do Norte, Natal, Brazil*

GORAN MUCIĆ • *Department of Biochemistry, Faculty of Chemistry, University of Belgrade, Belgrade, Serbia*

MARÍA FERNANDA PADRÓN • *Centre for Environmental, Biology and Chemistry Research, Facultad de Ciencias y Tecnología, Universidad de Carabobo, Valencia, Venezuela*

JÁNOS PÁLÓCZI • *Pharmahungary Group, Szeged, Hungary*

DANIEL PAN • *Department of Biological Sciences, State University of New York, Binghamton, NY, USA; Department of Systems Pharmacology and Translational Therapeutics, University of Pennsylvania Perelman School of Medicine, Philadelphia, PA, USA*

ISABELLA PANFOLI • *Biochemistry Laboratory, Pharmacy Department, University of Genova, Genova, Italy*

CHANG-SU PARK • *Department of Food Science & Technology, Catholic University of Daegu, Daegu, Republic of Korea*

HANDUNGE KUMUDU IRANI PERERA • *Department of Biochemistry, Faculty of Medicine, University of Peradeniya, Peradeniya, Sri Lanka*

MANU O. PLATT • *Wallace H. Coulter Department of Biomedical Engineering, Georgia Institute of Technology and Emory University, Atlanta, GA, USA*

NATALIJA POLOVIĆ • *Department of Biochemistry, Faculty of Chemistry, University of Belgrade, Belgrade, Serbia*

SHIVCHARAN PRASAD • *Department of Biotechnology, National Institute of Pharmaceutical Education and Research, Mohali, Punjab, India*

ZHE QU • *Department of Pathology and Anatomical Sciences, University of Missouri School of Medicine, Columbia, MO, USA*

JANA RADOSINSKA • *Institute for Heart Research Slovak Academy of Sciences, Bratislava, Slovak Republic; Institute of Physiology, Faculty of Medicine, Comenius University, Bratislava, Slovak Republic*

VIBHA RANI • *Department of Biotechnology, Jaypee Institute of Information Technology, Noida, Uttar Pradesh, India*

BRANKICA RAŠKOVIĆ • *Department of Biochemistry, Faculty of Chemistry, University of Belgrade, Belgrade, Serbia*

SILVIA RAVERA • *Biochemistry Laboratory, Pharmacy Department, University of Genova, Genova, Italy*

BAHAR S. RAZAVI • *Department of Agricultural Soil Science, University of Göttingen, Göttingen, Germany*

HERVÉ RÉMIGNON • *GenPhySE, Université de Toulouse, INRA, INPT, ENVT, Castanet Tolosan, France*

ZONGLI REN • *Vascular Surgery Research Laboratory, Division of Vascular and Endovascular Surgery, Brigham and Women's Hospital and Harvard Medical School, Boston, MA, USA*

ALINE RIMOLDI RIBEIRO • *Departmento de Parasitologia, Universidade Estadual de Campinas, Campinas, Brazil*

HUGO ALEXANDRE DE OLIVEIRA ROCHA • *Programa de Pós-graduação em Bioquímica, Centro de Biociências, Universidade Federal do Rio Grande do Norte, Natal, Brazil*

IPSITA ROY • *Department of Biotechnology, National Institute of Pharmaceutical Education and Research, S.A.S. Nagar, Mohali, Punjab, India*

KANIKA SHARMA • *Division of Structural Biology and Bioinformatics, CSIR-Indian Institute of Chemical Biology, Kolkata, India*
MARCELO SOUSA SILVA • *Programa de Pós-graduação em Bioquimica, Centro de BiocienciasUniversidade Federal do Rio Grande do Norte, Natal, Brazil; Global Health and Tropical Medicine, Instituto de Higiene e Medicina Tropical, Universidade Nova de Lisboa, Lisbon, Portugal; Departamento de Analises Clinicas e Toxicologicas, Centro de Ciencias da Saude, Universidade Federal do Rio Grande do Norte, Natal, Brazil*
GERGŐ SZŰCS • *Department of Biochemistry, Faculty of Medicine, University of Szeged, Szeged, Hungary*
ANNA TAN-WILSON • *Department of Biological Sciences, State University of New York, Binghamton, NY, USA*
MARÍA TAPIZQUENT • *Centre for Environmental, Biology and Chemistry Research, Facultad de Ciencias y Tecnología, Universidad de Carabobo, Valencia, Venezuela*
KEVIN K.W. WANG • *Program for Neurotrauma, Neuroproteomics & Biomarker Research, Department of Psychiatry, University of Florida, Gainesville, FL, USA; Department of Neuroscience, University of Florida, Gainesville, FL, USA; Department of Chemistry, University of Florida, Gainesville, FL, USA*
JEFF WILKESMAN • *Centre for Environmental, Biology and Chemistry Research, Facultad de Ciencias y Tecnología, University of Carabobo, Valencia, Venezuela*
KARL A. WILSON • *Department of Biological Sciences, State University of New York, Binghamton, NY, USA*
HIDETARO YASUMITSU • *Expert Laboratory for Life Environments (ELLE), Yokohama City University, Yokohama, Japan; Division of Pharmacy, Kamimach Hospital, Kochi, Japan*

Part I

Introduction

Chapter 1

Zymography Principles

Jeff Wilkesman and Liliana Kurz

Abstract

Zymography, the detection, identification, and even quantification of enzyme activity fractionated by gel electrophoresis, has received increasing attention in the last years, as revealed by the number of articles published. A number of enzymes are routinely detected by zymography, especially with clinical interest. This introductory chapter reviews the major principles behind zymography. New advances of this method are basically focused towards two-dimensional zymography and transfer zymography as will be explained in the rest of the chapters. Some general considerations when performing the experiments are outlined as well as the major troubleshooting and safety issues necessary for correct development of the electrophoresis.

Key words Electrophoresis, Enzymes, Zymography, Troubleshooting

1 Introduction

Zymography is the technique based on sodium dodecyl sulfate polyacrylamide gel electrophoresis (SDS-PAGE) which enables to detect enzymatic activity. The substrate may be copolymerized within the polyacrylamide gel matrix, but could also be submerged in a substrate solution or can be overlaid with another gel containing the substrate [1]. Samples submitted to zymography are generally prepared by the standard SDS-PAGE treatment buffer. However, nonreducing conditions must prevail at all times, viz., samples are not heated and there is no reducing agent [2-mercaptoethanol, dithiothreitol (DTT)] present. After the electrophoresis has been run, SDS is soaked out from the gel (zymogram) by incubation in a nonbuffered Triton X-100 (or other detergent, e.g., Tween-20, etc.). After that, an appropriate activation buffer is selected and the zymogram is incubated for a determined length of time and temperature, depending on the type of enzyme being assayed and the type of substrate being degraded. Finally, the zymogram is stained. Staining procedures may be direct, obtaining colored bands with a translucent background; or may be indirect, and areas of digestion are distinguished as pale bands under a dark background.

Jeff Wilkesman and Liliana Kurz (eds.), *Zymography: Methods and Protocols*, Methods in Molecular Biology, vol. 1626, DOI 10.1007/978-1-4939-7111-4_1, © Springer Science+Business Media LLC 2017

2 Classification

Different types of zymography exist, according to the type of enzyme. There are over 1000 different methods published for the detection of more than 400 different enzymes [2].

However three main types of zymography may be postulated [3, 4]:

- *In gel zymography* (IGZ): briefly, the SDS-PAGE is employed to separate the enzyme-containing samples and is later overlaid on an indicator gel. A variant of this type of zymography is the direct incorporation of the substrate to the gel matrix [4].
- In situ *zymography* (ISZ): based on zymography using SDS-PAGE, the sample—a tissue section or a cell preparation—is brought in contact with the substrate employing a photographic emulsion-containing gelatin (or a fluorescence-labeled protein substrate). After exposure, enzyme activity appears as white spots in a dark background (or as black spots in a fluorescent background). Therefore, exact localization of proteinase activity is achieved. As asset, endogenous enzymes in their biological contexts are visualized (*see part III of this book*).
- In vivo *zymography* (IVZ): relying on the use of protease-activated fluorogenic probes bearing different fluorophores, this type of zymography allows detection of diverse enzymes simultaneously. It has been used for the mapping of MMP activity patterns in intact organisms, e.g., zebrafish, as well as in roots from maize and lupine [5–7].

Based upon the type of detection performed, it is generally encountered as either a chromogenic or a fluorogenic reaction responsible for enzyme visualization on the gel. Some of the most common chromogenic reactions rely upon the following chemistry [2]:

- Reduction of tetrazolium salts (viz. NBT).
- Coupling diazonium salts (e.g., Fast Blue, Fast Red, etc.).
- Change in local pH (use of pH indicators for detection).
- Formation of orthophosphate or pyrophosphate.
- Formation of hydrogen peroxide (detection performed with redox dyes).
- Production of carbon dioxide.
- Use of chromogenic substrates (i.e., derivatives of *para*-nitrophenol or *para*-nitroaniline).

Besides chromogenic detection, fluorogenic detection is possible. The natural fluorochromes NADH and NADHP are widely used, as after oxidation they lose their fluorescence. Derivatives of *4*-methylumbelliferone are also widely used, especially for hydrolase detection [2].

Some variations of IGZ are described in the upcoming chapters. Worth mentioning here are electrophoretic *transfer zymography* [8]. Here, the enzyme is resolved in a polyacrylamide gel without protein substrate, and afterward the enzyme in the gel is transferred to a receptor gel that does contain the substrate. This step circumvents the issue of some enzymes exhibiting a lower mobility value when migrating in a substrate gel. In transfer zymography, the mobilities of the enzyme are comparable to those obtained by non-reducing SDS-PAGE.

Another variant is *two-dimensional zymography* (2DZ). In this case, the sample to be studied is first submitted to isoelectric focusing on IPG-strips. Later, the strip is submitted to conventional IGZ, gaining an extra dimension in the analysis, as now proteins are separated not only by molecular mass, but also according to their isoelectric point [9]. Not less important is another variation of IGZ called *reverse zymography,* which enables the detection of enzyme inhibitors. For this purpose, substrate and enzyme are both copolymerized within the gel matrix. After proper reactivation of the enzyme, the copolymerized substrate is evenly degraded except in those spots where an inhibitor is present preventing enzyme action. After staining, inhibitor location can be visualized as bands containing intact substrate [10].

3 Advantages of PAGZ

Zymography allows a number of advantages when analyzing unpurified enzyme preparations, namely [2, 4]:

- All kinds of biological samples may be analyzed, from cell and tissue extracts, to culture fluids, whole blood, plasma, serum as well as other complex body or lavage fluids, tissue sections, and even whole organisms [2, 3].
- Detection and identification of isozymes in crude cell extracts.
- Functional properties of isozymes directly in gel.
- Quantitative and qualitative information is given by means of estimating the molecular mass and the isoelectric point of the enzyme.
- Monitoring specific and non-specific enzymatic activities in complex biological and clinical samples.
- Discrimination of enzymes with similar and/or overlapping substrate specificities.
- Possible determination of subunit structure.

As can be seen, zymography provides general and specific and useful information in the broad area of pure and applied biochemistry. Clinical and diagnostic medicine has been more impacted by zymography. Topics like cancer, heart and vascular diseases, and immunological inflammatory diseases have standard zymography assays in their procedures. Zymography is also present in animal science, like the study of proteases in chicken, duck, and others [11] and agricultural and plant science [6, 7, 12], forensic science, etc. [2, 4].

In comparison with other biotechnological techniques, zymography remains a very simple and powerful procedure for the partition, identification, and characterization of gene products. Proteomic has a huge ally in zymography, given its advantages. 2DZ, as a result of 2DE and transfer zymography, gives valuable material to be further analyzed by mass spectrometry. The combination of flexible enzyme assay techniques with electrophoresis represents a powerful tool to cope with the rising demand of high-throughput screening coming from functional genomics, protein engineering, and combinatorial chemistry. Given the impact exerted by zymography, it is not difficult to see the importance to further continue the research in new and enhanced methodologies for enzymatic analysis.

4 Final Remarks

The choice of a given zymographic method may depend on highly diverse variables. Compatibility of reagents with the matrix and the enzyme, sensitivity, protein concentration, time-consuming protocols, quantification of results, band stability throughout time, costs, all these are elements to be considered before adopting a protocol. Beyond this, it may happen that even after thoughtful consideration, zymography fails in giving appropriate results. Here, we present some general conditions when the general procedure by some reason does not work. Given the amount of steps involved in the general zymography procedure, it is advised to analyze each step separately.

4.1 Troubleshooting

1. Support media: although the enzyme detection is not necessarily dependent on the support medium used, its choice is important as it will define the quality of the zymography. The most popular is polyacrylamide; still other media are also popular depending on the tested enzyme, viz., starch, cellulose acetate [13].
2. Gel staining: for some given enzymes, more than one method may be available. Choosing the most appropriate one may be difficult and can depend on variables such as length of the

procedure, chemicals involved, mode of application of the staining solution, and election of appropriate controls to test for positive and negative results. Staining solution must always be abundant. The amount of solution varies according to the size of the gel and the size of the tray where staining is taking place. This is particularly important, as some staining procedures are costly. Instead of submerging the gel in a tray containing the staining solution, an agar or a filter paper soaked with the solution can be overlaid. Direct application of drops of the solution onto the zymogram with a pippet is also an option sometimes [2].

3. Staining conditions: revise if a specific temperature is needed for reaction (generally 37 °C), and if luminic conditions are to be avoided. Keep track of the incubation time. Generally, staining reactions proceed in the first 10 min. However, there are cases where a 24–48 h incubation time is needed. This in turn implies how specific the enzyme is towards the substrate. For proteases, Coomassie blue is by far the most employed staining method; however, other stains may be used (protein silver stain, Congo red staining for cellulose, and Amido black stain). With the increasing versatility of gel scanners (including not only visible and UV lamps, but now also fluorescent detection), the range of zymography applications is boundless. New staining protocols are constantly being published (*see* Chapter 2), like the new RAMA stain [14].
4. Solution preparation: when working, effort must be made to prepare all solutions fresh. Indeed, it is possible to make stock solutions for gel polymerization and preserve them in cold. However, these solutions after 14 days tend to get contaminated by microorganisms, especially buffers. Many staining solutions must be prepared fresh and solution recycling is not advised in many cases. Of course, each experimenter must in some way find out if a determined solution may or may not be reused, and if, under which circumstances can it be preserved until its next use.
5. Running conditions: pre-running gels may be useful to eliminate possible non-polymerized elements from the gel that could potentially interfere in enzyme activation. Generally, it is advised to run the gel at 4 °C, especially if the enzyme is thermolabile. During electrophoresis, it has been suggested to run the gel at 20 mA per gel. If running condition is set at constant current, the voltage will increase with time reaching values eventually over 200 V. On the other hand, if the power supply is set to run at constant voltage (usually 150 V), the initial current will typically be around 35 mA and will drop with the pass of time, reaching 10 mA by the end of the run. Either way, it is suggested to select running conditions which will allow less stress to the enzyme, i.e., lower current and more running time.

6. pH conditions: this variable is hardly enough studied. Given the variation of pH with temperature, buffer concentration and pH range must be carefully controlled. The final pH of the staining solution must fulfill the criteria demanded by the pH optimum of the enzyme, the pH of the reaction taking place in the staining procedure, the pH of the buffer employed in gel preparation, and if there is a coupled reaction in the staining, then the pH optimum of the coupled reaction.
7. Buffers: besides checking the pH of the buffer at the real temperature where it is going to be used, its buffering capacities may also be affected if the concentration is too low. Generally, 50 mM buffer should be the least concentration used to assure proper buffering conditions.
8. Substrate: as there is a huge diversity of substrate, either natural or synthetic, it is wise to check its stability. Solubility is sometimes a problem. If incorporated to the matrix, it must be assured that the substrate will not migrate when electric field is applied. Substrate must be provided in a high purity state and its optimal concentration must be previously determined. Typical substrates for protease analysis are gelatin, casein, collagen, and fibrin [15].
9. Detergent removal: it is important to decide which detergent will be utilized in the washing buffer in IGZ [3]. Structure of the detergent's hydrophilic head group and its critical micelle concentration (CMC) affect the removal of SDS. Typically, 1 h incubation with a 1–2.5% (v/v) Triton X-100% is used. However, Tween-20 or Tween-80 may also be used satisfactorily.
10. Reactivation stage: after SDS removal, the enzyme is ready to be refolded and gain biological activity again. Besides checking pH and temperature conditions, the reactivation buffer selected should be also able to fulfill the requirement of any necessary cofactor. Depending on the enzyme, typical cofactors are divalent cations (Ca^{2+}, Mg^{2+}), NAD^{+}, etc.
11. Enzyme biological source: typical mistakes can be traced back to homogenate preparation. Related questions to be formulated are: how fresh is your sample? Where was it stored and for how long? Was it prepared in the appropriate buffers? Is the concentration high enough? Was a parallel enzyme test in test tube performed before electrophoresis?
12. Some frequently asked questions regarding failure in the procedure are:
    - Are all reagents in the staining solution present? Check the protocol and revise preparation of each solution. Many times, miscalculation occurs either in weighing or in dissolving.
    - Are the reagents stable in time? Check the stability of all reagents. Some of them are light-sensible and are spontaneously

hydrolyzed under inappropriate pH conditions. An inadequate storage can occur (left out the fridge, exposed to light, or contaminated).

- Was there a pH alteration? Measure pH of the staining solution and contrast with protocol requirements. Check pH of all solutions involved. Prepare fresh solutions if needed.
- Could there be enzyme inhibitors present in the solutions? Indeed, this is possible, especially if experiments are done in the field. Use always distilled water for experiments. Other enzyme inhibitors may be present in the gel matrix. Sometimes, reactivation of the enzyme after electrophoresis does not happen.

4.2 Safety Regulations

Many of the chemical reagents used during the zymography procedure are toxic, carcinogenic, and skin irritants. Therefore, it is wise to follow a strict security code assuming that all substances are potentially hazardous. The use of gloves throughout the entire experiment is mandatory. Dust mask shall be worn when handling chemicals, especially solids. Solution preparation may produce vapors; use the safety hood at all times. Do not discard the used solutions in the sink. Check first for the environmental risks of each of the chemicals and then proceed to discard appropriately. The ten principles of green chemistry shall be followed at all times. As many solutions will be prepared, label all the containers, indicating expiry date of the solutions.

Remember to keep all areas clean and tidy. Clean up immediately any spills on the bench. After completing experiments, discard gloves and wash hands thoroughly. Remember that the lab is not a kitchen! Do not keep food in the lab fridge.

Acknowledgments

We would like to thank Prof. Dr. J. Bubis and M. Calabokis (University Simón Bolívar) for their unconditional support during all this time. We thank Prof. Dr. W.E.G. Müller and H.C. Schröder, for opening the doors of their labs to host our research stay. We appreciate the support of the University of Carabobo.

References

1. Lantz MS, Ciborowski P (1994) Zymographic techniques for detection and characterization of microbial proteases. Methods Enzymol 235:563–594
2. Manchenko GP (2003) Handbook of detection of enzymes on electrophoretic gels, 2nd edn. CRC Press, USA, pp 1–39
3. Vandooren J, Geurts N, Martens E, Van den Steen PE, Opdenakker G (2013) Zymography methods for visualizing hydrolytic enzymes. Nat Methods 10:211–220
4. Wilkesman J, Kurz L (2009) Protease analysis by zymography: a review on techniques and patents. Recent Pat Biotechnol 3:175–184

5. Keow JY, Herrmann KM, Crawford BD (2011) Differential in vivo zymography: a method for observing matrix metalloproteinase activity in the zebrafish embryo. Matrix Biol 30:169–177
6. Sanaullah M, Razavi BS, Blagodatskaya E et al (2016) Biol Fertil Soils 52:505. doi:10.1007/s00374-016-1094-8
7. Spohn M, Carminati A, Kuzyakov Y (2013) Soil zymography A novel in situ method for mapping distribution of enzyme activity in soil. Soil Biol Biochem 58:275–280
8. Pan D, Hill AP, Kashou A, Wilson KA, Tan-Wilson A (2011) Electrophoretic transfer protein zymography. Anal Biochem 411:277–283
9. Wilkesman JG, Schröder HC (2007) Analysis of serine proteases from marine sponges by 2-D zymography. Electrophoresis 28:429–436
10. Snoek-van Beurden PA, Von den Hoff JW (2005) Zymographic techniques for the analysis of matrix metalloproteinases and their inhibitors. Biotechniques 38:73–83
11. Awde S, Marty-Gasset N, Wilkesman J, Rémignon H (2013) Proteolytic activity alterations resulting from force-feeding in Muscovy and Pekin ducks. Poult Sci 92:2997–3002
12. Rossano R, Larocca M, Riccio P (2011) 2-D zymographic analysis of Broccoli (*Brassica oleracea L. var. Italica*) florets proteases: follow up of cysteine protease isotypes in the course of post-harvest senescence. J Plant Physiol 168:1517–1525
13. Rothe GM (1994) Electrophoresis of enzymes, laboratory methods. Springer, New York
14. Yasumitsu H, Ozeki Y, Kanaly RA (2016) RAMA casein zymography: time-saving and highly sensitive casein zymography for MMP7 and trypsin. Electrophoresis 37:2959–2962
15. Kupai K, Szucs G, Cseh S, Hajdu I, Csonka C, Csont T, Ferdinandy P (2010) Matrix metalloproteinase activity assays: importance of zymography. J Pharmacol Toxicol Methods 61:205–209

Part II

Endopeptidase Zymography

Chapter 2

Serine Protease Zymography: Low-Cost, Rapid, and Highly Sensitive RAMA Casein Zymography

Hidetaro Yasumitsu

Abstract

To detect serine protease activity by zymography, casein and CBB stain have been used as a substrate and a detection procedure, respectively. Casein zymography has been using substrate concentration at 1 mg/mL and employing conventional CBB stain. Although ordinary casein zymography provides reproducible results, it has several disadvantages including time-consuming and relative low sensitivity. Improved casein zymography, RAMA casein zymography, is rapid and highly sensitive. RAMA casein zymography completes the detection process within 1 h after incubation and increases the sensitivity at least by tenfold. In addition to serine protease, the method also detects metalloprotease 7 (MMP7, Matrilysin) with high sensitivity.

Key words Casein zymography, CBB stain, RAMA stain, Electrophoresis, Serine protease, Proteinase, SDS-PAGE, MMP7

1 Introduction

Serine proteases are known to be one of the major enzymes spreading over all living things and to play important roles not only in their life but also in human health. Viral serine proteases are required for their propagation and aimed to be an important target for therapy [1]. Bacterial serine proteases are reported to be involved in their growth and virulence [2, 3]. Fungal serine proteases are known to be required for their life and induce allergy and inflammation in human body [4]. Plant serine proteases also play important roles in their life [5]. In relation to human life, snake venom is one of the most important proteases sources and has been studied and revealed to disturb coagulation and fibrinolysis in the prey systems [6, 7].

Human serine proteases have been studied for many years and revealed to play many important roles in normal and pathogenic processes including digestive system [8], defense [9], inflammation [10], allergy [11], cardiac function [12], and behavior [13].

Jeff Wilkesman and Liliana Kurz (eds.), *Zymography: Methods and Protocols*, Methods in Molecular Biology, vol. 1626, DOI 10.1007/978-1-4939-7111-4_2, © Springer Science+Business Media LLC 2017

To detect serine protease activities, protein substrate zymography is one of the major activity assays and frequently used. Casein and gelatin have been used as the protein substrates. Recently, several variations of zymography and new platforms for activity assay were reported [14–16]. However, since casein and gelatin zymography are able to detect many endopeptidase species at low cost and with decent sensitivity, they have been still used in many laboratories.

In casein zymography, samples are usually separated on a gel containing 1 mg/mL casein, followed then by a conventional CBB staining step [17, 18]. Although these conditions and procedures provide reproducible results, the zymography is time-consuming (usually overnight destaining) and its sensitivity is relatively low.

In that method, one of the major reasons for low sensitivity might be attributed to high concentration of substrate in a separation gel. Since proteases with weak activity might digest only small amount of substrate during the incubation period, i.e., within a protease-existing area, undigested substrates still might remain and then the residual substrate might be stained as substantially blue, and therefore, the area having weak protease activity might seem unclear and might be difficult to recognize as a band. When a lower concentration of substrate is employed, most of the substrate in a weak protease-existing area might be digested thoroughly. However, by using the low substrate concentration, the contrast between digested and undigested area might decrease, and therefore, all the active bands might become obvious.

To overcome the above-mentioned problems, low concentration substrate and high sensitive protein-staining methods (CGP stain and RAMA stain), which have recently been established by us [19, 20], are combined. Resulted new method of RAMA casein zymography, which employed RAMA stain in combination with low casein concentration, improves the detection sensitivities of trypsin and MMP7 by at least tenfold [21]. In addition, stain and destain processes complete within 1 h after incubation.

2 Materials

All the reagents used are reagent grade or analytical grade. The water used for preparation of reagents and for gel washing is Milli Q water (MQW: 18.2 mho). Importantly, all of the prepared aqueous solutions should be paid attention to for the contamination by microorganisms including bacteria and fungi. Once such microorganisms grow in a solution, it is contaminated with lot of proteases and artificial protease bands appear on the zymogram. RAMA casein zymography is sensitive enough to detect them. When slight amounts of precipitates appear or are noticed existing in your aqueous solution, discard it and prepare another fresh one.

2.1 Electrophoresis

1. Thirty percent of acrylamide stock solution for separation gel (a mixture of 29.2% acrylamide and 0.8% bisacrylamide) is prepared by dissolving them and stored at 4 °C in a bottle covered with aluminum foil for avoiding light (*see* **Note 1**).
2. Ten percent of acrylamide stock solution for stacking gel (a mixture of 10% acrylamide and 2.5% bisacrylamide) is prepared by dissolving them and stored at 4 °C, avoiding light (*see* **Note 2**).
3. Three percent of ammonium persulfate solution is prepared at the time of use (*see* **Note 3**).
4. 10× running buffer (250 mM Tris–HCl, 1.92 M glycine, 1% (w/v) SDS) is prepared by dissolving 30.3 g Tris, 144 g glycine, and 10 g SDS in MQW and up to 1000 mL (*see* **Note 4**).
5. 4× sample buffer [124 mM Tris–HCl pH 7.5, 50% glycerol, 4% SDS, 0.2% bromophenol blue (BPB)] is prepared by mixing 2 M Tris–HCl pH 7.5 with 1% BPB solution, glycerol, SDS, and MQW (*see* **Note 5**).
6. Casein substrate solution 10 mg/mL is prepared from bovine milk casein, which is dissolved in MQW with minimum amount of 0.1 M NaOH, dispensed in small aliquots, and stored at −20 °C until use (*see* **Note 6**).
7. Separation gel solution containing casein substrate (12.5% acrylamide, 375 mM Tris–HCl pH 8.8, 0.1% SDS, and 200 μg/mL casein) is prepared by mixing 2 M Tris–HCl pH 8.8 with 30% acrylamide stock solution and MQW, and the mixture is degassed and then added with 10 mg/mL casein substrate solution and 10% SDS (*see* **Note 7**).
8. Stacking gel solution (2.5% acrylamide, 125 mM Tris–HCl pH 6.8, and 0.1% SDS) is prepared by mixing 0.5 M Tris–HCl pH 6.8 with MQW and 10% acrylamide stock solution, and the mixture is degassed and then added with 10% SDS.

2.2 Protease Activity Assay

1. Twenty-five percent of Triton X-100 stock solution is prepared by dropwise adding 100% Triton X-100 into MQW with vigorous mixing (*see* **Note 8**).
2. Renaturation buffer (50 mM Tris–HCl pH 7.5, 2.5% Triton X-100) is prepared by mixing 2 M Tris–HCl pH 7.5 with 25% Triton X-100 stock solution and MQW. Prepare 200 mL buffer per gel (*see* **Note 9**).
3. Calcium chloride stock solution 1 M is prepared by dissolving powder $CaCl_2$ in MQW.
4. Incubation buffer (50 mM Tris–HCl pH 7.5, 1 mM $CaCl_2$) is prepared by mixing 2 M Tris–HCl pH 7.5 with 1 M $CaCl_2$ and MQW. Prepare 200 mL buffer per gel (*see* **Notes 10** and **11**).

2.3 Reagents for Detection

1. Coomassie Brilliant Blue R-250 (CBB) stock solution for RAMA stain (0.2% CBB and 60% methanol) is prepared as follows: First dissolve powder CBB R-250 in methanol, and after complete dissolution, add MQW as to final concentration (*see* **Note 12**).
2. Forty percent of ammonium sulfate stock solution for RAMA stain is prepared by dissolving powder ammonium sulfate in MQW.
3. Working solution of RAMA stain (0.05% CBB, 10% acetic acid, 15% methanol, 3% ammonium sulfate) should be freshly prepared by mixing 0.2% CBB stock solution with acetic acid, 40% ammonium sulfate stock solution, and MQW. Prepare 40 mL per gel (*see* **Note 13**).

3 Methods

All procedures are carried out at room temperature.

3.1 Substrate Gel

1. For appropriate preparation of separation gel solution containing casein and SDS, order of mixing solutions and degassing are important, as already mentioned in Subheading 2.1, **item 7**.
2. Add freshly prepared 3% ammonium persulfate and TEMED to the degassed mixture of separation gel solution (*see* **Note 14**).
3. The resulting mixture is immediately poured into gel apparatus, covered with small amounts of MQW for avoiding air, and incubated at room temperature until solidified (*see* **Note 15**).

3.2 Stacking Gel Preparation and Setting Up

1. As same as in Subheading 3.1, **step 1**, stacking gel also required degassing for polymerization, and moreover, different from separation gel, there is no air block off barrier of water.
2. Add freshly prepared 3% ammonium persulfate and TEMED to the stacking gel solution.
3. The resulting mixture is immediately poured into gel apparatus and is inserted with a silicon comb for sample loading (*see* **Note 16**).
4. Set up the electrophoresis apparatus.
5. Wash out any air bubbles from the bottom of gel and sample lanes.
6. Flush sample lanes with electrode buffer by using small syringe or pipette (*see* **Note 17**).

3.3 Sample Preparation

1. Mix protease samples with 4× sample buffer (*see* **Note 18**).
2. Prepare a standard protease sample used for a positive control (*see* **Note 19**).

3.4 Sample Loading

1. After preparation, immediately apply samples into sample lanes (*see* **Note 20**).
2. Before loading, confirm the loading position of each sample in a sample lane. Set a sample having large activity apart from the others with two-lane space and not near by the small activity (*see* **Note 21**).
3. When applied, load samples with smaller activity earlier. Do not overflow any aliquots from respective sample lane, nor push out any air bubbles into electrode buffer in a sample lane (*see* **Note 22**).

3.5 Electrophoresis

1. Start electrophoresis without pre-run at constant current 10 mA per gel (*see* **Notes 23** and **24**).

3.6 Renaturation

1. Prepare renaturation buffer 200 mL per gel and pour into a plastic vessel.
2. Wear plastic gloves, put the resolved gel into renaturation buffer with shaking on a reciprocal shaker (*see* **Note 25**).
3. Incubate the gel on a reciprocal shaker for at least 40 min with shaking at room temperature.
4. Discard renaturation buffer and add incubation buffer.

3.7 Incubation

1. After 10 min shaking in an incubation buffer, start incubation at 37 °C (*see* **Note 26**).
2. After 18 h incubation, shake the gel on a reciprocal shaker at room temperature for 10 min (*see* **Notes 27** and **28**).

3.8 Staining and Destaining

1. Pour 40 mL freshly prepared working solution of RAMA stain per gel into another clean plastic vessel and put it on a reciprocal shaker (*see* **Note 29**).
2. Start shaking and transfer an 18 h-incubated gel into the stain solution (*see* **Note 30**).
3. After 30 min stain with shaking, discard the solution and briefly rinse with roughly 20 mL MQW two times (*see* **Note 31**).
4. Add fresh 200 mL MQW and incubate for 15 min with vigorous shaking (*see* **Note 32**).

3.9 Scanning Image

1. Wear plastic glove, set a destained gel on an OHP sheet or clear transparent plastic sheet avoiding any air bubbles under the gel (*see* **Note 33**).
2. Set a gel with a sheet on a commercially available document scanner (Epson GT X-970 or similar one) and scan it.
3. Process obtained images with PowerPoint and/or other software for publication.

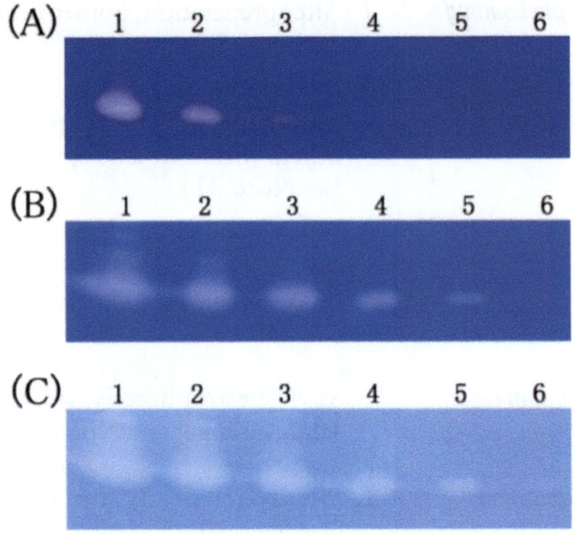

Fig. 1 Comparison of detection sensitivity among three methods such as conventional, RAMA, and CGP casein zymography using trypsin as the test protease. Purified bovine trypsin was separated on the casein-containing SDS-PAGE gels, renatured, incubated at 37 °C, and protease activities were detected with (**a**) conventional CBB stain, (**b**) RAMA stain, and (**c**) CGP stain. Trypsin was continuously diluted from left to right by tenfold in each lane (starting from 100 ng to 1 pg) and applied on the gels at three different concentrations of the substrate (casein), (**a**) 1 mg, (**b**) 200 μg, and (**c**) 67 μg. After separation, the gels were incubated at 37 °C for 18 h. In (**a**), the gel was stained with conventional CBB stain for 1 h and followed by an overnight destaining. In (**b**) and (**c**), gels were stained and destained for total time of 1 h with RAMA and CGP stain, respectively. All the stained gels were scanned at once, and their images were captured and compared simultaneously

4. As example, the detection sensitivity among three zymography methods such as conventional, RAMA, and CGP casein zymography was compared, and trypsin was chosen. As shown in Fig. 1, conventional method detected band at lane 4, RAMA strongly at lane 5 and weakly but significantly at lane 6, and CGP at lane 5. Therefore, sensitivity of RAMA casein zymography was revealed to be tenfold superior to conventional one. While CGP casein zymography demonstrated high sensitivity, its contrast was poor. Therefore, RAMA casein zymography was used for further experiments. In addition, the method was much faster than conventional one.
5. To evaluate the applicability of this method to the other class of protease, MMP7 was chosen. For MMP7, casein zymography is known to be more suitable than gelatin one [22, 23]. As shown in Fig. 2, conventional casein zymography faintly detected the band at lane 3. On the other hand, RAMA casein zymography significantly detected the band at lane 5.

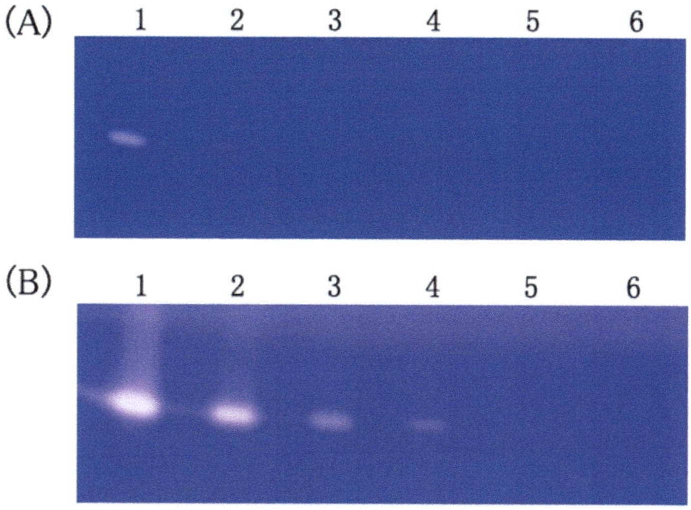

Fig. 2 High sensitive detection of MMP7 (Matrilysin) by RAMA casein zymography. Purified MMP7 was separated on the casein-containing SDS-PAGE gels, renatured, incubated at 37 °C, and detected with (**a**) conventional CBB stain, (**b**) RAMA stain. MMP7 was continuously diluted from left to right by tenfold in each lane (starting from 105 ng to 1 pg) and applied on two different concentrations of the substrate in the gels, (**a**) 1 mg, (**b**) 200 μg. After separation, the gels were incubated at 37 °C for 18 h. In (**a**), the gel was stained with conventional CBB stain for 1 h and followed by the overnight destaining. In (**b**), the gel was stained with RAMA stain for 30 min and washed with water for 15 min. Both stained gels were scanned at once, and their images were captured and compared simultaneously

Therefore, RAMA casein was revealed to be more sensitive by several tenfold than conventional one.

6. In conclusion, the method described in this chapter is simple, rapid, reproducible, low-cost, needed non-expensive equipments, and highly sensitive. Moreover, the strategy and method described herein, the low substrate concentration coupled with high sensitive staining, are also applicable to gelatin substrate zymography for high sensitive detection of metalloproteinases. Sensitivity is improved about tenfold (data not shown) [21].

4 Notes

1. Acrylamide and bisacrylamide are purchased at the highest grade and the powders are stored at 4 °C in light shielding bottles. To avoid wetting, do not open them until they reach room temperature.
2. Since bisacrylamide is difficult to dissolve at high concentration, make sure that all the powders are dissolved thoroughly and the solution is clear.

3. This solution should be prepared freshly and from power form, otherwise actual percentage of prepared gel might vary day by day. For storage of ammonium sulfate powder, temperature of 4 °C is not enough, because at that temperature it gradually gets wet and decays. Therefore, the powder should be dispensed into small aliquot in Eppendorf type tube with tight cap and stored at −20 °C avoiding light until use. The dissolved solution should be discarded after use.
4. For high sensitive detection, all the equipments in contact with this buffer including cylinders, bottles, stirrer bars, etc. should be extremely clean. Since the powerful detergent of SDS is contained in the solution at high concentration, it dissolves residual proteases sticking to the vessel wall derived from bacteria, fungi, or biological specimens into the solution and their activities are detected by zymography.
5. For sample handling and activity conservation, concentration of Tris and glycerol and pH of the buffer are modified from original report. To keep protease activity in the sample intact, any reducing reagents are omitted.
6. Since casein does not dissolve in MQW at the concentration of 10 mg/mL, small amount of alkali is required to dissolve it. However, too much alkali may disturb the separation dynamics of electrophoresis. Although it is possible to prepare a 10 mg/mL casein solution by using SDS instead of alkali, the sensitivity of SDS casein is slightly but significantly inferior to that of alkali. When compared among several caseins and casein-like proteins including skim milk and broking reagent, one from Merck demonstrated the best results.
7. Oxygen contained in the solution is known to prevent acrylamide polymerization and to result in the weakness of the gel. Therefore, degassing step is required for reproducibility of actual gel hardness and of protease resolution on a gel. An aspirator driven by running water is of low cost but unstable; on the other hand, a vacuum pump is stable and preferable.

   Since by the addition of SDS or casein into the solution, the resulted mixture is very easy to make bubbles and the removal of oxygen become insufficient, degassing should be performed without SDS and casein.
8. When concentrated Triton X-100 is mixed with MQW, it becomes solidified in the water and even under the vigorous stirring it takes a long time to be dissolved. However, 25% solution is easy to mix with aqueous solution. To perform a whole experiment in a short time, pre-preparation of 25% stock solution is required. To dissolve in a short time, it should be added into water dropwise. Too much addition into water at once results in stirrer bar stacking, and conversely, it sometimes results in overnight stirring.

9. Previously, even in our publication [24], NaCl is contained in a renaturation buffer as to final 0.1 M; however, it turned out to be not required for protease activity and is omitted.
10. Final concentration of calcium at 10 mM is reported; however, in my experiments it revealed to be rather inhibitory for the activity comparing to 1 mM.
11. For trypsin activity, pH at 7.5 is not optimum and at 8.0 is superior. However, former pH is suitable for many proteases tested including serine- and metalloproteases except for aspartic ones.
12. Since complete dissolving of CBB is required for high sensitivity and low background, powder CBB should be first dissolved with only methanol in a clear transparent glass bottle and stirred at least for 2 h. After confirmation by careful-visual-inspection for complete dissolution (if not dissolved thoroughly, insoluble ones are observed as the small granulated precipitates sticking on a vessel wall), the resulted dye solution is then mixed with MQW and stock at room temperature.
13. Since working solution of RAMA stain is not so stable, it should be used within several days after preparation. When precipitates are found to exist in a solution, not only in a prepared one, but also even in a fresh one, discard it and prepare another one fresh. Complete dissolving of dye is required for low background and uniform staining.
14. TEMED from several companies was tested. The one from Sigma-Aldrich demonstrated the reproducible results.
15. Although for blocking off air from a surface of an acrylamide solution isopropanol is also employed, it is inferior to water, because of its bad smell and environmental pollution after use. Since the relative density of gel solution is larger enough than that of water, interface between gel solution and water is stable against gentle addition of water. Pour water gently on it in a small amount, not dropwise but continuously.
16. A silicon comb should be very clean, otherwise polymerization of the gel surrounding comb results in incompletion. Several times of insert actions and/or re-arrangements of comb position might evoke air introduction into a solution and results in incomplete polymerization and short gel forming. One action of comb insertion and positioning is preferable for preparation of good sample lanes.
17. Without washing, in the sample lanes unpolymerized acrylamide and gel debris remain and they perturbate sample loading and separation on a gel.
18. Since this method is highly sensitive, most of the samples should be diluted to a large extent to adjust the intensity of

protease band. Protein at a low concentration is unstable and tends to stick to a vessel wall. Therefore, some manuals recommend an addition of carrier protein for sample stability; however, commercially available low-cost bulk proteins are sometimes contaminated with small amount of proteases, for example, BSA with serum proteases. When analyzed, sometimes faint but significant bands appear. Therefore, simple dilution with 4× sample buffer and immediate analysis are recommended. Once sample moves into substrate gel, it will be surrounded by substrate proteins and detergent and might be stabilized somehow.

19. Any sample containing known protease activity such as tissue extracts, blood sample, or commercially available protease are mixed together as the bulk, dispensed into small aliquots, and stored at −20 °C. At time of use, one of the control samples is mixed with 4× sample buffer and diluted to an appropriate concentration and can be used for as a positive control, which provides a reproducible activities on zymogram.
20. Since protein at a low concentration is time-dependently adsorbed to vessel wall, even in the presence of SDS, prepared samples should be loaded immediately.
21. Pre-experiments are required for rough estimation of the sample activities.
22. Since an air bubble spreads out small amounts of sample, if the sample has large amounts of protease activity, entire sample lanes might be contaminated by the protease activity. Therefore, the lane without any proteases applied should be required as the negative control.
23. In a previous publication, pre-run was recommended especially for casein zymography [25]. However, it took longer time and the resolution of the bands was not so good. Because interface between separation gel and stacking gel was partially perturbed by pre-run, the sharpness and resolution were lost.
24. Apparatus size is 10.1 (W) × 8.2 (L) cm. At constant current, quantity of heat produced during electrophoresis is constant, and it keeps the temperature almost same and prevents excess sample heating and loss of protease activity.
25. At first, put the resolved gel into renaturation buffer: do not start handling the gel before renaturation buffer is ready. The resolved gel tends to bend by itself and, by the contact with each other, some amount of protease in a gel is transferred even in a short time, and it results in the appearance of artificial protease bands. However, when contacted each other in a buffer under shaking conditions, such problems are almost negligible.

26. Since the gel percent is relatively high (12.5%), the shaking facilitates the equilibration of a gel with incubation buffer and increases the reproducibility of the experiments.
27. Longer incubation does not always improve the band intensity, partly because of its instability as a protease in a small amount.
28. By the shaking in a renaturation buffer, partially digested but still remaining caseins in a protease zone are washed out from gel before fixing and stained in a gel, and therefore, it enhances the contrast between clear protease zone and stained background.
29. Use one vessel for one gel.
30. For uniform staining, at the time of soaking a gel, shaking is required.
31. The rinsing step is required for efficient washing.
32. Usually 15-min washing is enough to provide sufficient contrast. Although longer washing provides better contrast, too much washing (3 h or more) may reduce the staining intensity of undigested area and result in low contrast.
33. By using a plastic sheet, handling of the fragile gel on a scanner such as changing orientation and position adjustment becomes easier.

Acknowledgment

The author is grateful to Mariko Kawasaki-Yasumitsu for her excellent support and also to members of Kamimachi Hospital. This work was supported by a 2010 Incentive Grant from Yokohama Academic Foundation.

References

1. Naggie S, Patel K, McHutchison J (2010) Hepatitis C virus directly acting antivirals: current developments with NS3/4A HCV serine protease inhibitors. J Antimicrob Chemother 65(10):2063–2069. doi:10.1093/jac/dkq284
2. Ruiz-Perez F, Nataro JP (2014) Bacterial serine proteases secreted by the autotransporter pathway: classification, specificity, and role in virulence. Cell Mol Life Sci 71(5):745–770. doi:10.1007/s00018-013-1355-8
3. Raju RM, Goldberg AL, Rubin EJ (2012) Bacterial proteolytic complexes as therapeutic targets. Nat Rev Drug Discov 11(10):777–789. doi:10.1038/nrd3846
4. Yike I (2011) Fungal proteases and their pathophysiological effects. Mycopathologia 171(5):299–323. doi:10.1007/s11046-010-9386-2
5. Antao CM, Malcata FX (2005) Plant serine proteases: biochemical, physiological and molecular features. Plant Physiol Biochem 43(7):637–650. doi:10.1016/j.plaphy.2005.05.001
6. Lu Q, Clemetson JM, Clemetson KJ (2005) Snake venoms and haemostasis. J Thromb Haemost3(8):1791–1799.doi:10.1111/j.1538-7836.2005.01358.x
7. Kini RM (2005) Serine proteases affecting blood coagulation and fibrinolysis from snake

venoms. Pathophysiol Haemost Thromb 34(4–5):200–204. doi:10.1159/000092424
8. Whitcomb DC, Lowe ME (2007) Human pancreatic digestive enzymes. Dig Dis Sci 52(1):1–17. doi:10.1007/s10620-006-9589-z
9. Zasloff M (2002) Trypsin, for the defense. Nat Immunol 3(6):508–510. doi:10.1038/ni0602-508
10. Kessenbrock K, Dau T, Jenne DE (2011) Tailor-made inflammation: how neutrophil serine proteases modulate the inflammatory response. J Mol Med (Berl) 89(1):23–28. doi:10.1007/s00109-010-0677-3
11. Smith PK, Harper JI (2006) Serine proteases, their inhibitors and allergy. Allergy 61(12):1441–1447. doi:10.1111/j.1398-9995.2006.01233.x
12. Wu Q, Kuo HC, Deng GG (2005) Serine proteases and cardiac function. Biochim Biophys Acta 1751(1):82–94. doi:10.1016/j.bbapap.2004.09.005
13. Almonte AG, Sweatt JD (2011) Serine proteases, serine protease inhibitors, and protease-activated receptors: roles in synaptic function and behavior. Brain Res 1407:107–122. doi:10.1016/j.brainres.2011.06.042
14. Vandooren J, Geurts N, Martens E, Van den Steen PE, Opdenakker G (2013) Zymography methods for visualizing hydrolytic enzymes. Nat Methods 10(3):211–220. doi:10.1038/nmeth.2371
15. Wilkesman J, Kurz L (2012) Advances in zymography techniques and patents regarding protease analysis. Recent Pat Biotechnol 6(2):106–114
16. Wilkesman J, Kurz L (2009) Protease analysis by zymography: a review on techniques and patents. Recent Pat Biotechnol 3(3):175–184
17. Lantz MS, Ciborowski P (1994) Zymographic techniques for detection and characterization of microbial proteases. Methods Enzymol 235:563–594
18. SJFaZ W (1995) The catabolism of extracellular matrix. In: MAHaJR H (ed) Extracellular matrix: a practical approach. Oxford University Press, New York, pp 261–287
19. Yasumitsu H, Ozeki Y, Kawsar SM, Toda T, Kanaly R (2010) CGP stain: an inexpensive, odorless, rapid, sensitive, and in principle in vitro methylation-free Coomassie Brilliant Blue stain. Anal Biochem 406(1):86–88. doi:10.1016/j.ab.2010.06.035
20. Yasumitsu H, Ozeki Y, Kawsar SM, Fujii Y, Sakagami M, Matuo Y, Toda T, Katsuno H (2010) RAMA stain: a fast, sensitive and less protein-modifying CBB R250 stain. Electrophoresis 31(12):1913–1917. doi:10.1002/elps.200900524
21. Yasumitsu H, Ozeki Y, Kanaly RA (2016) RAMA casein zymography: time-saving and highly sensitive casein zymography for MMP7 and trypsin. Electrophoresis 37(22):2959–2962. doi:10.1002/elps.201600346
22. Katsuno H, Shirakawa R, Miyazaki K, Ozeki Y, Yasumitsu H (2010) Production of active MMP7 in *E. coli* and its application for metalloproteinase inhibitors screening. Protein Pept Lett 17(5):568–572
23. Miyazaki K, Hattori Y, Umenishi F, Yasumitsu H, Umeda M (1990) Purification and characterization of extracellular matrix-degrading metalloproteinase, matrin (pump-1), secreted from human rectal carcinoma cell line. Cancer Res 50(24):7758–7764
24. Yasumitsu H, Miyazaki K, Umenishi F, Koshikawa N, Umeda M (1992) Comparison of extracellular matrix-degrading activities between 64-kDa and 90-kDa gelatinases purified in inhibitor-free forms from human schwannoma cells. J Biochem 111(1):74–80
25. Fernandez-Resa P, Mira E, Quesada AR (1995) Enhanced detection of casein zymography of matrix metalloproteinases. Anal Biochem 224(1):434–435

Chapter 3

Cysteine Protease Zymography: Brief Review

Jeff Wilkesman

Abstract

Cysteine proteases play multiple roles in basically all aspects of physiology and development. In plants, they are involved in growth and development and in accumulation and mobilization of storage proteins. Furthermore, they are engaged in signalling pathways and in the response to biotic and abiotic stresses. In animals and also in humans, they are responsible for senescence and apoptosis, prohormone processing, and ECM remodelling. When analyzed by zymography, the enzyme must be renaturated after SDS-PAGE. SDS must be washed out and substituted by Triton X-100. Gels are then further incubated under ideal conditions for activity detection. Cysteine proteases require an acidic pH (5.0–6.0) and a reducing agent, usually DTT. When screening biological samples, there is generally no previous clue on what peptidase class will be present, neither optimal proteolysis conditions are known. Hence, it is necessary to assess several parameters, such as incubation time, pH, temperature, influence of ions or reducing agents, and finally evaluate the inhibition profile. For detection of cysteine peptidase activity, the use of specific inhibitors, such as E-64, can be used to prevent the development of cysteine peptidase activity bands and positively confirm its presence. Here four different protocols to assess cysteine protease activity from different sources are presented.

Key words Cysteine protease, Zymography, Cathepsin

1 Introduction

Cysteine proteases were formerly known as thiol proteases. This group of enzymes is characterized by having at its active site a thiol group present in the Cys residue. Cysteine proteases have been isolated from plants, animals, and bacteria. Among the most representative examples are papain, actinidin, stem bromelain, ficin, cathepsins B and H, streptococcal proteinases, clostripain, and calpains. Typically, they all have relative small molecular masses, oscillating between 20,000 and 35,000. Some Cys proteases are glycosylated. The study of Cys proteases has become relevant, especially for modification of food proteins and for synthesis of biologically active peptides and their analogues.

The importance of cathepsins and other cysteine proteases from viruses and parasites is that they are targets for key diseases

Jeff Wilkesman and Liliana Kurz (eds.), *Zymography: Methods and Protocols*, Methods in Molecular Biology, vol. 1626,
DOI 10.1007/978-1-4939-7111-4_3, © Springer Science+Business Media LLC 2017

such as cancer, AIDS, osteoporosis, arthritis, atherosclerosis, as well as for parasitic diseases like amebiasis, leishmaniasis, malaria, Chagas disease, and African sleeping sickness [1].

To distinguish cysteine protease activity from other proteases, usually an inhibition assay must be performed in parallel. Typical inhibitors are leupeptin and E-64 (Table 1) [2].

Recently published information [3] reveals that when cysteine proteases—at least papain—are subjected to overlay zymography, enzyme activity is frequently lost. This activity loss is associated with the chemical lability of cysteine in the enzyme's active site and the presence of reactive oxidative species in the running buffer when using electrophoresis under acidic conditions.

Multiplex cathepsin zymography has been reported [4] as an effective assay based on SDS-PAGE, in order to quantify and identify levels of active cathepsins K, L, S, and V in cells and tissue. Among its benefits are: (a) antibodies are not required, hence the method is inexpensive and species-independent, (b) staining of the gel after non-reducing SDS-PAGE confirms cathepsin identity, (c) quantitative analysis is performed by densitometry, and (d) multiplexed detection enables distinction of active cathepsins K, L, S, and V in one cell or tissue extract, (e) cathepsin activity quantification is performed in short timeframes.

Here, we present a very brief review of some general methods to detect cysteine proteases.

Table 1
Some common inhibitors used for cysteine protease detection

Cysteine protease inhibitor	Optimal assay concentration	Remarks
N-Ethylmaleimide	Equimolar	Binds stoichiometrically to SH groups. Prepare fresh stock by dissolving 10 mg/mL water
Leupeptin	10–100 μM	Inhibits calpain, cathepsin B, H, and L, and papain. Prepare a 10 mM stock in water. Stock is stable for 6 months at −20 °C. Leupeptin also inhibits trypsin-like serine proteases (e.g., trypsin, chymotrypsin, pepsin, thrombin)
E-64	1–10 μM	Will not inhibit serine protease (excepting trypsin). A 1 mM stock is prepared in water and is stable at −20 °C. Preparation in 50% ethanol has also been reported [5]
Chymostatin	10–100 μM	It is not specific as it inhibits also chymotrypsin-like serine proteases as chymase cathepsins A, B, D, and G. A 10 mM stock is prepared in DMSO and is stable at −20 °C
Iodoacetic acid or iodoacetamide	1–150 mM	Covalent alkylation of the Cys residues at the active site. A 500 mM stock is freshly prepared in water

2 Materials

2.1 Protocol 1: Cysteine Protease Zymography

1. 11% polyacrylamide gels containing 0.1% gelatine.
2. 4% stacking gel.
3. Sample buffer (2×): 50 mM Tris–HCl pH 6.8, 10% (v/v) glycerol, 1% (w/v) SDS, 0.01% (w/v) bromophenol blue.
4. Activation buffer: 50 mM sodium citrate, 5 mM DTT, 5 mM $CaCl_2$, 1 mM $ZnCl_2$, pH 5.0.
5. Staining solution: 0.05% Coomassie R-250, 10% acetic acid, 40% methanol.
6. Destaining solution: 10% acetic acid, 40% methanol solution.

2.2 Protocol 2: Cathepsin Zymography

1. Non-reducing loading buffer 5×: 0.05% bromophenol blue, 10% SDS, 1.5 M Tris–HCl pH 6.8, 50% glycerol.
2. 12.5% SDS–polyacrylamide gels containing 0.2% gelatine.
3. 4.5% stacking gels.
4. Renaturing buffer: 65 mM Tris–HCl buffer pH 7.4 with 20% glycerol.
5. Activity buffer: 0.1 M sodium phosphate buffer pH 6.0, 1 mM EDTA. Add freshly 2 mM DTT just before use.
6. Coomassie stain solution: 10% (v/v) acetic acid, 25% (v/v) 2-propanol, 4.5% (w/v) Coomassie Blue R-250.
7. Destaining solution: 10% (v/v) 2-propanol, 10% (v/v) acetic acid.

2.3 Protocol 3: General Cysteine Zymography

1. 11% polyacrylamide gels containing 0.1% gelatine.
2. 4% stacking gel.
3. 2.5% Triton X-100.
4. Developing buffer: (a) for acidic cysteine protease: 0.1 M sodium citrate buffer pH 4.0, 0.05% Brij 35, 1 mM EDTA and 2 mM DTT; (b) for cysteine proteases with optimal neutral pH: 0.1 M sodium phosphate buffer pH 6.8, 0.05% Brij 35, 1 mM EDTA and 2 mM DTT.

2.4 Protocol 4: Specific Cysteine Zymography Employing Inhibitors

1. 12.5% SDS–polyacrylamide gels containing 0.2% gelatine.
2. 4.5% stacking gels.
3. 2.5% Tween 20.
4. Activation buffer: 50 mM acetic acid buffer pH 5.0; 2 mM cysteine.
5. Staining solution: 0.5% amido black, 30% methanol, 10% acetic acid.
6. Destaining solution: 30% methanol, 10% acetic acid.

7. Inhibitors: final assay concentration prepared in activation buffer. 20 mM EDTA, 25 mM pepstatin, 10 mM PMSF, and 10 μM E-64.

3 Methods

3.1 Protocol 1: Cysteine Protease Zymography

1. This protocol is based on the one proposed by Wagstaff et al. [5], where proteins from petal tissue were analyzed.
2. Prepare zymograms using 11% polyacrylamide gels containing 0.1% gelatin (type I from porcine skin) as substrate.
3. Use a 4% stacking gel as overlaid.
4. Load 5 μg protein per lane dissolved in sample buffer (2×).
5. Run gel under denaturing conditions at 180 V for 30 min.
6. The gels are renatured for 1 h in 2.5% Triton X-100.
7. Incubate gels overnight (15–18 h) at 37 °C in activation buffer. If protease specificity is going to be tested, include an appropriate inhibitor in the activation buffer (*see* **Notes 1** and **2**).
8. Stain gels in staining solution.
9. Destain in destaining solution until areas of protease activity are revealed as clear bands on a blue background.
10. Register the result by photographing or scanning.

3.2 Protocol 2: Cathepsin Zymography

1. This is a special case developed for cathepsin and based on Wilder et al. [6] when analyzing recombinant cathepsins from human sequences.
2. Add non-reducing loading buffer 5× to samples prior loading.
3. Resolve equal amounts of cell or tissue protein in 12.5% SDS–polyacrylamide gels containing 0.2% gelatin at 4 °C.
4. Remove gels carefully from the electrophoresis unit.
5. Renature enzymes in renaturing buffer. Repeat for a total of three washes, 10 min each with gentle orbital shaking, at room temperature (~22 °C).
6. Incubate gels in activity buffer for 30 min at room temperature (~22 °C). It is possible to vary the pH value tested altering the pH buffer or the chemical nature of the buffer (*see* **Note 3**).
7. Exchange activity buffer for fresh one (of the same pH) and incubate further for 18–24 h (overnight) at 37 °C.
8. Discard activity buffer and rinse gels once with deionized water and incubate for 1 h in Coomassie stain solution.
9. Destain in destaining solution.

3.3 Protocol 3: General Cysteine Zymography

1. Based on the method of Saitoh et al. [7], it is performed as a usual SDS-PAGE.
2. After SDS-PAGE, remove SDS by washing gel twice in 2.5% Triton X-100 with gentle orbital shaking at room temperature (~22 °C) for 15 min each.
3. Rinse gel briefly with distilled water twice.
4. Incubate gel at 37 °C in 15 ml developing buffer for 24 h (*see* **Note 4**).

3.4 Protocol 4: Specific Cysteine Zymography Employing Inhibitors

1. According to Grudkowska et al. [8], proceed to reactivate proteins after electrophoretical separation by incubating gels twice in 2.5% Tween 20 for 30 min each.
2. Rinse gels briefly three times with distilled water.
3. Incubate for 16 h (overnight) in 50 mM acetic buffer pH 5.0 containing 2 mM cysteine, with gentle shaking at 25 °C.
4. Rinse gels with water as before (**step 2**).
5. Stain for 3 h in staining solution [0.5% amido black, 30% methanol, 10% acetic acid].
6. Destain gels in 30% methanol and 10% acetic acid, until clear bands are visible on a dark blue background.
7. To detect proteinase specificity, use inhibitors. Prepare inhibitors to the indicated assay concentration: 20 mM EDTA (for metalloproteinases), 25 mM pepstatin (for aspartic proteinases), 10 mM PMSF (for serine proteinases), and 10 μM E-64 for inhibition of cysteine proteinases [9]. Other inhibitors may be used (*see* Table 1).
8. Incubate protein extracts for 2 h in 50 mM acetic buffer pH 5, containing the appropriate inhibitor, prior to electrophoresis.
9. After electrophoreses, place the gels (or gel strips if individual lanes are going to be tested) in the presence of the inhibitor. Gelatin digestion must be carried out in the presence of the inhibitors.
10. Again, incubate for 16 h (overnight) in 50 mM acetic buffer pH 5.0 containing 2 mM cysteine plus the inhibitor, with gentle shaking at 25 °C.
11. Rinse gels with water as before (**step 2**).
12. Stain for 3 h in staining solution [0.5% amido black, 30% methanol, 10% acetic acid].
13. Contrast the results. The presence of the specific inhibitor will diminish the intensity of the band signal corresponding to cysteine protease (*see* **Notes 5–7**).

4 Notes

1. In order to determine the specificity of the bands corresponding to cysteine protease, a protease inhibitor must be used. For this purpose, include in the overnight activation buffer either (a) 2 μM leupeptin or (b) 100 μM E-64 in 50% ethanol. Always perform a control without inhibitor. Alternatively, IAA can be added in the sample buffer before submitting the sample to electrophoresis. As this is covalent modification, there should be no need to add more IAM at the activation buffer. However, tests should be run before to check the inhibitor efficiency.
2. The use of other inhibitors is also recommended, e.g., PMSF, EDTA, and pepstatin, in order to rule out any other concomitant protease with similar *Rf*.
3. For different pH conditions, use, e.g., 0.1 M sodium acetate buffer pH 4, or sodium phosphate buffer pH 6, 7or 8. However, be aware that if using phosphate buffer, calcium and magnesium ions present will precipitate in the respective phosphate salts.
4. Take into consideration that for acidic cysteine protease, use 0.1 M sodium citrate buffer pH 4.0, but for cysteine proteases with an optimal neutral pH, use 0.1 M sodium phosphate buffer pH 6.8. Other buffers may also be employed.
5. Care must be taken when analyzing inhibition results. Some cysteine proteases may present unspecific inhibition towards other type of inhibitors.
6. Take into consideration that some specific cysteine protease inhibitors are dissolved in solvents other than water, like DMSO or ethanol. A control must be run to interpret correctly the result, as it may turn that the solvent alone may inhibit the enzyme as well.
7. Quantification of the inhibition process can be performed via densitometry. A calibration curve must be previously generated employing a commercial standard cysteine protease, with and without the presence of the inhibitor. Intensity of band is proportional to the activity. It is suggested to test with different enzyme concentrations and different inhibitor concentration until optimal band signals intensities are achieved on the gel. For this purpose, staining must be first standardized.

Acknowledgments

The author thanks Dr. Rebeca Giamate and Dr. Liliana Kurz from the University of Carabobo for technical and emotional support during the writing and editing of this chapter.

References

1. Verma S, Dixit R, Pandey KC (2016) Cysteine proteases: modes of activation and future prospects as pharmacological targets. Front Pharmacol 7:107
2. Smith C, Gates R. Protease inhibition and detection. Life Science. Biofiles 4(2). http://www.sigmaaldrich.com/content/dam/sigma-aldrich/flashapps/biofiles-movie/pdf/BioFiles_4.2_v1.pdf. Accessed 27 Sep 2016
3. Vootukuri Reddy S, Philpott MP, Trigiante G (2016) Retaining in-gel zymographic activity of cysteine proteases via a cysteine-supplemented running buffer. Electrophoresis 37:2644–2648
4. Dumas JE, Platt MO (2013) Systematic optimization of multiplex zymography protocol to detect active cathepsins K, L, S, and V in healthy and diseased tissue: compromise among limits of detection, reduced time, and resources. Mol Biotechnol 54:1038–1047
5. Wagstaff C, Leverentz MK, Griffiths G, Thomas B, Chanasut U, Stead AD, Rogers HJ (2002) Cysteine protease gene expression and proteolytic activity during senescence of *Alstroemeria* petals. J Exp Bot 53:233–240
6. Wilder CL, Park KY, Keegan PM, Platt MO (2011) Manipulating substrate and pH in zymography protocols selectively distinguishes cathepsins K, L, S, and V activity in cells and tissues. Arch Biochem Biophys 516:52–57
7. Saitoh E, Yamamoto S, Okamoto E, Hayakawa Y, Hoshino T, Sato R, Isemura S, Ohtsubo S, Taniguchi M (2007) Identification of cysteine proteases and screening of cysteine protease inhibitors in biological samples by a two-dimensional gel system of zymography and reverse zymography. Anal Chem Insights 2:51–59
8. Grudkowska M, Lisik P, Rybka K (2013) Two-dimensional zymography in detection of proteolytic enzymes in wheat leaves. Acta Physiol Plant 35:3477–3482
9. Zhang N, Jones BL (1995) Characterization of germinated barley endoproteolytic enzymes by two-dimensional gel electrophoresis. J Cereal Sci 21:145–153

Chapter 4

Aspartic Protease Zymography Case Study: Detection of Fungal Acid Proteases by Zymography

Gavin Kernaghan and Michael Mayerhofer

Abstract

This chapter describes a method for the production and characterization of fungal acid proteases. Protease production is induced by growth on BSA media over a pH gradient and protein levels are monitored over time with the Bradford assay. Once protein is depleted, the media is purified and proteases are characterized by gelatin zymography using acrylamide and buffers at near-neutral pH. Maintaining pH levels below those found in traditional zymographic systems avoids the potential loss of activity that may occur in aspartic proteases under alkaline conditions.

Key words Neutral zymography, Fungal enzymes, Protein utilization, SDS-PAGE, Laemmli gels, Bradford assay

1 Introduction

Fungi exhibit external digestion and therefore rely heavily on extracellular enzymes to access nutrient resources. Fungal cells secrete a wide variety of extracellular proteases for the degradation of proteins, including aspartic, cysteine, serine, and metalloproteases [1–3]. Both the production and activity of proteases are optimal at specific pH values, which vary among protease types and fungal species [4].

Zymography is a useful tool for characterizing extracellular proteases, as it allows for direct visualization of enzymatic activity. Standard zymography utilizes SDS-PAGE with the Laemmli buffering system [5]. This involves basic Tris-glycine running buffer and basic Tris-chloride gel buffer, resulting in pH levels up to 9.5 [6]. However, many fungal proteases (e.g., aspartic proteases) exhibit optimal activity at low pH (as low as 2.0) [7] and can become unstable and susceptible to structural modification or irreparable inactivation under the alkaline conditions found within a Laemmli gel [8, 9]. Therefore, a neutral to acidic environment should be maintained in order to retain the enzymatic activity of

Jeff Wilkesman and Liliana Kurz (eds.), *Zymography: Methods and Protocols*, Methods in Molecular Biology, vol. 1626, DOI 10.1007/978-1-4939-7111-4_4, © Springer Science+Business Media LLC 2017

aspartic proteases. Although this has been achieved through the addition of compounds such as potassium acetate [10], this results in a system that may be acidic, but not buffered at the desired pH, and therefore subject to fluctuation.

More recently, a modification to protein electrophoresis has been developed in which the standard Tris-glycine gel buffer is replaced with bis-Tris and the standard Tris-chloride running buffer is replaced with MES-Tris or MOPS-Tris [11]. This results in a near-neutral pH system and represents a general improvement in protein electrophoresis as resolution is increased through reductions in protein–acrylamide interactions and disulfide formation. The neutral system also reduces the breakdown of buffer components and increases gel shelf life. This system is easily adapted to zymography, reducing the problems related to aspartic proteases under alkaline conditions.

Here we describe a method in which near-neutral electrophoresis is adapted for the zymography of fungal aspartic proteases. The method involves the induction of fungal proteases and subsequent analysis by zymography under near-neutral conditions. Cultures are grown on protein media (BSA) over a range of pH levels until the protein is exhausted (based on Bradford assay). Media is then purified and analyzed using zymography with gels and buffers at near-neutral pH.

2 Materials

2.1 Acid Protease Production

1. Fungal cultures: obtain fungal isolates from culture collections or by direct isolation using media appropriate for the fungal group (*see* **Note 1**).
2. Water agar medium: 1.5% agar in distilled H_2O (dH_2O).
3. BSA medium (1 L): 0.3 g BSA (96% lyophilized powder, Sigma), 0.03 g yeast extract, 1 g $MgSO_4{\cdot}7H_2O$, 2 g KH_2PO_4, 1 mg $FeSO_4$, 1 mg $ZnSO_4{\cdot}7H_2O$, 0.1 mg $MnSO_4{\cdot}H_2O$, 0.1 mg H_3BO_3, 0.1 mg $CoCl_2{\cdot}6H_2O$, 0.1 mg $CuSO_4{\cdot}5H_2O$ (*see* **Note 2**).
4. Basic lab material: Petri dishes, stericup filter units (0.22 μm filter), automatic pipetter with a 0.22 μm filter, sterile 1 and 10 mL pipettes, sterilized 18 × 150 mm culture tubes, loose fitting culture tube closures (e.g., Kap-Uts from Bellco), clear, flat-bottom 96-well microplates.
5. Laminar flow hood or biosafety cabinet.
6. Incubator for fungal cultures.
7. Transfer tubes (Spectrum Labs).
8. Microplate reader (spectrophotometer) capable of reading absorbance at 580 nm.
9. Bradford assay reagent (Coomassie Brilliant Blue-G reagent).

2.2 Near-Neutral Zymography

1. Centrifuge capable of spinning 15 mL tubes at 18,400 × *g* and 4 °C.
2. Centrifugal filters with 10 kDa cutoff (e.g., Microsep or Amicon).
3. Mini-protean protein electrophoresis cell (BioRad) and power supply.
4. Gelatin solution: 1% Gelatin in dH_2O.
5. Resolving gel: 2.05 mL dH_2O, 2.85 mL 1.25 M bis-Tris buffer (pH 6.7), 4 mL 30% acrylamide:bis-acrylamide (29:1), 100 μL 10% sodium dodecyl sulfate (SDS) in dH_2O, 1 mL 1% gelatin in dH_2O, 20 μL *N,N,N′,N′*-tetramethyl-ethylenediamine (TEMED), 7.5 mg ammonium persulfate (APS).
6. Stacking gel: 2.88 mL dH_2O, 1.42 mL 1.25 M bis-Tris buffer (pH 6.7), 0.65 mL 30% acrylamide:bis-acrylamide (29:1), 50 μL 10% SDS in dH_2O, 25 μL TEMED, 7.5 mg APS.
7. Running buffer: 250 mM 3-(*N*-morpholino)propanesulfonic acid (MOPS), 250 mM Tris(hydroxymethyl)aminomethane (Tris), 0.1% SDS in dH_2O, stored at 4 °C.
8. Loading buffer: 125 mM Tris–HCl (pH 6.8), 8% SDS, 50% glycerol, 0.02% bromophenol blue.
9. 10–250 kDa protein standard (Precision Plus; Bio Rad).
10. Wash buffer: 2.5% Triton X-100.
11. Development buffer: 200 mM NaCl, 5 mM $CaCl_2$, 0.02% Brij L23 solution, in 50 mM sodium citrate–citric acid buffer.
12. Staining Solution: 45% methanol, 10% glacial acetic acid, 45% dH_2O, 0.1% amido black.
13. Destaining solution: 50% methanol, 2% glacial acetic acid, 48% dH_2O.
14. Flatbed scanner (300–600 dpi).

3 Methods

3.1 Growth of Fungal Cultures and Optimization of Protease Utilization

1. Plate isolates onto water agar medium (*see* **Note 3**).
2. Allow cultures to grow to a point at which there is enough mycelium to transfer. The optimum growth temperature will depend on the group of fungi under investigation.
3. Autoclave empty culture tubes for 15 min at 121 °C. Enough tubes are required to accommodate all replicates of each isolate at each pH level of interest (*see* **Note 4**). Label each tube with respect to isolate, replicate, and pH.
4. Prepare liquid BSA media (*see* **Notes 5** and **6**).

5. Set up a range of media pH levels by dividing medium into batches and titrating each batch to the desired pH with 0.1 M HCl or NaOH (*see* **Note 7**).
6. Filter-sterilize each batch of pH-adjusted BSA medium with 0.22 μm filters using a glass vacuum filter holder or disposable Stericup filter units (*see* **Notes 8** and **9**).
7. Using sterile pipettes and an automatic pipetter with a 0.22 μm filter, add 10 mL of each pH-adjusted media to culture tubes labeled with respect to fungal isolate, pH level, and replicate number. Attach loose fitting closures to culture tubes. If quantification of protein utilization is desired, retain a small amount of each batch (media at each pH level) for constructing standard curves in **step 11**.
8. In a laminar flow hood or biosafety cabinet, use sterile Transfertubes to transfer each fungal isolate from water agar to the corresponding culture tubes containing pH-adjusted liquid BSA media.
9. Test the liquid cultures for protein content by removing a small aliquot (e.g., 0.1 mL) of BSA media from each culture using sterile pipettes and a filtered pipetter in a laminar flow hood or biosafety cabinet.
10. Add 6 μL of each aliquot to each of three wells of a clear flat-bottom 96-well microplate (*see* **Note 10**). Also add 6 μL dH_2O to each of three wells to assess background absorbance levels.
11. If protein quantification is desired, prepare a protein standard curve using the fresh BSA media retained at **step 8** (*see* **Note 11**).
12. Add 56 μL Coomassie protein assay reagent (*see* **Note 12**).
13. Incubate plate at room temperature for 15 min.
14. Place plate in spectrometer and read absorbance at 580 nm.
15. Repeat the Bradford assay on a regular basis (*see* **Note 13**) until it indicates that the BSA has been completely utilized. Zymograms are run when two or more consecutive Bradford readings are not significantly above the background spectrometer readings (*see* **Notes 14** and **15**).

3.2 Near-Neutral Zymography

1. Filter the fungal hyphae from the BSA media using a coarse filter. Sterile conditions are not necessary (*see* **Note 16**).
2. Centrifuge 7.5 mL of each filtered media sample for 10 min at 18,400 × *g* in order to remove residual fragments of fungal hyphae.
3. Concentrate 5 mL of the resulting supernatant with centrifugal filters with a 10 kDa cutoff at 7600 × *g* for 20 min at 4 °C.
4. Keep filtrate in the refrigerator or on ice until use in zymography.

5. Set up glass gel plates for electrophoresis cell.
6. Prepare a fresh 1% gelatin solution (5 mL).
7. Mix components of the stacking and resolving gels separately, leaving out APS and TEMED until just prior to pouring gel (*see* **Note 17**).
8. Dissolve 7.5 mg APS and 20 μL TEMED in the resolving gel mixture.
9. Pipette 3–3.5 mL of the liquid resolving gel between the gel plates (*see* **Note 18**).
10. Pipette a layer of 70% ethanol on top of the resolving gel (*see* **Note 19**).
11. Let polymerize (approximately 1 h) and blot excess ethanol.
12. Dissolve 7.5 mg APS and 25 μL TEMED in the stacking gel mixture.
13. Pipette the stacking gel mixture between the glass plates until full.
14. Add an 8- or 12-well comb and allow to polymerize (approximately 30 min).
15. Place completed gel plates into electrophoresis cell sitting in an ice bath.
16. Add cold running buffer.
17. Gently remove comb (*see* **Note 20**).
18. Pipette a small amount of running buffer in and out of the wells to remove any interfering gel material.
19. Mix each sample 3:1 with 4× loading buffer (*see* **Note 21**).
20. Load the protein standards (10 μL) and samples (*see* **Note 22**).
21. Run gel at 125 volts until the dye front is almost at the bottom of the gel and the protein standards are clearly separated (*see* **Note 23**).
22. Remove the gel plates from electrophoresis box and separate them carefully (*see* **Note 24**).
23. Wash three times in 2.5% Triton X-100 for 15 min each time.
24. Wash in development solution for 15 min (*see* **Note 25**).
25. Incubate in development solution overnight at 37 °C.
26. Transfer the gel from the development buffer to the staining solution for 40–60 min and shake gently at room temperature.
27. Remove and place in the destaining solution for 15 min.
28. Repeat this step until bands of clearing become visible (Fig. 1).
29. Scan gel using a flatbed scanner at 300–600 dpi.

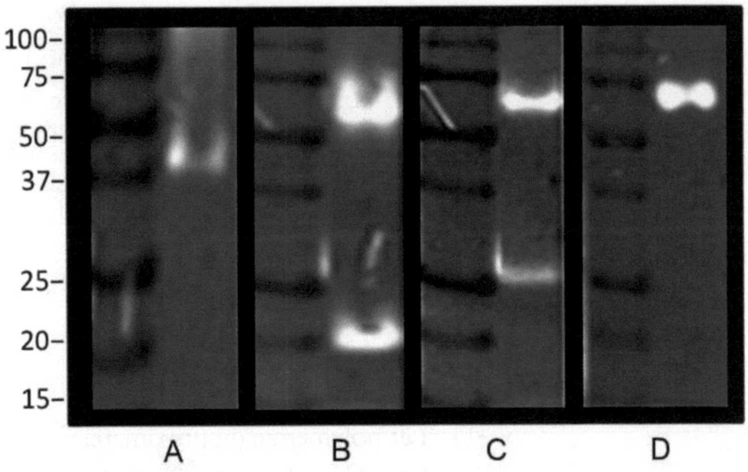

Fig. 1 Examples of fungal aspartic proteases from four fungi characterized by near-neutral zymography. (**a**) *Irpex lacteus* (42 kDa); (**b**) *Meliniomyces variabilis* (60 and 20 kDa); (**c**) *Phialocephala fortinii* (63 and 26 kDa); (**d**) *Umbelopsis isabellina* 70 kDa. Each zymogram includes a 10–250 kDa protein standard

30. Analyze gel images for protease size (kDa) and intensity using software such as Gel-Pro Analyzer or Image J.
31. To confirm the aspartic nature of the proteases on the gel, the specific inhibitor Pepstatin A (1 mM) can be incubated with the sample prior to loading on the gel, or added to the development solution after running the gel, or both. Loss of individual bands after the addition of Pepstatin A indicates an aspartic protease (*see* **Note 26**).

4 Notes

1. Most culturable filamentous fungi can be maintained on standard malt or yeast extract–malt extract agar (YMA) plates. Many yeasts can be maintained on yeast extract peptone dextrose (YPD) plates.
2. For convenience, the solutions of $FeSO_4$, $MnSO_4 \cdot H_2O$, H_3BO_3, $ZnSO_4 \cdot 7H_2O$, $CoCl_2 \cdot 6H_2O$, and $CuSO_4 \cdot 5H_2O$ can be kept as a 1000× nutrient stock solution, and added to the basal media at 1 mL/L.
3. Passage through water agar avoids carryover of nutrients from rich media such as YMA. Ideally, all Petri dishes should contain the same amount of water agar in order to reduce weighing errors if biomass determination after growth in liquid BSA is desired. We fill 9 mm diameter Petri dishes with 25 mL media.
4. We grow the fungi over a range of pH values to optimize growth and protease production as only small amounts of protease may

be produced under conditions of suboptimal pH. For example, if no prior information is available on pH optima for the fungi of interest, each isolate could be grown at pH 2–7 (in one unit increments).

5. For each isolate, 10 mL of medium is required for each replicate culture at each pH level. For example, each isolate would require 180 mL given three replicates, each at pH 2–7. As we also use a small amount of media for creating a standard curve at each pH, we would prepare at least 250 mL for each fungal isolate.
6. Add BSA in small amounts to avoid clumping.
7. We found it most effective to work with the largest volume of media possible. After initially determining the pH of the bulk media, titrate it all to the closest desired pH value, then split the solution as necessary and titrate each part up or/and down to achieve the desired pH range.
8. Filter sterilization should be done in a laminar flow hood or biosafety cabinet to avoid contamination.
9. We find that filter sterilizing the complete media is quite efficient and also avoids any potential pH changes caused by autoclaving. However, it may be more economical to autoclave the media prior to adding the BSA and yeast extract, and then adding these to the cooled media using a 0.22 μm syringe filter.
10. We are describing the Bradford assay developed for microplate readers. A single cuvette spectrometer could also be used, but would be time-consuming. The 96-well format also lends itself well to make each measurement in triplicate, allowing outlier readings (often caused by bubbles in Coomassie reagent) to be discarded.
11. The Bradford assay is somewhat pH-sensitive, so it is best to create protein standard curves using fresh media at the same pH as the growth media being tested. To prepare a standard curve, pipette 12 μL of media into the first well. Then add 6 μL of distilled water to wells 2–10. Then transfer and mix 6 μL from well 1 to well 2, and 6 μL from well 2 to well 3. Continue until well 10, from which 6 μL is removed and discarded.
12. Pipette Coomassie protein assay reagent directly on top of the samples of media. Do not mix, as this may create bubbles that may interfere with the absorbance readings.
13. The time interval for sequential measurement of the media protein depends on the fungi being tested. At optimal pH, some fungi may completely deplete the protein very rapidly and should be measured every day or two. Other fungi may take 2 weeks or more.

14. At the point at which the BSA has been exhausted, it no longer has the potential to interfere with the enzymes on the zymograms. The rate of protein loss also indicates which cultures (pH levels) are producing the highest levels of protease and are therefore most likely to result in successful zymograms. We find it effective to graph protein loss over time in order to determine which cultures are most active, and when to assay proteases.
15. Once the optimal pH for protease production has been determined, isolates could optionally be grown on 1% BSA media at that pH. The higher concentration of BSA induces more protease production and results in greater clearing of the protein background during zymography, although it will take longer for the cultures to exhaust the BSA.
16. We find that organza fabric works well as a coarse filter. Fungal biomass can also be dried and weighed at this point if data on biomass is desired.
17. The recipe given is for two gels. If only one is required, the other can be wrapped in plastic and stored in the refrigerator.
18. From 3 to 3.5 mL of the resolving gel mixture should be about 6 cm high between the glass plates, leaving space for the stacking gel.
19. Ethanol, distilled water, 0.1% SDS, or butanol can be added to obtain a flat surface at the top of the resolving gel.
20. Gently remove the comb by wiggling as the stacking gel tends to stick to it.
21. A 4× sample buffer helps to maximize the amount of sample loaded in the well. For example, 75 μL of sample plus 25 μL of loading buffer will be enough to load two 50 μL wells.
22. We recommend loading two wells of each sample to maximize the chance of bands with good resolution.
23. If only high molecular weight proteases are present, the gel could be run until the first or second bands of the protein standard are run off the gel.
24. The flat end of a scientific spatula works well for prying apart the plates.
25. The pH of the development solution can be modified in order to determine the pH optimum and range of individual proteases based on band intensity. The pH of the development solution can be brought as low as 2.5 by adjusting the ratio of sodium citrate to citric acid.
26. Similarly, the specific inhibitors E-64 (1 mM), PMSF (1 mM), and EDTA (10 mM) can be used to indicate serine proteases, cysteine proteases, or metallo-proteases, respectively.

References

1. Pavlukova EB, Belozersky MA, Dunaevsky YE (1998) Extracellular proteolytic enzymes of filamentous fungi. Biochemistry (Mosc) 63: 899–928
2. Monod M, Capoccia S, Léchenne B et al (2002) Secreted proteases from pathogenic fungi. Int J Med Microbiol 292:405–419
3. Yike I (2011) Fungal proteases and their pathophysiological effects. Mycopathologia 171: 299–323
4. Mayerhofer MS, Fraser E, Kernaghan G (2015) Acid protease production in fungal root endophytes. Mycologia 107:1–11
5. Laemmli UK (1970) Cleavage of structural proteins during the assembly of the head of bacteriophage T4. Nature 227:680–685
6. Hachmann JP, Amshey JW (2005) Models of protein modification in Tris–glycine and neutral pH Bis-Tris gels during electrophoresis: effect of gel pH. Anal Biochem 342:237–245
7. Leake JR, Read DJ (1990) Proteinase activity in mycorrhizal fungi. I: The effect of extracellular pH on the production and activity of proteinase by ericoid endophytes from soils of contrasted pH. New Phytol 115:243–250
8. Piper DW, Fenton BH (1965) pH stability and activity curves of pepsin with special reference to their clinical importance. Gut 6:506
9. Kulkarni A, Gaikwad S, Rao M (2008) pH induced structural alterations in an aspartic protease from *Vigna radiata* indicating an alkali induced molten globule state. Int J Biol Macromol 43:373–376
10. Burton KS, Smith JF, Wood DA, Thurston CF (1997) Extracellular proteinases from the mycelium of the cultivated mushroom *Agaricus bisporus*. Mycol Res 101:1341–1347
11. Updyke TV, Engelhorn SC (2000) System for pH-neutral stable electrophoresis gel. US patent 6162338 A

Chapter 5

Detection of Aspartic Proteinase Activities Using Gel Zymography

Handunge Kumudu Irani Perera

Abstract

Gel zymography is a two-stage process where the proteins from the test sample are first separated by electrophoresis followed by the detection of the activity of hydrolytic enzymes. Many zymography procedures use sodium dodecyl sulfate (SDS) polyacrylamide gels copolymerized with an appropriate substrate. The procedure described here uses native polyacrylamide gel electrophoresis (PAGE) in the absence of both SDS and substrate. In order to visualize aspartic proteinase activity, the gel is impregnated in bovine hemoglobin at pH 3.0 for 15 min after the electrophoresis procedure. Subsequently, the gel is incubated in a humid container in the absence of hemoglobin for 1 h at 37 °C. At the end, the gel is stained with amido black and destained. Clear areas against a dark background corresponding to aspartic proteinase activities can be detected.

Key words PAGE, Zymography, Aspartic proteinases, Hemoglobin

1 Introduction

Gel zymography is a two-stage process where the proteins from the test sample are first separated by electrophoresis and then the activity of hydrolytic enzymes is detected. Qualitative or quantitative information on hydrolytic enzymes [1] is collected based on the visualization of the breakdown of a substrate [2]. Zymography is mostly used to detect the activities of the proteolytic enzymes [3]. It is a method used for screening, identification, and characterization of proteinases [4]. Rather than visualizing all proteins separated on the gel using a general staining method, zymography recognizes a particular group of enzymes which remain active after the electrophoresis. Zymography is a flexible procedure which provides room for the investigator to change the substrate, pH, temperature, ion concentration as well as to incorporate inhibitors according to the type of enzyme which is under investigation [1].

The method elaborated here describes a gel zymography procedure designed to recognize activities of aspartic proteinases

Jeff Wilkesman and Liliana Kurz (eds.), *Zymography: Methods and Protocols*, Methods in Molecular Biology, vol. 1626, DOI 10.1007/978-1-4939-7111-4_5, © Springer Science+Business Media LLC 2017

which belong to one catalytic class of proteinases [4]. One specific feature of the aspartic proteinases is having acidic pH optima. They are usually inactive at neutral pH, unlike most other enzymes.

Many zymography procedures use sodium dodecyl sulfate (SDS) polyacrylamide gel electrophoresis which separates proteins based on their molecular weight [4, 5]. The procedure described here uses native polyacrylamide gel electrophoresis (PAGE) in the absence of SDS. Native polyacrylamide gel electrophoresis (PAGE) separates proteins in their native conformation under non-denaturing and non-reducing conditions [6]. A key determinant of the protein separation by PAGE is the net charge of the protein as governed by its amino acid composition and pH of the medium during electrophoresis [6]. Advantage of preserving the active state during separation of a protein with the native PAGE method is used in the described procedure to demonstrate the proteinase activity. PAGE method can resolve proteins of the same molecular mass as long as their net charges are considerably different [7].

Generally, zymography procedures use resolving gels copolymerized with a substrate [4, 5]. Gels copolymerized with gelatin have long been used to detect matrix metalloproteinase activities [1]. In the present technique, the substrate solution at the optimum pH was added to the gel after the protein separation is completed. Hemoglobin, which is the most common substrate used to measure aspartic proteinase activity in biological samples [8], is used in this procedure. After allowing time to react with the substrate, the gel is stained with amido black and destained [9]. The proteinase activities can be seen as clear areas against a dark background due to the breakdown of protein substrate [1]. Activity due to multiple enzymes can be detected using this procedure as they will be visualized separately.

2 Materials

2.1 Polyacrylamide Gel

1. *Resolving gel buffer*: 1.5 M Tris–HCl, pH 8.8. Weigh 45.43 g of Tris and add into a 250 mL graduated beaker. Add 200 mL distilled water and mix the contents using a stir bar. Adjust the pH to 8.8 using HCl (*see* **Note 1**). Make the volume up to 250 mL with distilled water. Store at 4 °C.
2. *Stacking gel buffer*: 0.5 M Tris–HCl, pH 6.8. Weigh 15.14 g Tris and add into a 250 mL graduated beaker. Add 200 mL distilled water and mix the contents using a stir bar. Adjust the pH to 6.8 using HCl (*see* **Note 1**). Make the volume up to 250 mL with distilled water. Store at 4 °C.
3. *Acrylamide (30%), bisacrylamide solution*: Acrylamide 29.2%, bisacrylamide 0.8%. Weigh 73 g of acrylamide and 2 g bisacrylamide into a beaker containing 100 mL of distilled

water (*see* **Note 2**). Mix and make up to 250 mL with distilled water. Filter and store at 4 °C in a dark bottle.

4. *10% ammonium persulfate (APS)*: 10% APS. Weigh 100 mg of ammonium persulfate in a microfuge tube and add 1 mL of distilled water. Mix and store at 4 °C after covering with aluminum foil (*see* **Note 3**).
5. *Tetramethylethylenediamine (TEMED)*: Store at 4 °C.
6. *PAGE running buffer (×10)*: 250 mM Tris, 1.92 M glycine. Weigh 30 g Tris, 144 g glycine and dissolve in 1 L distilled water (*see* **Note 4**).
7. *PAGE running buffer (×1)*: 25 mM Tris, 192 mM glycine. Add 130 mL of running buffer (×10) to distilled water and make up to 1300 mL (*see* **Note 4**). Store at 4 °C.
8. *PAGE sample buffer (×3)*: 0.5 M Tris–HCl (pH 6.8) (93.75 mM), glycerol (37.5%), and bromophenol blue (0.015%). Add 187.5 μL 0.5 M Tris–HCl (pH 6.8), 375 μL glycerol, and 100 μL of 0.15% (150 mg in 100 mL distilled water) bromophenol blue into a microfuge tube. Add distilled water up to 1 mL. Store at 4 °C.

2.2 Zymography Components

1. *Formate buffer*: 100 mM formate buffer. Add 5.75 mL formic acid (80% solution) into 900 mL distilled water. Adjust to pH 3.0 with concentrated NaOH solution. Dilute up to 1 L with distilled water. Store at 4 °C.
2. *Bovine hemoglobin solution (Hb)*: 1% bovine Hb, pH 3.0. Prepare fresh (20 mL/gel). Add 200 mg Hb in 20 mL distilled water. Mix with a magnetic stirrer for about 1 h. Check the pH using a pH paper and get it to pH 3.0 by adding HCl drops (*see* **Note 5**).
3. *Amido black stain*: 0.1% Amido black, 10% acetic acid, 30% methanol. Weigh 100 mg Amido black 10 B. Add 7 mL acetic acid and 30 mL methanol. Make up to 100 mL in distilled water, mix with a magnetic stirrer to dissolve. Store at room temperature.
4. *Destain*: Methanol 20 mL (20%), glacial acetic acid 10 mL (10%), and distilled water up to 100 mL (70%). Store at room temperature.

3 Methods

It is extremely important to minimize the loss of activity of enzymes until the zymography is done (*see* **Notes 6–9**). Furthermore, having an idea about the optimum pH, activity of the enzyme, and whether the protein is negatively charged at pH 8.5 will facilitate the selection of optimum conditions for zymography (*see* **Notes 10** and **11**).

3.1 Preparation and Fractionation of Porcine Ovarian Extracts

1. Ovaries of pigs (6–8 months age) are obtained from an abattoir during routine operations. Ovaries are cleaned and brought to the laboratory in dry ice. All procedures are conducted at 4 °C.
2. Crude ovarian extract is obtained by homogenizing the ovaries in 50 mM phosphate-buffered saline, pH 7.5. Extract is centrifuged to remove debris and the supernatant is dialyzed overnight in 20 mM phosphate buffer.
3. This extract is centrifuged again at 15,000 × *g*.
4. Supernatant is fractionated using DEAE-52 cellulose equilibrated with 20 mM Tris–HCl (pH 8.5). The bound proteins are eluted with a 0–1 M linear NaCl gradient.
5. Active fractions are subjected to Sephacryl S-200 column, equilibrated with 20 mM phosphate buffer pH 7.5, containing 0.2 M NaCl. Crude extract and the partially purified fractions are then subjected to gel zymography.

3.2 Polyacrylamide Gel Electrophoresis Under Non-Denaturing Conditions

1. Standard protocol of PAGE is used [9–12]. Single percentage (7.5%) 1 mm thickness mini gels are casted. Tris glycine buffer system is used.
2. Assemble the gel cassette according to the manufacturer's recommendations (*see* **Note 12**). Place the comb into the cassette. Using a marker pen, place a mark on the glass plate 1 cm below the lower edge of the comb to highlight the level to which the resolving gel is poured. Remove the comb (*see* Fig. 1a).
3. Prepare the required volumes of 7.5% resolving gel (*see* **Notes 13** and **14**) and stacking gel in a small beaker or a capped tube (*see* Table 1), without adding APS and TEMED (*see* **Note 15**). Add 10% APS and TEMED into the resolving gel, swirl gently to mix and pipette (or gently pour) the resolving gel into the gel cassette up to the marked line. Immediately, add a few drops of distilled water gently over the resolving gel using a pipette to make a thin layer (*see* **Note 16**) and leave aside without disturbing for approximately 30 min to set.
4. Once the resolving gel is set (*see* **Note 17**), add 10% APS and TEMED to the stacking gel, gently swirl (or invert if it is a tube) to mix, discard the distilled water layer, and pipette the stacking gel over the resolving gel. Gently insert the comb (thickness 1.0 mm 10 well) immediately after adding the stacking gel. Care should be taken not to trap air under the comb while inserting the comb. Allow approximately 30 min without disturbing to set (*see* **Note 17**).
5. Once the gel is set (*see* **Note 17**), mix 20 μL sample (*see* **Note 18**) with 10 μL sample buffer just before the sample loading (*see* **Notes 14** and **19**).
6. Load the prepared samples into the wells (*see* **Note 20**) using gel loading tips. Electrophoresis is carried out at

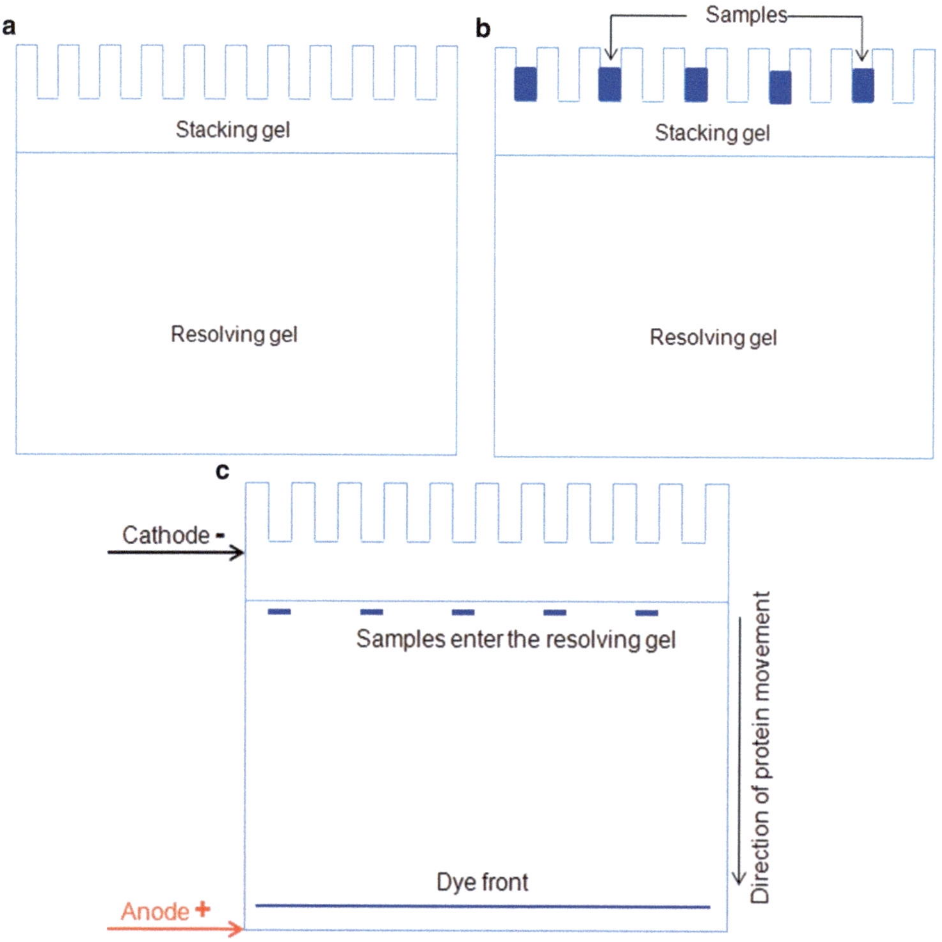

Fig. 1 Polyacrylamide gel electrophoresis. (**a**) Preparation of the gel without SDS and the substrate, (**b**) loading the samples without SDS and reducing agents, (**c**) electrophoresis at 4 °C

4 °C under constant current of 20 mA until the dye front reaches the lower end of the gel or as desired (*see* **Note 21**) (*see* Fig. 1b and c).

3.3 Detection of Enzyme Activity Using Zymography

1. After electrophoretic separation, enzyme activity was detected, based on previously published methods [8, 13, 14]. Once the electrophoresis is finished, disconnect the power supply. Invert the glass plate with the gel, gently loosen the gel from the glass plate from one edge using a spatula, and float it in a small container (*see* **Notes 22** and **23**) with the formate buffer. Gently swirl and discard the buffer, add formate buffer again, and leave the gel to equilibrate with the buffer for 10 min (*see* **Note 24**). Discard the formate buffer.
2. Incubate the gel with 1% bovine hemoglobin at pH 3.0 for 15 min. Subsequently, wash the gel once gently with the formate buffer (pH 3.0) and incubate for 1 h at 37 °C in a humid container (*see* **Notes 25** and **26**) (*see* Fig. 2a and b).

Table 1
Composition of the gel components

Components	Resolving gel (7.5%)[a]	Stacking gel (4%)
Water	3.7 mL	3.05 mL
1.5 M Tris–HCl (pH 8.8)	1.875 mL	–
0.5 M Tris–HCl (pH 6.8)	–	1.25 mL
Acrylamide (30%), bisacrylamide (0.8%)	1.875 mL	660 μL
Add the following reagents immediately before dispensing into the gel cassette		
10% ammonium persulfate (APS)	37.5 μL	50 μL[b]
Tetramethyl ethylenediamine (TEMED)	3.75 μL	5 μL[b]

[a]Volumes given are sufficient to prepare 1 gel
[b]APS and TEMED are added to the stacking gel mixture only when the resolving mini gel is set

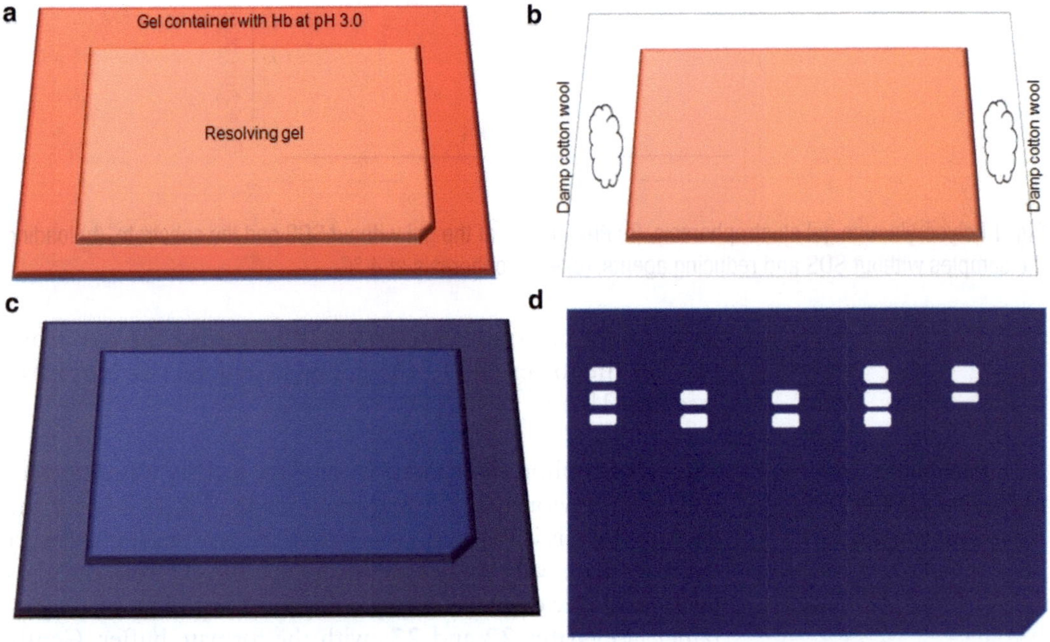

Fig. 2 Zymography. (**a**) Addition of the substrate (bovine hemoglobin at pH 3.0), (**b**) incubation of the gel with the bound substrate in a humid chamber for hydrolysis to occur, (**c**) gel is stained with amido black, (**d**) Aspartic proteinase activities are detected after destaining the gel

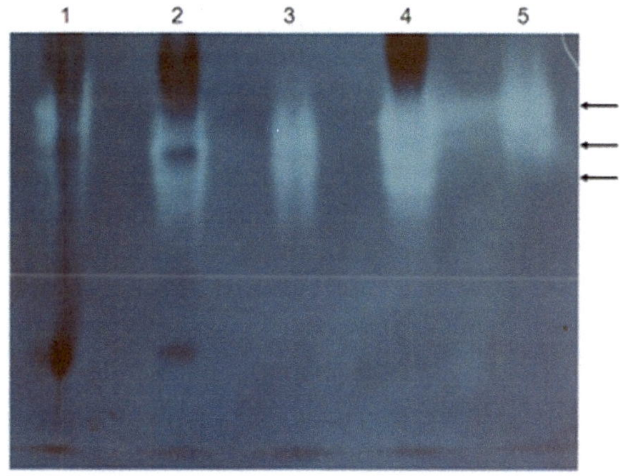

Fig. 3 Detection of multiple aspartic proteinase activities from a tissue extract. *Arrows* show the multiple aspartic proteinase activities (*see* Ref. 9)

3. Add the amido black stain to the gel container and keep on a shaker for 30 min. Remove the stain (*see* **Note 27**) (*see* Fig. 2c), add destain, and leave overnight (*see* **Note 28**).
4. Keep the gel container on a light box and visualize the clear areas in a dark background (*see* **Note 28**) (*see* Figs. 2d and 3).

4 Notes

1. As the pH of the Tris buffer changes significantly with the temperature, pH should be adjusted after storing the buffer at 4 °C as the electrophoresis is conducted at 4 °C. It is better to avoid concentrated HCl during the final adjustments of pH.
2. Caution: Acrylamide and bisacrylamide monomers are neurotoxic! Wear a mask and gloves and use a fume hood during preparation. Prevent any spills during weighing. Acrylamide, bisacrylamide reagent, is also available commercially.
3. APS stored at 4 °C can be used for a few times within a couple of weeks.
4. Do not adjust the pH of the ×10 and ×1 running buffer. Prepare the ×1 running buffer fresh and store at 4 °C. SDS is not added to the running buffer.
5. Prepare the hemoglobin solution fresh immediately after starting the electrophoresis and keep it stirring using a magnetic stir bar to dissolve. This may take approximately 1 h. The substrate was brought to the optimum pH (pH 3.0) of the aspartic protease.
6. It is extremely important to take precautions to preserve the enzyme activity until the end of the reaction with the substrate

in zymography. Avoid heat, detergents, foaming, and over-dilution to minimize loss of activity. All procedures involving the samples such as collection, transport, extraction, and electrophoresis should be conducted at 4 °C.

7. Avoid bringing the sample near the optimum pH of the proteinase during extraction and storage, to avoid activation of the proteinase which might cause auto-degradation of the protease. A buffer at pH 7.4 would be fine for aspartic proteases.
8. Protease inhibitors should not be added when cell lysates are prepared.
9. Tissue extracts should be preferably stored at −80 °C until the preparation of the polyacrylamide gel is complete. Avoid repeated thawing of the frozen samples.
10. It is necessary to assess the optimum pH of the aspartic proteinase and the activity prior to starting the procedure as this information is necessary to select the pH and the incubation period for the zymography.
11. Since the protein migration during PAGE depends on the net charge, it is important to know that the protein under investigation is negatively charged during the separation (most of the proteins are negatively charged at pH 8.5). In case that the p*I* of the protein is above 8.5, the two electrodes should be reversed when connecting to the power supply to prevent a backward movement. However, reported p*I*s of aspartic proteinases are less than 8.5. Aspartic proteinases identified from the porcine ovarian extracts showed a movement towards the anode indicating that their p*I*s are less than 8.5.
12. After assembling the gel cassette, fill the cassette with distilled water using a wash bottle and leave for about 5 min to ensure there is no leaking. Discard distilled water before pipetting the gel solution into the cassette.
13. Some proteases may give better results with 10% resolving gels. To get the desired percentage, adjust only the volumes of acrylamide, bisacrylamide solution, and distilled water (*see* Table 1).

    Gel percentage = (Volume of acrylamide, bisacrylamide solution/Total volume) × 30.
14. Sodium dodecyl sulfate and β-mercaptoethanol should be omitted from all reagents.
15. TEMED and APS are the polymerizing agents and therefore are added only when the resolving and stacking gels are about to pour into the cassette. As the stacking gel is added only when the resolving gel is set, delay the addition of TEMED and APS until such time.

16. Addition of a thin layer of distilled water will leave a smooth and even edge on the resolving gel and prevent drying during the period of setting.
17. Approximately 5.6 mL resolving gel is poured into the gel cassette with a 1 mm spacer for each mini gel. Resolving and stacking gels were prepared with extra volume (7.5 mL resolving gel) to retain extra gel in the container after pouring into the gel cassette to monitor whether the gel is set. Do not handle the gel cassette to check for gel setting as it might cause formation of uneven gels.
18. Total protein concentration of the samples can be determined by the standard Bradford method [15] with bovine serum albumin as the standard.
19. Do not boil the sample after adding sample buffer. SDS is not added to the sample buffer.
20. It is better to load the samples well apart to leave sufficient space to view the catalytic activity (*see* Fig. 1b). The capacity of a well is 30 μL, if a 10 well comb with a 1 mm thickness is used. Better is not to use more than 90% of the maximum volume of a well to prevent overflow of the sample to adjacent wells. Add sample buffer alone onto the empty wells.
21. If the proteases have a lesser mobility, the duration of electrophoresis can be prolonged after the dye front leaves the gel.
22. Handle the 7.5% gel with care as the gel is fragile. Before removing the gel from the glass plate, mark one end of the gel to identify the lanes and separate the stacking gel. As a practice, I make a small cut at the lower end of the right side (towards the last well) (*see* Fig. 2c and d) of the gel.
23. Use a transparent container with a lid which is little larger than the gel. Use one container for each gel. A volume of 20 mL of buffer, substrate, stain, and destain is sufficient for each gel.
24. At the end of the electrophoresis, wash and equilibrate the gel in formate buffer (pH 3.0) in order to bring the gel to the acidic pH required to activate the aspartic proteases.
25. Keep two or three damp cotton wool plugs (or paper towels) along the sides of the gel container to prevent the gel from drying during the incubation (*see* Fig. 2b). Keep the container closed during the incubation. An incubator or a water bath can be used (take precautions to prevent water leaking into the container if a water bath is used).
26. The duration of incubation in the humid container after removing the substrate can be adjusted to suit the desired results. Incubation period will depend on the activity of the proteases. Even though a longer incubation increases the sensitivity, it might result in broader bands due to diffusion of

proteins, decreasing the resolution. Broader bands might mask multiple proteinase activities found in close proximity, while a shorter incubation may not reveal sufficient visibility of clear areas. Duration of proteolysis should be strictly controlled if the data are to be used in a quantitative manner.

27. Amido black stain can be reused for several times.
28. Partial destaining will occur within a much shorter time with visible results. I prefer to use the gel for viewing and imaging after an overnight destaining. Clear areas observed with zymography correspond with the aspartic proteinase bands with different net charges (*see* Figs. 2d and 3). Make images for the future reference.

References

1. Kandapur R (2013) Zymography: enzymes in action. Sci Int 1(4):70–75
2. Vandooren J, Geurts N, Martens E, Van den Steen PE, Opdenakker G (2013) Zymography methods for visualizing hydrolytic enzymes. Nat Methods 10(3):211–220
3. Wilkesman J, Kurz L (2012) Advances in zymography techniques and patents regarding protease analysis. Recent Pat Biotechnol 6(2):106–114
4. d'Avila-Levy CM, Santos ALS, Cuervo P, de Jesus JB, Branquinha MH (2012) Applications of zymography (substrate-SDS-PAGE) for peptidase screening in a post-genomic era. In: Magdeldin S (ed) Gel electrophoresis—advanced techniques. ISBN: 978-953-51-0457-5. InTech. Available from http://www.intechopen.com/books/gel-electrophoresis-advanced-techniques/applicationsofzymography-substrate-sds-page-for-peptidase-screening-in-a-post-genomic-era
5. Pan D, Hill AP, Kashou A, Wilson KA, Tan-Wilson A (2011) Electrophoretic transfer protein zymography. Anal Biochem 411(2):277–283
6. Arndt C, Koristka S, Bartsch H, Bachmann M (2012) Native polyacrylamide gels. Methods Mol Biol 869:49–53
7. A guide to polyacrylamide gel electrophoresis and detection. http://www.bio-rad.com/webroot/web/pdf/lsr/literature/Bulletin_6040.pdf. Accessed 10 Feb 2016
8. Barrett AJ (2013) In: Rawlings ND, Woessner JF (eds) Handbook of proteolytic enzymes. Academic Press. Elsevier, London
9. Perera HKI, Fernando PHP, Athauda SBP (2015) Zymographic detection of aspartic proteinase activities in porcine ovarian extracts. Int J Biochem Res Rev 7(4):166–174
10. Ornstein L, Davis BJ (1964) Disc electrophoresis-I: background and theory. Ann N Y Acad Sci 121:321–349
11. Laemmli UK (1970) Cleavage of structural proteins during the assembly of the head of bacteriophage T4. Nature 227:680–685
12. Ausubel FM, Brent R, Kingston RE, Moore DD, Seidman JG, Smith JA, Struhl K (eds) (2003) Current protocols in molecular biology. John Wiley & Sons, New York
13. Takahashi K, Matsumoto K, Nishii W, Muramatsu M, Kubota K, Hachioj HI et al (2009) Comparative studies on the acid proteinase activities in the digestive fluids of nepenthes, cephalotus, dionaea, and drosera. Carniv Pl Newsl 38:75–82
14. Furihata C, Kawachi T, Sugimura T (1972) Premature induction of pepsinogen in developing rat gastric mucosa by hormones. Biochem Biophys Res Commun 47:705–711
15. Bradford M (1976) A rapid and sensitive method for the quantification of microgram quantities of protein utilizing the principle of protein-dye binding. Anal Biochem 72:248–254

Chapter 6

MMP Activity Detection in Zymograms

Péter Bencsik, Monika Bartekova, Anikó Görbe, Krisztina Kiss, János Pálóczi, Jana Radosinska, Gergő Szűcs, and Péter Ferdinandy

Abstract

Matrix metalloproteinases (MMP) belong to a distinguished class of zinc-dependent endopeptidases. Zymography is a semi-quantitative tool for determining the activity of different MMP isoenzymes in a variety of biological samples. In substrate gel zymography, protein samples of different origin (tissue, cell lysates, plasma/serum, perfusates, other liquids) are separated in sodium dodecyl sulfate (SDS) polyacrylamide gels containing copolymerized substrate (gelatin, casein, elastin, etc.), and after incubation-enabling substrate cleavage by MMPs, MMP activities are detected after the gel staining as transparent bands against a dark-blue background. In situ zymography is a histological modification of substrate zymography in frozen sections, allowing detection of the localization of the MMP activities within the tissue. Here, we describe detailed experimental protocols of all abovementioned techniques and provide examples for several sample measurements.

Key words Matrix metalloproteinase activity, Substrate zymography, Gelatin, Casein, In situ zymography

1 Introduction

1.1 Matrix Metalloproteinases in Health and Disease

Matrix metalloproteinases (MMPs) are enzymes that belong to the family of zinc-dependent endopeptidases and are known to play a crucial role in the dynamic processing of the extracellular matrix (ECM) facilitating the degradation of matrix material. On the other hand, MMPs have been also shown to be present intracellularly, thereby influencing physiological as well as pathological intracellular signal transduction processes and the contractile machinery. Therefore, MMPs are interesting drug targets for several pathologies (*see* for a review: [1]). The family of MMPs include close to 30 members (*see* for reviews: [2, 3]); many of them shown to be activated due to different physiological as well as pathological situations in different tissues. Some of the important family members together with their characteristic features are shown in Table 1. MMPs are synthesized as zymogens and can be activated by proteolytic cleavage of an amino-terminal domain, by oxidative/nitrosative-induced conformational

Jeff Wilkesman and Liliana Kurz (eds.), *Zymography: Methods and Protocols*, Methods in Molecular Biology, vol. 1626,
DOI 10.1007/978-1-4939-7111-4_6,

Table 1
The family of MMPs and their characteristic features

MMP codes	Alternative names	Molecular weight (kDa)		Substrates	Pathologies
		Z[a]	A[a]		
MMP-1	Interstitial collagenase	57	52	Gelatin	Atherosclerosis, melanoma, heart failure
MMP-2	Gelatinase A, type IV collagenase	75/72	64	Gelatin, elastin	Myocardial infarction, heart failure, gastritis, rheumatoid arthritis
MMP-3	Stromelysin-1	57	45	Gelatin, elastin, casein	Brain injury, neurodegeneration
MMP-7	Matrilysin	28	19	Gelatin, elastin, casein	Tumor-induced osteolysis, colon cancer
MMP-8	Neutrophil collagenase	75	57	Gelatin	Coronary artery disease, angina
MMP-9	Gelatinase B	92	86	Gelatin, elastin	Myocarditis and subsequent dilated cardiomyopathy, ulcerative colitis
MMP-10	Stromelysin-2	57	44	Gelatin, elastin, casein	Lung cancer
MMP-11	Stromelysin-3	51	44	Gelatin, elastin, casein	Tumor progression, breast carcinomas
MMP-12	Macrophage metalloelastase	54	22	Gelatin, elastin, casein	Granulomatous skin diseases, inflammatory disorders
MMP-13	Collagenase-3	65	48	Gelatin	Breast carcinomas
MMP-14	MT1-MMP	66	54	Gelatin, casein	Tumor growth by activating MMP-2
MMP-15	MT2-MMP	76	N/A	Fibronectin Laminin	Obesity, preeclampsia, ovarian carcinoma
MMP-16	MT3-MMP	64	52/30	Gelatin, casein	Breast cancer
MMP-17	MT4-MMP	71	67	N/A	Preeclampsia
MMP-19	RASI-1	57	N/A	Gelatin	Rheumatoid arthritis
MMP-20	Enamelysin	54	42.5		Amelogenesis imperfecta

(continued)

Table 1 (continued)

MMP codes	Alternative names	Molecular weight (kDa) Z[a]	A[a]	Substrates	Pathologies
MMP-21	N/A	N/A	N/A	N/A	Melanoma, ovarian and colon carcinomas
MMP-22	N/A	N/A	42	N/A	N/A
MMP-23	N/A	N/A	N/A	N/A	Breast cancer
MMP-24	MT5-MMP	N/A	N/A	N/A	Brain tumors
MMP-25	MT6-MMP, leucolysin	56	38	pro-MMP-2	Inflammatory hyperalgesia
MMP-26	Endometase	29	19	Gelatin	Lung cancer

[a]Z and A indicates the zymogen or active form(s) of MMPs, respectively

change (without a change in molecular weight), or by phosphorylation [2, 4]. Activities of MMPs are tightly regulated by their endogenous tissue inhibitors (TIMPs) [5].

MMPs play an important role in many physiological and pathological processes, including embryogenesis, wound healing, inflammation, cardiovascular diseases, and tumor development or progression [2, 6]. Increased activities of different MMPs have been reported to be connected with different pathological situations such as ischemia-reperfusion injury [7], myocardial contractile dysfunction [8], heart failure [9], arthritis [10], neurodegenerative disorders [11], cancer invasion and metastasis [12], liver cirrhosis [13], fibrotic lung disease [14], periodontal disease [15] as well as with responses to some invasive interventions like anthracycline treatment [16, 17] or chest irradiation [18] used in the cancer therapy. On the other hand, inhibition of MMP activities has been shown to be connected with some kinds of tissue protection such as ischemic preconditioning [19, 20] or flavonoid-induced cardioprotection [17]. Moreover, pharmacological inhibition of MMP activities has been shown to be cardioprotective in animal models of acute myocardial infarction [21, 22] and has also been shown to be altered in coronary artery disease patients [23]. Being able to detect MMPs at early stages of the disease is opening a perspective to use MMPs as diagnostic markers. MMPs have been well-investigated in clinical studies of cardiovascular diseases: MMP-2 and -9 in Chagas cardiomyopathy [24] and MMP-9 in ST-segment elevation myocardial infarction [25]. Protein expression and activation of MMP-2 and MMP-9 has clinical relevance and prognostic value in patients with colorectal cancer [26]. Fecal MMP-9 is a useful tool for the differential diagnosis of diarrheic disorders and in the noninvasive evaluation of disease activity and mucosal healing in ulcerative colitis [27].

Regarding the abovementioned facts, determination of MMP activities belongs to very useful methodologies in biomedical research and is of high clinical importance since it seems to be a powerful diagnostic and/or therapeutic tool for the detection or follow-up of the abovementioned pathologies.

1.2 MMP Activity Detection in Biological Samples

MMP activities can be determined by zymography in different kinds of biological samples such as heart, brain, liver, lung tissues, blood vessels, or in isolated or cultured cell lineages (*see* Figs. 1, 2, and 3). Zymography can be performed as a substrate zymography, in which the substrate of the certain MMP is incorporated, copolymerized in a sodium dodecyl sulfate (SDS) polyacrylamide gel, and MMPs are separated according to their molecular weights.

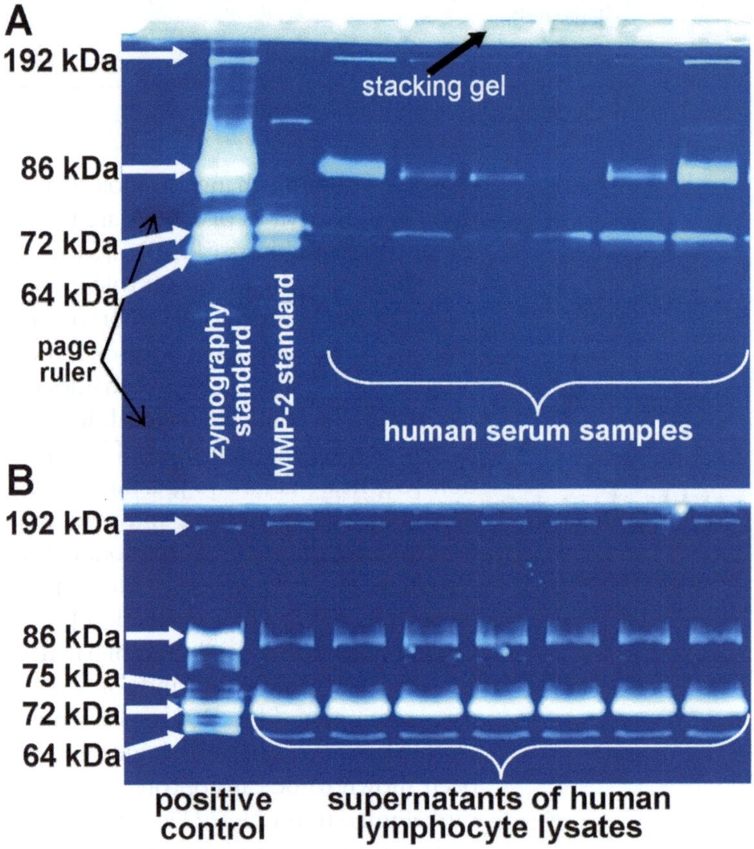

Fig. 1 Representative gelatin zymograms performed from human samples. (*Panel A*) Human serum samples from patients with coronary artery disease. Some of the patients have increased MMP-2 and/or MMP-9 activity as indicated by the zymographic intensity at 72 or 86 kDa, respectively. (*Panel B*) Gelatin zymogram from human isolated lymphocytes. 72 kDa MMP-2 activity is markedly visible; however, other activities at 64 kDa (MMP-2) and at 86 kDa (MMP-9) can also be observed. Weak gelatinolytic activity signals can be seen in both panels at 192 kDa, which may indicate the dimerized form of MMP-9 [28]

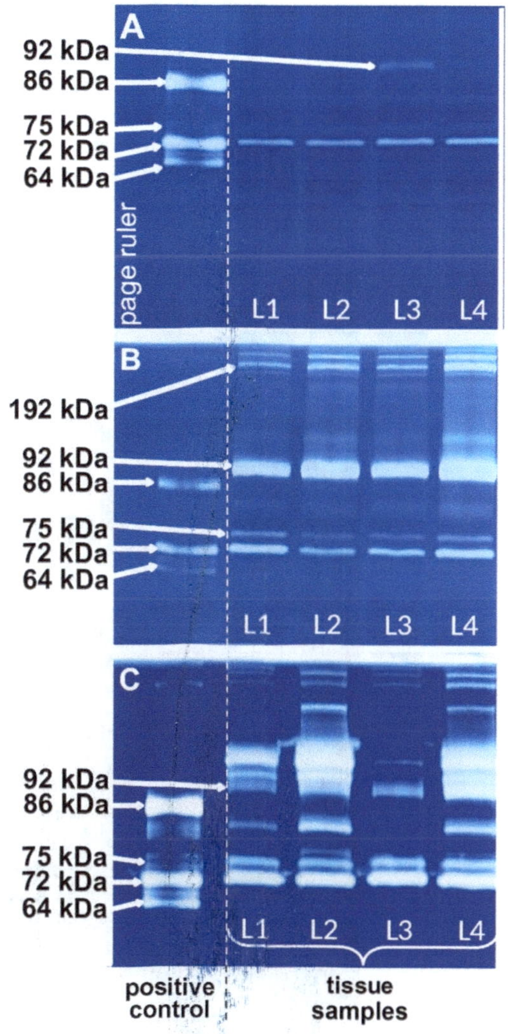

Fig. 2 Representative gelatin zymograms performed from homogenates of murine tissues (*L1–4*: *lanes 1–4*; 15 μg protein was loaded into each wells). *Panel A* represents mouse heart homogenates, which expresses mainly the 72 kDa isoform of MMP-2. Weak signal for MMP-9 can be observed; however, it indicates inappropriate removal of blood from heart samples. (*Panel B*) Lung samples from mice subjected to chronic tobacco smoking (*L2* and *4*) and their controls (*L1* and *3*). A markedly increased intensity can be observed at 92 kDa (MMP-9). (*Panel C*) Samples derived from the aorta of transgenic mice. *L1* and *2*: ApoB100LDL−/− mice, control and Chlamydia pneumoniae (Cpn) infected; *L3* and *4*: ApoE−/−, control and Cpn infected

The activity of MMPs is detected by the absence of gelatin in the gel, which can be visualized by transilluminating the gel. The activity of the certain MMP is proportional with the intensity and the thickness of the corresponding band on zymogram, which can be evaluated electronically by using different software after scanning the gels.

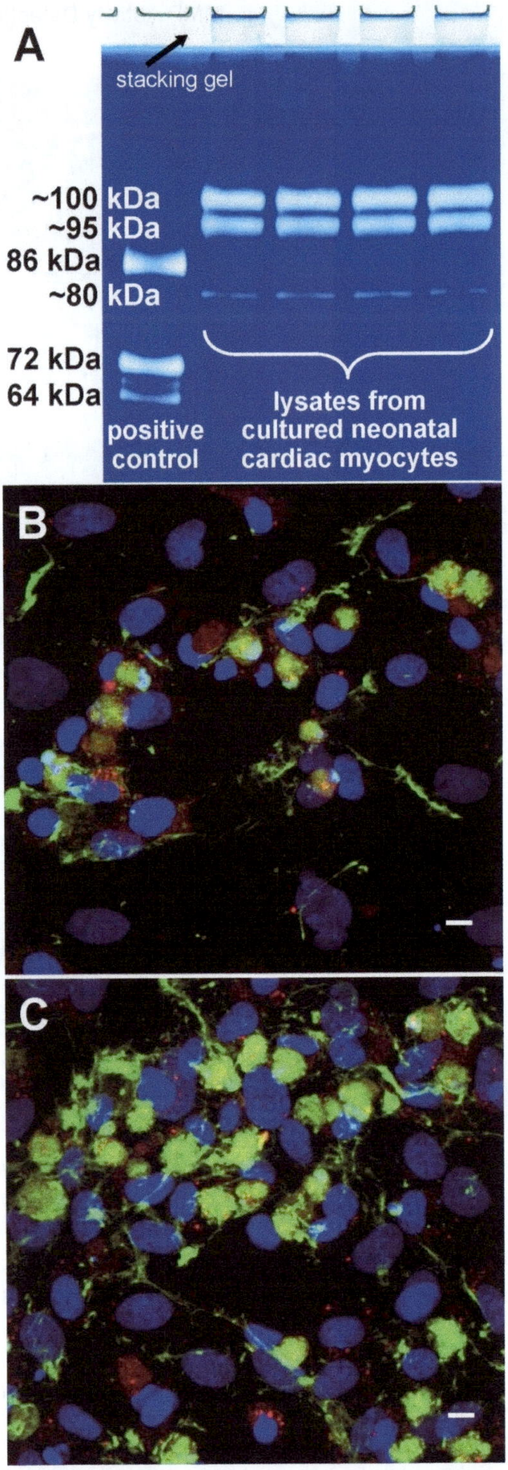

Fig. 3 Examples of MMP zymography in isolated and/or cultured cells. (*Panel A*) Gelatin zymogram from cultured neonatal cardiac myocytes after resuspension. Gelatinolytic activities can be detected at completely different molecular weights than that of heart homogenates (for comparison *see* Fig. 2a). (*Panel B, C*) In situ gelatin zymography in fixed neonatal cardiac myocyte culture in normoxic conditions (*Panel B*) and subjected to simulated ischemia/reoxygeation (*Panel C*). *Panel C* shows an increased gelatinolytic activity (represented as *green fluorescence*—DQ™ fluorescent gelatin; *red fluorescence*: MMP-2 immunostaining by rhodamine-labeled goat anti-mouse antibody, *blue fluorescence*: cell nuclei by Hoechst 33342 staining) in cardiac myocytes, which indicates the presence and activation of MMP-2 during simulated ischemic stress. Scale bars = 20 μm

For the detection of MMP activity in situ in different cell or tissue cultures, in situ zymography is a suitable method (Fig. 3). In in vivo systems, natural inhibitors of MMPs (tissue inhibitors of MMPs, TIMPs) are presented [29]. Therefore, when a study is designed for detecting MMPs activity in a certain physiological or pathological condition, one should calculate not only with the activation, but the inhibition of MMPs by TIMPs as well. For this reason, reverse zymography has been developed, which allows detection of TIMPs activities in gel zymograms. However, this chapter is limited for showing the opportunities to detect MMP activities in biological samples, thus here we do not provide detailed description on the available techniques for detecting TIMPs activity (for TIMP measurements, *see* for review [30]).

1.3 Types of Zymography Substrates

MMPs cleave different substrates; therefore, the substrate, which is copolymerized in the gel, should be determined according to the MMP isoform.

Gelatin zymography is predominantly used for measurements of activities of MMP-2 and MMP-9 as these two MMPs exert strong ability to cleave gelatin as a substrate and are commonly called "gelatinases."

In casein zymography, casein is copolymerized into the polyacrylamide gel as a substrate for MMP cleavage. Casein zymography is used for estimation of proteolytic activity of MMP-7 due to its ability to cleave casein.

The most rarely applied type of zymography is elastin zymography, in which soluble elastin is copolymerized in the gel and, beside the activity of elastases, it can suitably show elastinolytic activity for such MMPs, which cleave basically other substrates like gelatin (e.g., MMP-2).

2 Materials

2.1 Sample Preparation

1. *For pulverized tissue samples*: Homogenization buffer: 500 mL double distilled water (ddH_2O), 0.335 g (50 mM) Tris base, 1 mL (0.5%) Triton X-100. In 500 mL beaker dissolve compounds in 500 mL ddH_2O. Adjust to pH 7.4 with 1 M HCl. Aliquot into 15 mL Falcon tubes. Store at −20 °C for 1 year. Homogenator: Pellet Pestle Motor, Centrifuge (e.g., Hettich Universal 320R), BCA Protein Assay Kit for protein measurement.
2. *For organ perfusates*: Concentrating tubes: Amicon Ultra-4 30 kDa centrifugal filter unit with Ultracel-30 membrane for perfusate concentration.
3. *For cell culture lysates*: Phosphate Buffered Saline tablets (PBS): In 200 mL beaker dissolve 1 PBS tablet in 100 mL ddH_2O. Adjust

to pH 7.2 with 1 mM NaOH prepared freshly; Homogenization buffer (*see* above **item 1**); Amicon Ultra 10 kDa concentrating tubes. Centrifuge (e.g., Hettich Universal 320R)

2.2 Gelatin Zymography

1. 30% Acrylamide/0.8% Bisacrylamide, Store: at +4 °C for 1 year.
2. Separating Gel Solution—1.5 M Tris–HCl, pH 8.8, Store at +4 °C for 1 year.
3. 2% Gelatin Solution: 100 mg gelatin (type A, from porcine skin; stored between 20–25 °C) dissolved in 4.5 mL ddH_2O.
4. 10% (w/v) SDS (sodium dodecyl sulfate) solution. Weigh 10 g SDS and dissolve in 100 mL ddH_2O (Storage between 20–25 °C for 1 year).
5. 10% (w/v) ammonium persulfate (APS) solution: Dissolve 100 mg of APS in 1 mL ddH_2O. Storage: at +4 °C for 1 month
6. TEMED. Storage: at +4 °C for 1 year.
7. Stacking Gel Solution: 0.5 M Tris–HCl/SDS, pH 6.8. Storage: at +4 °C for 1 year.
8. ELFO Buffer (25 mM Tris–HCl, 192 mM glycine, 0.1% SDS, pH 8.3). Storage: at +4 °C for 1 year. Or self-prepared: 28.83 g glycine, 6.0 g Tris base, 2.0 g SDS. Dissolve Tris base and glycine in 1000 mL of ddH_2O. Bring solution to 1950 mL with ddH_2O. Add SDS. Bring solution to 2000 mL total volume with ddH_2O. Storage: at 4 °C for 1 year.
9. Non-reducing loading (sample) buffers: use commercial "Blue" Zymogram Sample Buffer or commercial "Pink" non-reducing lane marker.
10. For positive control, use "zymography standard," containing a mixture of purified and activated MMP-2 and MMP-9; or MMP-2 standard for gelatin and elastin zymography (Fig. 1a). For casein zymography, human, recombinant active MMP-7 can be used.
11. Renaturation buffer: Renaturation buffer (Bio-Rad). Store: at +4 °C for 1 year. Add 1× mL Renaturation buffer (Bio-Rad) + 9× mL ddH_2O. Storage: prepare freshly.
12. Development buffer: Development buffer (Bio-Rad) Storage: at +4 °C for 1 year. Add 1× mL Development buffer + 9× mL ddH_2O. Storage: prepare freshly.
13. For negative control, use 10 mM ethylene glycol tetraacetic acid (EGTA, binds Ca^{2+} ions, which is obligatory for activity of MMPs). Dissolve 381 mg EGTA in 90 mL ddH_2O + 10 mL development buffer.
14. Coomassie Brilliant Blue (0.05%): 250 mg Coomassie Brilliant Blue G-250, 125 mL methanol, 50 mL glacial acetic acid,

325 mL ddH_2O. Dissolve 250 mg Coomassie Brilliant Blue in the mixture of 125 mL methanol and 50 mL glacial acetic acid and dilute it by adding 325 mL ddH_2O. Storage: between 20–25 °C.

15. Destaining solution: 40 mL methanol, 80 mL acetic acid, 880 mL ddH_2O. Storage: between 20–25 °C

2.3 Casein Zymography

All materials and procedures are identical with gelatin zymography except gelatin solution, which is substituted with casein.

1. Casein solution: 100 mg casein (storage: between 20–25 °C) dissolved in 4.5 mL phosphate buffer.
2. Phosphate buffer (126 mM).

2.4 Elastin Zymography

All materials and procedures are identical with gelatin zymography except gelatin solution, which is substituted with elastin.

1. Elastin solution: 54 mg soluble elastin from bovine neck ligament dissolved in 4.5 mL ddH_2O.

2.5 Preparation of Substrate for Electrophoresis

1. *Gelatin solution*: Add 100 mg gelatin to 4.5 mL ddH_2O. Gently heat and mix solution until it dissolves (beaker will be warm to touch, max. 40 °C). If it has cooled down, add 0.5 mL 10% (w/v) SDS aqueous solution to reach final desired volume and concentration. Prepare freshly (*see* **Note 1**).
2. *Casein solution*: Dissolve 100 mg casein in 4.5 mL phosphate buffer. Stir until casein dissolves. Preparation of phosphate buffer: mix 7.12 g Na_2HPO_4 dissolved in 400 mL ddH_2O and 1.56 g NaH_2PO_4 dissolved in 100 mL ddH_2O, set at pH 7.4. Add 0.5 mL 10% w/v SDS aqueous solution to reach final desired volume. Alternatively, dissolve 30 mg casein in 2 mL of 75 mM Tris–HCl, pH 8.8.
3. *Elastin solution*: Add 54 mg κ-elastin to 4.5 mL ddH_2O. Stir until elastin dissolves. Add 0.5 mL 10% w/v SDS aqueous solution to reach final desired volume.

2.6 In Situ Zymography

The EnzCheck Gelatinase/Collagenase Assay Kit from Invitrogen was used. This assay includes the following reagents:

1. DQ gelatine from pig skin: five vials (1 mg DQ substrate lyophilized from 1 mL of PBS in each vial).
2. 10× Reaction buffer: 50 mL.
3. 1,10-phenanthroline monohydrate: 30 mg powder in a vial.
4. Collagenase type IV from *Clostridium histolyticum*: 500 U collagenase powder in a vial.

3 Methods

3.1 Sample Preparation

3.1.1 Pulverized Tissue Sample

1. Weigh out 30–50 mg heart (lung, pancreas, aorta, spleen) tissue powder into a liquid-nitrogen-frozen 1.5 mL Eppendorf tube. Avoid thawing. It can be stored at −80 °C for 2 years.
2. Thaw an appropriate amount of homogenization buffer (*see* **Note 2**).
3. Add 4× volume homogenization buffer to the sample (e.g., 30 mg sample and 120 μL buffer).
4. Homogenize the mixture by Pellet Pestle Motor for 3 × 10 s (*see* **Note 3**).
5. Centrifuge the homogenate at 4 °C for 10 min at 10,000 × *g*, and collect the supernatant, and store at −80 °C for maximum 1 month.
6. Measure protein concentration by a BCA kit. Usually 20× dilution of tissue homogenates is required.

3.1.2 Preparing Perfusate Samples

1. In case of perfusate sample, use Amicon Ultra 30 kDa concentrating tubes to concentrate 4 × 3 mL perfusate sample to 50–100 μL.
2. Pour 3 mL perfusate in the concentrating insert of an Amicon tube.
3. Put concentrating insert into the tube and close it. Spin samples at 7500 × *g* for 20 min, 4 °C.
4. Remove the concentrating insert and discard the flow-through from the tube. Reinsert the concentrating insert.
5. Pour 3 mL perfusate in the insert, recap, and spin it for 20 min. Repeat **steps 3–4** twice more.
6. Pipette out the concentrated sample from the insert into an Eppendorf tube.
7. Measure protein concentration by BCA kit. Usually 3× dilution of perfusate concentrates is required.

3.1.3 Cell Culture Lysates

1. For cell culturing *see* Ref. 31 (culturing neonatal cardiac myocytes).
2. Remove treating solutions and wash cells in 2 mL PBS two times and then remove PBS.
3. Scrape cells from two wells of a 6-well plate in 200 μL zymography homogenization buffer (two wells together are 400 μL and they are collected into one tube after washing both wells). In case of 25 cm^2 flask use 400 μL buffer; in case of 75 cm^2 flask, use 1 mL buffer.
4. Keep Eppendorf tubes with the suspensions on ice and take into −80 °C freezer or concentrate them freshly.

3.2 Concentration of Samples

1. Keep samples on ice.
2. Homogenize the mixture by ultrasonic homogenizer 2× 5 s on ice.
3. Centrifuge cell homogenates at 5000 × *g* for 10 min at 4 °C.
4. Collect supernatant.
5. Centrifuge the supernatant in Amicon Ultra 10 kDa concentrating tubes to increase the sample protein concentration (4000 × *g* for 30–50 min at −4 °C).
6. Put 50–100 μL samples into −80 °C freezer in two aliquots (one aliquot for determination of protein concentration).
7. Measure protein concentration by BCA kit. Usually 3× dilution of cell concentrates is required.

3.3 Preparation of Separating Gel

1. Assemble electrophoresis unit (*see* **Note 4**).
2. Insert a comb between the gel-casting glasses and mark desired level of separating gel on the front glass (at the bottom of the teeth of the comb).
3. Mix 30% acrylamide/0.8% bisacrylamide solution with Tris–HCl pH 8.8, gelatin solution, and ddH_2O (Table 2; *see* **Notes 5** and **6**).
4. Add 10% APS solution and TEMED to the mix quickly (Table 2).

Table 2
Preparation of separating gel between 0.75–1.5 mm thickness

	Final acrylamide concentration in the separating gel (%)					
			8.0			
Stock solutions[a]	**7.0**	**7.5**	**Small gel (15 mL)**	**Large gel (22.5 mL)**	**9.0**	**10.0**
30% acrylamide/0.8% bisacrylamide (mL)	3.5	3.75	4.0	6	4.5	5.0
1.5 M Tris–HCl, pH 8.8 (mL)	3.75	3.75	3.75	5.62	3.75	3.75
ddH_2O (mL)	6.25	6.0	5.75	8.62	5.25	4.75
Gelatin/casein/elastin solution (20/20/12 mg/mL, 1% w/v SDS; mL)	1.5	1.5	1.5	2.25	1.5	1.5
10% w/v Ammonium Persulfate Solution (APS; μL)	50	50	10	75	50	50
TEMED (μL)	10	10	10	15	10	10

[a] *See* **Notes 8** and **9**

5. Swirl to get homogenous gel. Avoid bubbling. Use immediately as polymerization process has begun (*see* **Note 7**).
6. Using a pipette, pour a small amount into sandwich plates and watch for leakage. In the absence of leakage, continue filling up to 1 mm above line.
7. Gently add butanol (with a 27G-needle connected to a 10-mL syringe) along the surface of the gel to remove bubbles (*see* **Note 8**).
8. Allow gel to polymerize (approximately 20 min at 25 °C; *see* **Note 9**). Use this time to prepare stacking gel (without adding TEMED and 10% APS).
9. A layer of H_2O on the surface of the gel will be visible when polymerization is completed. Drain this layer from the unit with a small stripe of blotting/filter paper.

3.4 Preparation of Stacking Gel

1. Mix 30% acrylamide/0.8% bisacrylamide solution with Tris–HCl, pH 6.8 and ddH_2O (Table 3).
2. Add 10% SDS, 10% APS, and TEMED quickly (Table 3).
3. Swirl to mix. Avoid bubbling. Use immediately as polymerization process has begun.
4. Place comb in units and then use pipettes to pour stacking gel.
5. Allow gels to polymerize (approximately 15 min at 25 °C). Use this time to make sample calculation.

3.5 Sample Calculation

1. According to the results of protein measurement, sample loading mass and volume should be calculated.
2. From a tissue homogenate, 50 μg protein per lane should be loaded. Since we load 15 μL per lane, this means that the final protein concentration of sample needs to be 50 μg/15 μL.
3. In case you want to load a sample only once, it is enough to prepare 2× volume of one load (30 μL), which means that we

Table 3
Preparation of 5 mL stacking gel

Stock solutions	Volume
30% acrylamide/0.8% bisacrylamide	1 mL
0.5 M Tris–HCl pH 6.8	1.25 mL
ddH_2O	3.05 mL
10% SDS	50 μL
10% w/v Ammonium Persulfate Solution (APS)	25 μL
TEMED	8 μL

should add 6 μL (1/5 part) "pink" or 20 μL (2/3 part) "blue" loading buffer.

4. The remaining 24 or 10 μL, respectively, should contain 2 × 50 μg = 100 μg protein. Therefore, volume of the sample will be: V_1 = 100 μg/protein concentration of your sample.
5. Then we should add ddH_2O to dilute samples. The required volume of ddH_2O is: $V_2 = 30 - V_1$.

Taken together:	
Loading volume:	15 μL
Loaded protein:	50 μg
Prepared volume:	30 μL (2× loading volume)
Prepared (sample) mass:	2× loaded protein (100 μg)
"Pink" loading buffer:	1/5 of prepared volume (6 μL) OR
"Blue" loading buffer:	2/3 of prepared volume (20 μL)
Sample volume:	V_1 = 2× loaded protein (2× 50 μg)/ sample protein concentration (μg/μL)
ddH_2O:	V_2 = Prepared volume (30 μL) – V_1

3.6 Sample Loading and Running Gels

1. Prepare and cool down ELFO buffer: 50 mL ELFO + 450 mL ddH_2O (avoid bubbles, mix it gently) (*see* **Note 10**).
2. Mark gel lanes for sample loading. Do not use the two outside lanes.
3. When gel is polymerized, remove combs by pulling it straight up.
4. Remove gel plates and snap onto electrode assembly (*see* **Note 11**).
5. Fill up the lower and the upper buffer container with tank buffer.
6. In case of casein zymography: pre-run electrophoresis at 40 mV for 15 min at 4 °C before the samples are loaded into the wells. Another possibility is to load sample buffer into one well and pre-run electrophoresis at 4 °C until it reaches the bottom of the gel. Afterwards continue with **step 7**, but keep gels at 4 °C (*see* **Note 12**).
7. Load samples

   For identification of different isoforms of detected MMPs, positive controls (e.g., zymography standard containing human MMP-2 and -9 or MMP-2 standard) should be used. Page ruler is also useful to detect the different size of bands. Leave at least one lane loaded with sample(s) for negative control (*see* in Subheading 3.6).

8. Connect electrodes properly (red to red, black to black) and set to 90 V.
9. Let the samples run until loading buffer (running front) reaches the bottom of the gel (*see* **Note 13**).
10. Use this time to prepare renaturation and development buffers.

3.7 Washing and Incubating Gels

1. Set dry incubator at 37 °C.
2. Disassemble the casting apparatus.
3. Cut down lanes for negative control (*see* **Note 14**).
4. Wash gels for 40 min in 200 mL renaturation buffer between 20–25 °C (*see* **Notes 15** and **16**).
5. Place gels into 200 mL freshly prepared development buffer (*see* **Note 16**). For negative control, incubate lane(s) separately in development buffer substituted with 10 mM EGTA solution.
6. Incubate gels in dry incubator at 37 °C for 20–40 h (*see* **Note 17**).

3.8 Staining Gels and Preparation for Evaluation of MMP Activity

1. Immerse gels into 0.05–0.1% Coomassie Brilliant Blue solution. Place on a shaker for 1 h (*see* **Note 18**).
2. Incubate gels in destaining solution and place on a shaker (at least for 60 min; *see* **Notes 19** and **20**).
3. Gelatinolytic activities should be detected as transparent bands against the blue background of Coomassie Brilliant Blue-stained gelatin (*see* Figs. 1–3).
4. Scan the gel in transilluminator mode with a special gel scanner or gel-documentation system.
5. Evaluate MMPs activity by using a gel/film evaluation software (e.g., Quantity one, Bio-Rad).

3.9 In Situ Zymography

The present description provides details for performing in situ zymography for MMP-2 in cardiac myocytes (Fig. 3); however, techniques for showing other MMP activities in different tissue sections or cell cultures in situ also exist and a description for general in situ zymography is available in Ref. 32.

1. Culture neonatal rat cardiac myocytes [31] or other cell types in 24-well tissue culture plate at the density of 10^5 cells/well for 3 days.
2. Replace the growth medium (DMEM—Dulbecco's Modified Eagle Medium—supplemented with Glu, AB/AM, 1% FBS)

with a "stress" solution containing DQ substrate at 40 μg/mL concentration. In case of control group, replace the medium of the cells with a control solution containing DQ substrate at the abovementioned concentration.

3. Subject cells to circumstances according to your aim/project, which may induce intracellular MMP-2 activation.
4. For negative control samples, use ilomastat (at 0.5 μM final concentration), or other non-specific MMP inhibitor (e.g., 1,10-phenanthroline, between 1 and 5 mM final concentration, SB-3CT between 1 and 10 nM final concentration).
5. Subsequently, replace "stress" solution with growth medium containing DQ substrate at 40 μg/mL concentration (250 μL volume must be applied onto the cells).
6. Replace the medium, and wash cells twice with Dulbecco's-PBS.
7. Rinse cells in 3.7% paraformaldehyde in PBS between 20–25 °C for 15 min.
8. Wash cells twice with PBS.
9. Rinse coverslips with mounting medium and view fluorescent signal under fluorescent microscope.
10. If it is necessary, you can combine in situ zymography with immunocytochemistry. In this case, after the fixation you should continue with an appropriate immunostaining protocol.

4 Notes

1. Gelatin is very sensible. Make sure that gelatin dissolves completely (clear, transparent solution without any opalescent particle). Avoid gelatinization: prolonged cooling leads to gel formation. When SDS is added, avoid precipitation. If gelatin precipitates, try to heat again. If precipitates do not disappear from the solution, prepare a new gelatin solution. Precipitated gelatin does not polymerize homogenously in the gel, which may lead to equivocal results.
2. Avoid reducing agents (e.g., dithiothreitol; DTT) or protease inhibitors (e.g., phenylmethanesulfonyl fluoride; PMSF) in the homogenization buffer. They may reduce or inhibit MMP activities, thereby may lead to false results.
3. Avoid ultrasonic homogenizer. It can destroy the native structure of MMPs; therefore, enzymes may lose their activities.
4. 7.5–10% polyacrylamide gels are recommended for gelatin zymography. For casein zymography, we recommend to use

10–15% gels according to the MW of active MMP-7 (19–21 kDa; 28–30 kDa pro-MMP-7).

5. Mix 15.0 mL of separating gel. This volume is sufficient for 1 small unit (containing ten lanes; i.e., two gels), or mix 22.5 mL for a triple-wide unit (containing 30–34 lanes).
6. In order to avoid leakage, ensure that spacers, comb, and glass plates are aligned properly.
7. Make sure that gel is horizontal and there is no difference in the levels of the two edges.
8. Avoid "shooting" of butanol. Butanol should cover the whole surface of the gel uniformly.
9. TIP: leave pipette tip in the remnant of separating gel; when it is polymerized, you will be able to lift it with the pipette.
10. Calculate the necessary volume of ELFO buffer according to the buffer tank.
11. Eliminate bubbles under the gel, they may disturb gel running.
12. Casein migrates in gel during electrophoresis. Due to its low molecular weight (23 kDa), the zone containing casein can obscure MMP-7 (latent form: 29 kDa, active form: 20 kDa) after staining. Therefore, pre-run of casein-embedded gel is recommended before classical zymogram procedure in order to get excess of casein out of the gel. The amount of remaining casein is sufficient for detection of MMP activities [33].
13. Average time for gel running: 1.5–2 h.
14. Avoid gel rupture. Cut the different corner of the gels to be able to identify the gels later (e.g., cut the bottom left corner of gel #1, and both the top and bottom left corners for gel #2). (Ensure that the gel is oriented correctly so that you don't accidentally cut the right side corner.)
15. Before preparation of renaturation and development buffers, make sure that the buffers do not contain any visible contamination (e.g., fungal particles). Buffers should be clear and transparent.
16. Make sure that gels immerse in the buffers and are not attached to the wall of the dish.
17. Even before staining, gelatinolytic activity can be visible. Hold the gel up against a dark background to visualize it, if not, longer incubation should be applied. A pilot zymography is recommended to run for setting up the optimal incubation time. For example, 20-h incubation is recommended in case of human plasma, rat heart, and rodent lung samples, while 40-h incubation is required for mouse heart samples.

18. Staining can be longer, if it is necessary. The gel should be dark blue, protein ladder should be invisible.
19. Use clear dish and change destaining solution after 5–10 min, if it becomes bluish.
20. Leave the gel in destaining solution until the stacking gel becomes completely destained (transparent) again (*see* Figs. 1 and 3). It can last even for 12 h.

Acknowledgments

This work (P. Bencsik, A. Görbe) was supported by the János Bolyai Research Scholarship of the Hungarian Academy of Sciences and has received funding from the European Union's Horizon 2020 research and innovation programme under grant agreement No. 698297 Acronym: Infarnosys. P. Ferdinandy was a Szentágothai Fellow of the National Excellence Program of Hungary (TÁMOP-4.2.4.A/2-11/1-2012-0001).

References

1. Dormán G, Cseh S, Hajdú I, Barna L, Kónya D, Kupai K et al (2010) Matrix metalloproteinase inhibitors: a critical appraisal of design principles and proposed therapeutic utility. Drugs 70(8):949–964
2. Kupai K, Szucs G, Cseh S, Hajdu I, Csonka C, Csont T et al (2010) Matrix metalloproteinase activity assays: importance of zymography. J Pharmacol Toxicol Methods 61(2):205–209
3. Apte SS, Parks WC (2015) Metalloproteinases: a parade of functions in matrix biology and an outlook for the future. Matrix Biol 44–46:1–6
4. DeCoux A, Lindsey ML, Villarreal F, Garcia RA, Schulz R (2014) Myocardial matrix metalloproteinase-2: inside out and upside down. J Mol Cell Cardiol 77:64–72
5. Nagase H, Woessner JF Jr (1999) Matrix metalloproteinases. J Biol Chem 274:21491–21494
6. Massova I, Kotra LP, Fridman R, Mobashery S (1998) Matrix metalloproteinases: structures, evolution, and diversification. FASEB J 12(12):1075–1095
7. Wang W, Schulze CJ, Suarez-Pinzon WL, Dyck JR, Sawicki G, Schulz R (2002) Intracellular action of matrix metalloproteinase-2 accounts for acute myocardial ischemia and reperfusion injury. Circulation 106:1543–1549
8. Gao CQ, Sawicki G, Suarez-Pinzon WL, Csont T, Wozniak M, Ferdinandy P et al (2003) Matrix metalloproteinase-2 mediates cytokine-induced myocardial contractile dysfunction. Cardiovasc Res 57(2):426–433
9. Spinale FG (2002) Matrix metalloproteinases: regulation and dysregulation in the failing heart. Circ Res 90(5):520–530
10. Gruber BL, Sorbi D, French DL, Marchese MJ, Nuovo GJ, Kew RR et al (1996) Markedly elevated serum MMP-9 (gelatinase B) levels in rheumatoid arthritis: a potentially useful laboratory marker. Clin Immunol Immunopathol 78(2):161–171
11. Mukherjee A, Swarnakar S (2015) Implication of matrix metalloproteinases in regulating neuronal disorder. Mol Biol Rep 42(1):1–11
12. Li M, Yamamoto H, Adachi Y, Maruyama Y, Shinomura Y (2006) Role of matrix metalloproteinase-7 (matrilysin) in human cancer invasion, apoptosis, growth, and angiogenesis. Exp Biol Med (Maywood) 231(1):20–27
13. Kurzepa J, Mądro A, Czechowska G, Kurzepa J, Celiński K, Kazmierak W et al (2014) Role of MMP-2 and MMP-9 and their natural inhibitors in liver fibrosis, chronic pancreatitis and non-specific inflammatory bowel diseases. Hepatobiliary Pancreat Dis Int 13(6):570–579
14. Craig VJ, Zhang L, Hagood JS, Owen CA (2015) Matrix metalloproteinases as therapeutic targets for idiopathic pulmonary fibrosis. Am J Respir Cell Mol Biol 53(5):585–600
15. Meschiari CA, Marcaccini AM, Santos Moura BC, Zuardi LR, Tanus-Santos JE, Gerlach RF

(2013) Salivary MMPs, TIMPs, and MPO levels in periodontal disease patients and controls. Clin Chim Acta 421:140–146

16. Ivanova M, Dovinova I, Okruhlicova L, Tribulova N, Simoncikova P, Bartekova M et al (2012) Chronic cardiotoxicity of doxorubicin involves activation of myocardial and circulating matrix metalloproteinases in rats. Acta Pharmacol Sin 33(4):459–469
17. Bartekova M, Simoncikova P, Fogarassyova M, Ivanova M, Okruhlicova L, Tribulova N et al (2015) Quercetin improves postischemic recovery of heart function in doxorubicin-treated rats and prevents doxorubicin-induced matrix metalloproteinase-2 activation and apoptosis induction. Int J Mol Sci 16(4):8168–8185
18. Barancik M, Okruhlicova L, Fogarassyova M, Bartekova M, Slezak J (2013) Mediastinal irradiation modulates myocardial and circulating matrix metalloproteinases. Exp Clin Cardiol 18S:37A–40A
19. Cheung PY, Sawicki G, Wozniak M, Wang W, Radomski MW, Schulz R (2000) Matrix metalloproteinase-2 contributes to ischemia-reperfusion injury in the heart. Circulation 101(15):1833–1839
20. Lalu MM, Csonka C, Giricz Z, Csont T, Schulz R, Ferdinandy P (2002) Preconditioning decreases ischemia/reperfusion-induced release and activation of matrix metalloproteinase-2. Biochem Biophys Res Commun 296(4):937–941
21. Bencsik P, Kupai K, Giricz Z, Görbe A, Pipis J, Murlasits Z et al (2010) Role of iNOS and peroxynitrite – matrix metalloproteinase-2 signaling in myocardial late preconditioning in rats. Am J Physiol Heart Circ Physiol 299(2):512
22. Bencsik P, Paloczi J, Kocsis GF, Pipis J, Belecz I, Varga ZV et al (2014) Moderate inhibition of myocardial matrix metalloproteinase-2 by ilomastat is cardioprotective. Pharmacol Res 80:36–42
23. Bencsik P, Sasi V, Kiss K, Kupai K, Kolossváry M, Maurovich-Horvat P et al (2015) Serum lipids and cardiac function correlate with nitrotyrosine and MMP activity in coronary artery disease patients. Eur J Clin Investig 45(7):692–701
24. Bautista-López NL, Schulz R (2014) Matrix metalloproteinases 2 and 9 as diagnostic tools in Chagas cardiomyopathy. Int J Cardiol 177(1):46–47
25. Zhang Y, Lin P, Jiang H, Xu J, Luo S, Mo J et al (2015) Extensive serum biomarker analysis in patients with ST segment elevation myocardial infarction (STEMI). Cytokine 76(2):356–362
26. Mook OR, Frederiks WM, Van Noorden CJ (2004) The role of gelatinases in colorectal cancer progression and metastasis. Biochim Biophys Acta 1705(2):69–89
27. Annaházi A, Molnár T, Farkas K, Rosztóczy A, Izbéki F, Gecse K et al (2013) Fecal MMP-9: a new noninvasive differential diagnostic and activity marker in ulcerative colitis. Inflamm Bowel Dis 19(2):316–320
28. Roomi MW, Kalinovsky T, Rath M, Niedzwiecki A (2014) Effect of a nutrient mixture on matrix metalloproteinase-9 dimers in various human cancer cell lines. Int J Oncol 44(3):986–992
29. Arpino V, Brock M, Gill SE (2015) The role of TIMPs in regulation of extracellular matrix proteolysis. Matrix Biol 44–46:247–254
30. Troeberg L, Nagase H (2007) Analysis of TIMP expression and activity. Methods Mol Med 135:251–267
31. Gorbe A, Giricz Z, Szunyog A, Csont T, Burley DS, Baxter GF et al (2010) Role of cGMP-PKG signaling in the protection of neonatal rat cardiac myocytes subjected to simulated ischemia/reoxygenation. Basic Res Cardiol 105(5):643–650
32. Yan SJ, Blomme EA (2003) In situ zymography: a molecular pathology technique to localize endogenous protease activity in tissue sections. Vet Pathol 40(3):227–236
33. Fernandez-Resa P, Mira E, Quesada AR (1995) Enhanced detection of casein zymography of matrix metalloproteinases. Anal Biochem 224(1):434–435

Chapter 7

Characterization of Novel Collagenolytic Proteases

Goran Mucić, Brankica Rašković, and Natalija Polović

Abstract

Collagenolytic proteases have many potential applications in different areas of science, industry, and medicine. The determination of the activity of such proteins is paramount to their application. Here, we describe methods which can be applied to determine the activity and some basic characteristics of potential collagenases.

Key words Collagenase, Collagen, Caseinolytic activity, Collagenolytic activity, BAPNA assay

1 Introduction

Proteases which cleave the triple helix of native collagen are called collagenases. Mammalian collagenases can be divided into three subgroups, metallocollagenases (matrix metalloproteinases), serine collagenases, and cysteine collagenases (e.g., human cathepsins), while collagenases from different sources are found to belong to different groups such as plant cysteine proteases, invertebrate serine and cysteine proteases, and bacterial metallocollagenases. Metallocollagenases contain zinc in their active site and sometimes require calcium for stability and optimum activity [1, 2]. Many studies have been conducted regarding metallocollagenases from mammalian and bacterial sources. Serine collagenases are found in various marine species; they, like mammalian and bacterial metalloproteases, possess collagenolytic activity; however, they too possess a broad proteolytic activity [3].

Enzymes, such as collagenases, have many applications like meat tenderizers, isolation and cultivation of mammalian cells, and elimination of scar tissue; products of their enzymatic hydrolysis of collagen can be used in lotions, as dietary supplements, etc. Considering various applications of collagenases, it is natural that there is a high interest in such proteins [4–20]. Detection and characterization of these proteins is important not only for research purposes, but for medical reasons as well. Matrix metalloproteinases are commonly detected via zymography in clinical practice since these enzymes are involved in tumor invasion and metastasis [5].

Jeff Wilkesman and Liliana Kurz (eds.), *Zymography: Methods and Protocols*, Methods in Molecular Biology, vol. 1626,
DOI 10.1007/978-1-4939-7111-4_7, © Springer Science+Business Media LLC 2017

2 Materials

2.1 SDS-PAGE and Gelatin SDS-PAGE

1. Monomer solution: 30% acrylamide, 0.8% bisacrylamide. Dissolve 58.4 g of acrylamide and 1.6 g of bisacrylamide in 100–120 mL of distilled water and, after dissolving, add distilled water to 200 mL. Filter the solution through filter paper and store at 4 °C.
2. Running gel buffer: 1.5 M Tris–HCl buffer pH 8.8. Dissolve 36.3 g of Tris in 150 mL of distilled water. Use 4 M HCl to bring pH value to 8.8 and add distilled water to 200 mL. Store at 4 °C.
3. Stacking gel buffer: 0.5 M Tris–HCl buffer pH 6.8. Dissolve 6.0 g of Tris in 80 mL of distilled water. Use 4 M HCl to bring pH value to 6.8 and add distilled water to 100 mL. Store at 4 °C.
4. SDS solution: 10% (w/v) solution in water. Dissolve 10.0 g of SDS in a total of 100 mL distilled water. Store at room temperature.
5. Ammonium persulfate: 10% (w/v) solution in water. Dissolve 0.1 g of APS in a total of 1 mL distilled water (*see* **Note 1**).
6. *N*,*N*,*N*′,*N*′-Tetramethyl-ethylenediamine (TEMED).
7. Solution for overlaying running gel during polymerization: n-butanol saturated with distilled water. Add distilled water to 100 mL of n-butanol until stable layer of water is formed (*see* **Note 2**).
8. SDS-PAGE buffer: 0.025 M Tris–HCl, 0.192 M glycine, 0.1% SDS, pH 8.3. Dissolve 3.0 g of Tris, 14.4 g of glycine, and 1.0 g of SDS in a total of 1000 mL distilled water. No need for pH adjustment.
9. Solution buffer (5× concentrate): 60 mM Tris buffer pH 6.8, 25% glycerol, 2% SDS, 14.4 mM 2-mercaptoethanol, and 0.1% bromphenol blue. Mix 1.2 mL 0.5 M Tris buffer pH 6.8, 5.0 mL glycerol, 2 mL 10% SDS, 0.5 mL 2-mercaptoethanol (*see* **Note 3**), 1.0 mL 1% bromphenol blue, and add water to 10 mL. Store at −20 °C.
10. Staining solution: 0.1% CBB R-250, 50% methanol, 10% acetic acid. Dissolve 0.5 g of CBB R-250 in 100 mL of methanol until it is completely without particles. Add 10 mL of acetic acid and water to 200 mL.
11. Destaining solution: 50% methanol, 10% acetic acid. Mix 500 mL methanol with 100 mL acetic acid and add water to 1000 mL.
12. Gel storage solution: 7% acetic acid. 7 mL of acetic acid in a total of 100 mL solution.

2.2 Collagenolytic and Gelatinolytic Assay (Ninhydrin Method)

1. The ninhydrin reagent is made by mixing equal volumes of ninhydrin solution [4% (w/v) ninhydrin in propylene glycol] and 200 mM citrate buffer, 0.16% (w/v) $SnCl_2$, pH 5.0. Incubate 10 min at 100 °C, after which cool to room temperature and dilute with 400 μL 50% (v/v) 1-propanol.

2.3 Effect of pH on Collagenolytic Activity—Ninhydrin Method

For both the ninhydrin method and the gelatin SDS-PAGE, prepare the following buffers:

1. 50 mM Na citrate buffer pH 3.0–6.0.
2. 50 mM Na phosphate buffer pH 6.5 and 7.0.
3. 50 mM Tris–HCl buffer pH 8.0 and 9.0.
4. 50 mM Glycine buffer pH 9.5.

3 Methods

3.1 PAGE and Zymography

1. Prepare the samples by mixing with the adequate volume of 5× solution buffer and incubating at 95 °C for 5–10 min. For gelatin SDS-PAGE incubate the samples mixed with 5× solution buffer at 60 °C for 30 min.
2. Prepare running and stacking gel solutions by mixing previously prepared solutions according to Tables 1 and 2, respectively (*see* **Note 4**).
3. Cast running gel within a gel cassette. Allow space for stacking gel and overlay immediately with n-butanol solution.
4. Polymerization of the gel is usually completed after 20–30 min. Pour off n-butanol layer and wash the surface of the gel 3× with distilled water.

Table 1
Solution volumes for 10 mL of running gel

Solution	7.5% running gel	10% running gel	12% running gel
Monomer solution (mL)	2.50	3.33	4.00
Running gel buffer (mL)	2.50	2.50	2.50
[a]Distilled water (mL)	4.84	4.00	3.33
TEMED (μL)	5	5	5
SDS solution (μL)	100	100	100
APS solution(μL)	50	50	50

[a]For gelatin SDS-PAGE add 0.3% gelatin solution instead of water

Table 2
Solution volumes for 4% stacking gel

Solution	5 mL	10 mL
Monomer solution (mL)	0.67	1.33
Stacking gel buffer (mL)	1.25	2.50
Distilled water (mL)	3.00	6.00
TEMED (μL)	2.5	5
SDS (μL)	50	100
APS (μL)	25	50

5. Add stacking gel solution. Place the well-forming comb into the stacking gel. Polymerization is usually completed after 20–30 min. Remove the comb and wash the wells 3× with distilled water.
6. Fill the upper and bottom tank with SDS-PAGE buffer and load samples into the wells.
7. Connect the apparatus to the power pack and set to 80 V at the beginning until samples concentrate and reach the running gel. Then increase the voltage to 150–180 V and run until bromophenol blue dye reaches the end of the gel.
8. For the SDS-PAGE, separate the cassette and place the running gel into the staining solution for constant mixing for 30 min [21].
9. For the gelatin SDS-PAGE, separate the cassette and place the running gel into distilled water. Wash the gel for 30 min (3× 10 min) to remove SDS. Place the gel into buffer with optimal pH for collagenase activity and incubate overnight at room temperature. Stain the gel with staining solution for 30 min [22].
10. Destain the gels with destaining solution until clearly visible blue bands appear on transparent background.
11. Gel can be stored in 7% acetic acid solution.

3.2 Collagenolytic and Gelatinolytic Assay (Ninhydrin Method)

1. Slowly dissolve collagen or gelatin in 50 mM TBS pH 8.1 (or buffer with pH optimum for collagenase activity) at 37 °C for 15 min to a concentration of 5 and 20 mg/mL respectively.
2. Add 10 μL of the sample to 100 μL of the substrate (collagen or gelatin) and incubate for 5 h at 37 °C with gentle shaking (*see* **Note 5**).
3. Add an equal volume (110 μL) of cold 20% (w/v) solution of PEG 6600.
4. Incubate with 110 μL 20% (w/v) PEG 6600 at 4 °C for 1 h.

5. Centrifuge at 12,000 × *g* for 30 min at 4 °C.
6. Add 20 μL of the supernatant to 200 μL ninhydrin reagent (*see* **Note 6**).
7. Measure the absorbance at 570 nm, compare it with the absorbance of the blank (ninhydrin reagent and buffer). One arbitrary unit of collagen or gelatin digestion activity (CDU and GDU, respectively) was defined as the amount of enzyme that releases peptides from collagen or gelatin equivalent in ninhydrin color to 1 mmol of leucine in 5 h. Calibration curve with different concentrations of leucine needs to be constructed [23].

3.3 Effect of Inhibitors and Metal Ions on Collagenolytic Activity

1. Prepare the inhibitor and metal ion solutions in adequate concentration in order to fully inhibit sample collagenase.
2. Preincubate the sample with the various inhibitors or metal ions at room temperature for 30 min.
3. To test the effect of the inhibitors and metal ions on enzymatic activity, run the preincubated samples on gelatin SDS-PAGE.
4. After electrophoresis slice the gel into strips containing each sample and then incubate the slice in a buffer containing the respective inhibitor or ion overnight.
5. After incubation stain the gel strips with staining solution (Subheading 2.1). The inhibition is expressed as a percentage of intensity of the proteolytic bands in the presence of inhibitor compared to a control [24].

3.4 Collagen Cleavage Assay

1. Add 15 μL of 0.1 mg/mL sample to 60 μL of 0.5 mg/mL collagen in buffer with pH optimum for collagenase activity and incubate for an adequate period of time at 37 °C (*see* **Note 5**).
2. Prepare similar reaction mixtures and preincubate with adequate concentration of inhibitors.
3. Run the samples on 7.5% SDS-PAGE gel. Observe collagen degradation products [2].

3.5 Effect of pH on Collagenolytic Activity—Ninhydrin Method

1. Prepare 2% gelatin in the various buffers.
2. Incubate the sample in the various 2% gelatin solutions for 5 h, after which determine the enzyme activity as described in Subheading 2.2 (starting from **item 3**).

3.6 Effect of pH on Collagenolytic Activity—Gelatin SDS-PAGE

1. Resolve collagenase on gelatin SDS-PAGE as described in Subheading 2.1.
2. After electrophoresis (and after washing the gel 3× with distilled water) slice the gel into strips containing sample triplicates and then incubate each strip in a previously prepared buffers of various pH values overnight.

3. After incubation stain the gel strips with CBB. The relative activity is expressed as a percentage of intensity of the most intensive proteolytic bands.

3.7 Effect of pH on Collagenolytic Stability—Ninhydrin Method

1. Prepare the following buffers: 50 mM Na citrate buffer pH 3.0–6.0, 50 mM Na phosphate buffer pH 6.5 and 7.0, 50 mM Tris–HCl buffer pH 8.0 and 9.0 and 50 mM glycine buffer pH 9.5.
2. Preincubate collagenase in previously prepared buffers of various pH values for 30 min.
3. Prepare 2% gelatin solution in 0.5 M TBS pH 8.1 (or buffer with pH optimum for collagenase activity).
4. Determine the enzyme activity as described in Subheading 2.2 (starting from the **item 2**).

3.8 Effect of pH on Collagenolytic Stability—Gelatin SDS-PAGE

1. Prepare the following buffers: 50 mM Na citrate buffer pH 3.0–6.0, 50 mM Na phosphate buffer pH 6.5 and 7.0, 50 mM Tris–HCl buffer pH 8.0 and 9.0 and 50 mM Glycine buffer pH 9.5.
2. Preincubate collagenase in previously prepared buffers of various pH values for 30 min.
3. Resolve collagenase on gelatin SDS-PAGE as described in Subheading 2.1.
4. After electrophoresis (and after washing the gel 3× with distilled water) incubate the gel in the buffer optimal for collagenase activity overnight.
5. After incubation stain the gel with CBB. The relative activity is expressed as a percentage of intensity of the most intensive proteolytic bands.

3.9 Effect of Temperature on Collagenolytic Activity

1. To determine the optimal temperature for enzyme activity, incubate the reaction mixtures at various temperatures (20–80 °C) for 10 min.
2. Determine the enzyme activity as described in Subheading 2.2.
3. The residual activity was expressed as a percentage of maximal activity.

3.10 Effect of Temperature on Collagenolytic Stability

1. To determine the thermostability of the enzyme, preincubate the reaction mixture at a given temperature between 20 and 90 °C for 30 min, and then determine the enzyme activity at 37 °C after incubation as described in Subheading 2.2.
2. The residual activity was calculated as the ratio between the enzymatic activity, observed at the end of each incubation run, and that at the beginning, and expressed as a percentage of maximal activity [24].

4 Notes

1. It is recommended to prepare this solution immediately before usage.
2. Mix the solution before usage and wait for the layers to separate. Use upper layer.
3. For electrophoresis under nonreducing conditions add equal volume of water instead of 2-mercaptoethanol.
4. Solution of APS needs to be added immediately before casting the gel.
5. The amounts of collagenase and substrate need to be optimized.
6. Ninhydrin reagent needs to be made immediately before usage.

References

1. Birkedal-Hansen H (1995) Proteolytic remodeling of extracellular matrix. Cell Biol 7:728–735
2. Kim M, Hamilton SE, Guddat LW, Overall CM (2007) Plant collagenase L unique collagenolytic activity of cysteine proteases from ginger. Biochim Biophys Acta 1770:1627–1635
3. Spiriti J, Van der Vaart A (2010) Mechanism of the calcium-induced trans-cis isomerization of a non-prolyl peptide bond in *Clostridium histolyticum* collagenase. Biochemistry 49:5314–5320
4. Daboor SM, Ghaly AE, Brooks SL, Dave D (2010) Extraction and purifiacation of collagenase enzymes: a critical revies. Am J Biochem Biotechnol 6:239–263
5. Hu X, Beeton C (2010) Detection of functional matrix metalloproteinases by zymography. J Vis Exp (45):1–4
6. Grant GA, Eisen AZ (1980) Substrate specificity of the collagenolytic serine protease from *Uca pugilator*: studies with noncollagenous substrates. Biochemistry 19:6089–6095
7. Van Wormhoudt A, Le Chevalier P, Sellos D (1992) Purification, biochemical characterization and N-terminal sequence of a serine-protease with chymotrypsic and collagenolytic activities in a tropical shrimp, *Penaeus vannamei* (Crustacea, Decapoda). Comp Biochem Physiol B 103:675–680
8. Foegeding EA, Laric DK (1986) Tenderization of beef with bacterial collagenase. Meat Sci 18:201–214
9. Kanth SV, Venba R, Madhan B, Chandrababu NK, Sadulla S (2008) Studies on the influence of bacterial collagenase in leather dyeing. Dyes Pigments 76:338–347
10. Takashima A (2001) Establishment of fibroblast cultures. Curr Protoc Cell Biol. Chapter 2, Unit2.1:2.1.1–2.1.12
11. Mazda T, Makino K, Yabe R, Nakata K, Fujisawa K, Ohshima H (1995) Use of standardized protease enzymes for antibody screenin of blood donor samples with the microplate system AutoAnalyzer. Transfus Med 5:45–50
12. Hansbrough JF, Achauer B, Dawson J, Himel H, Luterman A, Slater H, Levenson S, Salzberg CA, Hansbrough WB, Dore C (1995) Wound healing partial-thickness burn wounds treated with collagenase ointment versus silber sulfadizine cream. J Burn Care Rehabil 16:241–247
13. Tallis A, Motley TA, Wunderlich RP, Dickerson JE Jr, Waycaster C, Slade HB (2013) Clinical and economic assessment of diabetic foot ulcer debridement with collagenase: results of a randomized controlled study. Clin Ther 35:1805–1820
14. Sgniukiv AM, Rudenskaya GN, Isaev VA, Baydakov AV, Zhantlev RD, Korsunovskaya OS, Ageeva LV, Starikova NV (2006) A comparative study of collagenase complex and new homogeneous collagenase preparations for scar treatment. J Drug Deliv Sci Technol 16:285–292
15. Yagishita D, Ajijola OA, Vaseghi M, Nsair A, Zhou W, Yamakawa K, Tung R, Mahajan A, Shivkumar K (2013) Electrical homogenization of ventricular scar by application of collagenase: a novel strategy for arrhythmia therapy. Circ Arrhythm Electrophysiol 6:776–783
16. Kin T, Johnson PR, Shapiro AM, Lakey JR (2007) Factors influencing the collagenase digestion phase of human islet isolation. Transplantation 83:7–12

17. McCarthy RC, Breite AG, Green ML, Dwulet FE (2011) Tissue dissociation enzymes for isolating human islets for transplantation: factors to consider in setting enzyme acceptance criteria. Transplantation 91:137–145
18. Gelbard M, Hellstrom WJ, McMahon CG, Levine LA, Smith T, Tursi J, Kaufman G, Goldstein I (2013) Baseline characteristics from an ongoing phase 3 study of collagenase Clostridium histolyticum in patients with Peyronie's disease. J Sex Med 10:28222831
19. Jin B, Alter HJ, Zhang ZC, Shih JW, Esteban JM, Sun T, Yang YS, Qiu Q, Liu XL, Yao L, Wang HD, Cheng LF (2005) Reversibility of experimental rabbit liver cirrhosis by portal collagenase administration. Lab Investig 85:992–1002
20. Kim SK, Mendis E (2006) Bioactive compounds from marine processing byproducts a review. Food Res Int 39:383–393
21. Laemmli UK (1970) Cleavage of structural proteins during the assembly of the head of bacteriophage T4. Nature 227:680–685
22. Gavrovic-Jankulovic M, Polovic N, Prisic S, Jankov R, Atanaskovic-Markovic M, Vuckovic O, Cirkovic Velickovic T (2005) Allergenic potency of kiwi fruit during fruit development. Food Agric Immunol 16:117–128
23. Zhang Y, Fun Y, Zhou S, Kang L, Li C (2013) A straightforward ninhydrinbased method for collagenase activity and inhibitor screening of collagenase using spectrophotometry. Anal Biochem 437:46–48
24. Raskovic B, Bozovic O, Prodanovic R, Niketic V, Polovic N (2014) Identification, purification and characterization of a novel collagenolytic serine protease from fig (*Ficus carica* var Brown Turkey) latex. J Biosci Bioeng 118:622–627

Chapter 8

Zymography as a Research Tool in the Study of Matrix Metalloproteinase Inhibitors

Zongli Ren, Juanjuan Chen, and Raouf A. Khalil

Abstract

Matrix metalloproteinases (MMPs) are proteolytic enzymes that degrade various components of the extracellular matrix (ECM) and play a role in tissue remodeling. Changes in MMPs have been observed in cancer, connective tissue disorders, and vascular disease, and both endogenous tissue inhibitors of MMPs (TIMPs) and synthetic MMP inhibitors (MMPIs) have been evaluated as modulators of MMP activity in various biological systems. Zymography is a simple technique that is commonly used to assess MMP activity and the efficacy of MMPIs. Also, reverse zymography is a modified technique to study the activity of endogenous TIMPs. However, problems are often encountered during the zymography procedure, which could interfere with accurate assessment of MMP activity in control specimens, and thus make it difficult to determine the pathological changes in MMPs and their responsiveness to MMPIs. Simplified protocols for preparation of experimental solutions, tissue preparation, regular and reverse zymography procedures, and zymogram analysis are presented. Additional helpful tips to troubleshoot problems in the zymography technique and to enhance the quality of the zymograms should make it more feasible to determine the changes in MMPs and assess the efficacy of MMPIs in modulating MMP activity in various biological systems and pathological conditions.

Key words Blood vessels, Uterus, Extracellular matrix, Matrix metalloproteinase, TIMP

Abbreviations

A/C	Acrylamide/bis-acrylamide
ECM	Extracellular matrix
MMP	Matrix metalloproteinase
MMPI	MMP inhibitor
MT-MMP	Membrane-type-MMP
RUPP	Reduction in uterine perfusion pressure
TIMP	Tissue inhibitor of MMP
Zn^{2+}	Zinc ion

Jeff Wilkesman and Liliana Kurz (eds.), *Zymography: Methods and Protocols*, Methods in Molecular Biology, vol. 1626, DOI 10.1007/978-1-4939-7111-4_8, © Springer Science+Business Media LLC 2017

1 Introduction

Matrix metalloproteinases (MMPs) are structurally and functionally related Ca^{2+}-dependent and Zn^{2+}-containing endopeptidases that degrade the extracellular matrix (ECM) and connective tissue proteins [1–3]. Since its discovery, the MMP family has grown to 28 members in vertebrates, at least 23 in humans and 14 in blood vessels [2]. Based on their activity toward specific substrates and the organization of their different domains, MMPs are classified into collagenases, gelatinases, stromelysins, matrilysins, membrane-type (MT)-MMPs, and other MMPs [4–6]. MMPs are secreted as inactive or latent pro-MMPs which undergo proteolytic activation by other MMPs or other proteases before they can degrade ECM proteins [7] (Fig. 1). Activated MMPs play a role in vascular remodeling and angiogenesis [1], the uterine and vascular changes associated with pregnancy and preeclampsia [8], and many pathological conditions such as neoplasm, connective tissue disorders, and vascular disease.

Because there are no specific activators of MMPs, MMP inhibitors (MMPIs) are often used to test the role of MMPs in different processes. Also, in many tissues, MMP activity is modulated by endogenous tissue inhibitors of metalloproteinases (TIMPs) [9, 10] (Fig. 1). TIMPs include four homologous members; TIMP-1, -2, -3, and -4 [11–13]. While all TIMPs can inhibit all MMPs, the efficacy of MMP inhibition varies with each TIMP. For example, TIMP-1 is a poor inhibitor of membrane-type 1-MMP (MT1-MMP), MT3-MMP, MT5-MMP, and MMP-19 [12]. Also, TIMP-1 and -3 interact with pro-MMP-9, while TIMP-2, -3, and -4 interact with pro-MMP-2 [12]. Although TIMPs could restrict ECM deposition, their ultimate effect on ECM turnover depends on the TIMP/MMP ratio in the tissue. Other endogenous MMPIs include α2-macroglobulin, a glycoprotein consisting of four identical subunits that is found in blood and tissue fluids and acts as a general proteinase inhibitor. Most endopepidases are inhibited by being entrapped within the macroglobulin and the complex is then cleared by endocytosis via a low-density lipoprotein receptor-related protein-1 [14].

In addition to endogenous MMPIs, several classes of synthetic MMPIs have been developed and evaluated as diagnostic and therapeutic tools in cancer, autoimmune, and vascular disease [15] (Table 1). Early investigations on MMPIs focused on developing compounds that contain a group that chelates the catalytic Zn^{2+} ion and a backbone that mimics the natural peptide substrate of MMPs [16]. The first-generation hydroxamate-based MMPIs have the Zn^{2+}-binding group hydroxamate and the basic backbone of collagen [17]. Batimastat, a low molecular mass hydroxamate derivative, was the first MMPI to enter clinical trials [18], but the results were

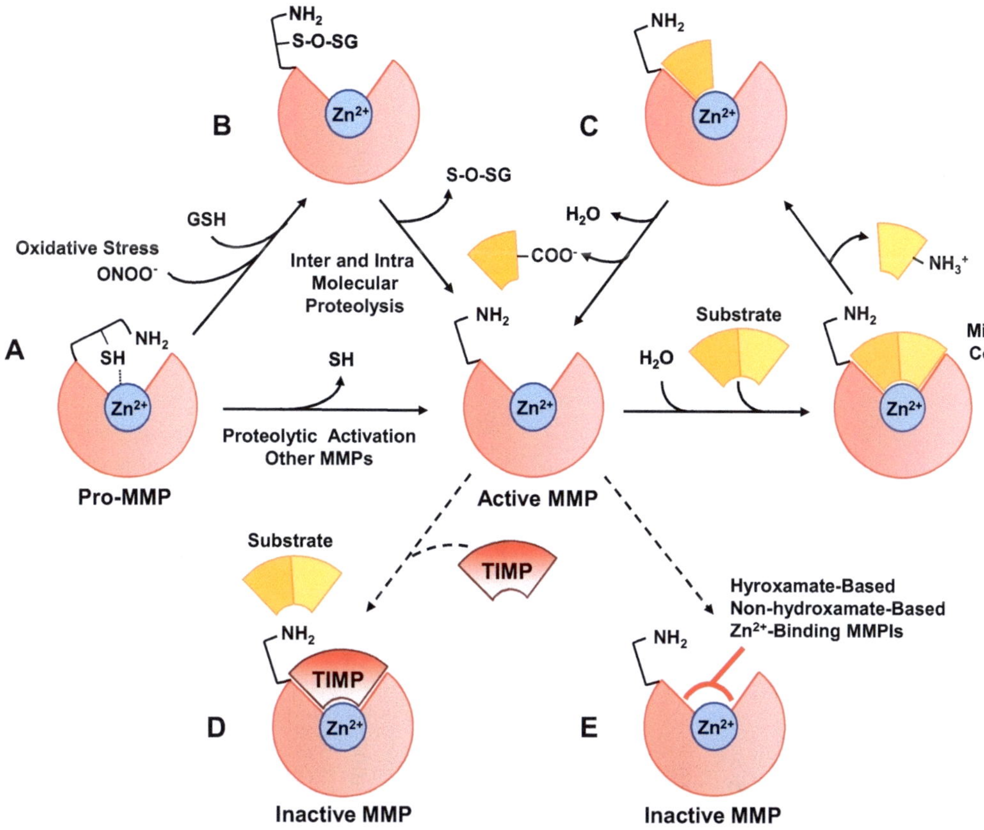

Fig. 1 Mechanisms of MMP activation, MMP–substrate interaction, and MMP inhibition. Full-length pro-MMP can be activated in two ways. (**a**) Proteolytic activation of MMPs by MT-MMP/TIMP or other proteases occurs by removal of the cysteine switch motif-SH autoinhibitory propeptide region resulting in a truncated active MMP. (**b**) In the presence of oxidative stress, reactive O_2 species such as peroxinitrite ($ONOO^-$) and cellular glutathione (GSH), the critical cysteine residue in the propeptide region, undergoes S-glutathiolation, leading to the release of cysteine binding to the catalytic Zn^{2+} ion and active enzyme. (**c**) Active MMP interacts with its substrate through a series of biochemical reactions. Using H^+ from free H_2O, the substrate carbonyl binds to Zn^{2+}, forming a Michaelis complex. The Zn^{2+}-bound H_2O performs a nucleophilic attack on the substrate, resulting in the release of an H_2O molecule, breakdown of the substrate, and the release of active MMP to be ready for interaction with another substrate. (**d**) TIMP interacts with MMP in a manner similar to that of substrate substituent, further contributing to expelling Zn^{2+}-bound H_2O and preventing substrate degradation. (**e**) Zn^{2+}-binding MMPIs act as anchor that is locked in the active site and direct the backbone of the inhibitor into the target substrate-binding pockets resulting in inactive MMP. *Dashed lines* indicate inhibition

disappointing due to the metabolically labile nature of the hydroxamate Zn^{2+}-binding group, the metabolic inactivation and chelation of metal ions of other metalloproteins, and the serious side effects such as musculoskeletal pain experienced by patients [19].

Non-hydroxamate-based Zn^{2+}-binding MMPIs such as carboxylates, hydrocarboxylates, sulphydryls, phosphoric acid derivatives, and hydantoins are more stable and do not have the limitations associated with hydoxamate-based MMPIs. Rebimastat is a broad-spectrum MMPI that has a thiol Zn^{2+}-binding group

Table 1
Representative synthetic MMPIs and their IC_{50} or K_i toward specific MMPs

MMPI[a] (other name)	MMP specificity (IC_{50} or K_i) (nM)										
	MMP-1	2	3	7	8	9	10	11	13	14	16
Hydroxamate-based Batimastat (BB-94)	3	4	20	6	10	1					
Marimastat (BB-2516)	5	6	200	20	2	3				1.8	
Ilomastat (GM6001)	0.4	0.4	27		0.1	0.2				5.2	
CGS-27023-A (MMI-270)	33	20	43		8	8			6		
MMI-166		0.4			400	90				100	
ABT-770	4600	3.7	42	>10,000		120					
PD-166793	6.1	47	12	7.2		7.9			8	240	
Prinomastat (AG3340)	8.3	0.05	0.3	54		0.26			0.03	0.33	
Cipemastat (Ro 32-3555)	3.0	154			4.4	59			3.4		
Non-hydroxamate Rebimastat (BMS-275291)	9	39	157	23		27				40	

Tanomastat (BAY 12-9566)		11	134			301			1470		
Mechanism-Based SB-3CT	206	14	15	96		600					
Pyrimidine-based Ro 28-2653	16,000	7–246	1200		15	12–23				96	91
Phosphorous-based RXP03	2000	55		20,000	4	41	45	6.2	16	90	
Tetracyclines Doxycycline	>400,000	56,000	32,000	28,000	26,000–50,000				2000–50,000		
Metastat (COL-3;CMT-3)	34 μg/mL				48 μg/mL				0.3 μg/mL		
Antibody-based DX-2400 (Ki)										0.8	
REGA-3G12						+++					

[a]*MMPI* MMP inhibitor, *CMT* chemically modified tetracycline, IC_{50} half-maximal inhibitory concentration, K_i inhibition constant

[20, 21]. Tanomastat has a thioether Zn^{2+}-binding group and a biphenyl deep-pocket-binding segment and is well-tolerated, but may show variable efficacies and outcomes depending on the timing of administration [22].

Mechanism-based MMPIs bind to the MMP active site and cause covalent enzyme modification. SB-3CT, a thiol-based inhibitor that contains a diphenylether deep-pocket-binding scaffold, is a mechanism-based selective inhibitor of MMP-2 and -9 through a process involving slow-binding inhibition similar to that of TIMP-1 and -2 [16, 23, 24]. SB-3CT directly binds the catalytic Zn^{2+} ion of MMP-2 and changes the conformation around the Zn^{2+} active site to that of the proenzyme [25].

Pyrimidine-based MMPIs include Ro 28-2653, an orally bioavailable MMPI that inhibits MT1-MMP, MT3-MMP, and MMP-2, -8, and -9 [26, 27]. Phosphorous-based MMPIs have phosphinate as Zn^{2+}-binding group and include 582311-81-7 and the MMP11-selective inhibitor RXP03 [28, 29]. Tetracyclines, such as minocycline and doxycycline, have innate MMPI capacity. Doxycycline is the only MMPI approved by the United States Food and Drug Administration and is indicated for periodontal disease [30]. Metastat (COL-3) is a chemically modified tetracycline that has been tested in a Phase I clinical trial in patients with refractory metastatic cancer [31].

Because of the high structural homology of the MMP catalytic site, most of the early MMPIs show broad-spectrum effects on different MMPs. In order to reduce off-target effects of MMPIs, investigations have shifted from targeting the catalytic site to alternative sites in the MMP molecule. MMPs have unprimed subsites S1, S2, and S3 on the left side of the Zn^{2+} ion and primed S1′, S2′, and S3′ on the right side of Zn^{2+} ion [32, 33]. The S1′ pocket is the main substrate recognition subsite and is the most variable among different MMPs in terms of amino acid sequence and pocket depth (shallow, intermediate, and deep) [16, 34–36]. These variabilities in MMP structure have been utilized to design more specific MMPIs. For example, extending the P1′ substituent (the group in MMPI or substrate that binds to the S1′ pocket of MMP) was used to enhance selectivity of MMP-13 over the highly homologous MMP-2 by taking advantage of the steric limitations of the shorter S1′ loop of MMP-2 [37]. However, identifying alternative MMP-specific sites could be challenging as they are scattered in different locations on the surface and even hidden inside the MMP molecule. Combining structural spectroscopic analyses, nuclear magnetic resonance, and protein crystallography with computational prediction of binding sites have helped to reveal these hidden sites and made it possible to design novel molecular effectors and therapeutic agents [38, 39]. For example, peptide-based MMPIs interact with secondary binding sites on MMPs and thereby have greater selectivity [40]. Also, phage display peptide libraries have been used to identify selective MMP-2, MMP-9, and MT1-MMP inhibitors that are effective in vivo [41, 42].

Antibody-based MMPIs have been designed to achieve high selectivity and potency [43]. For example, combining a human antibody phage display library with automated selection and screening strategies resulted in the identification of the highly selective antibody-based MMP14 inhibitor DX-2400 [44, 45]. Also, the neutralizing monoclonal antibody REGA-3G12 is a selective inhibitor of MMP-9 directed against the catalytic domain, but not the fibronectin or Zn^{2+}-binding domains [46, 47]. Another strategy for generating inhibitory antibodies that effectively target the in vivo activity of dysregulated MMPs is mimicking the mechanism used by TIMPs [48].

While several MMPIs have been developed, only one MMPI is approved by the Food and Drug Administration [1]. This could be due to the numerous side effects of MMPIs and their lack of specificity in various MMP assays. In order to determine the role of MMPs in pathological conditions and the efficacy of various MMPIs, it is critical to have reliable methods for measuring MMP activity. Zymography is a simple technique first described in 1980 [49] and is now widely used to measure MMP activity in various systems. The zymography technique is based on separation of proteins by nonreducing sodium dodecyl sulfate polyacrylamide gel electrophoresis (SDS–PAGE). Special polyacrylamide gel is made by inducing acrylamide polymerization in the presence of the specific substrate of the MMP(s) of interest. For instance, using gelatin as a substrate, zymography has been used to assess the activity of MMP-2 (gelatinase A) and MMP-9 (gelatinase B) [50]. During electrophoresis, MMPs are activated in a nonproteolytic manner by SDS [51]. Following electrophoretic separation and a "renaturation" step, the gel is incubated at 37 °C in a Ca^{2+}- and Zn^{2+}-containing buffer optimized for measuring MMP activity toward the specific substrate (Fig. 2).

If performed carefully and analyzed accurately, zymography can be a very valuable technique to measure MMP activity in various biological systems including plasma, cells, culture media, and tissue extracts [52, 53]. Zymography has also been used to measure the changes in MMP activity in vascular remodeling and vascular disorders [3]. Also, reverse zymography is a modified technique to study the activity of endogenous TIMPs. However, problems encountered during the zymography procedure may lead to inaccurate assessment of MMP activity in control specimen, and thus make it difficult to determine the pathological changes in MMP activity and their responsiveness to MMPIs.

In this chapter, we will discuss the materials needed for a good zymography experiment, the preparation of experimental buffers, sample preparation, running and developing the gels during the regular and reverse zymography procedures, and zymogram analysis

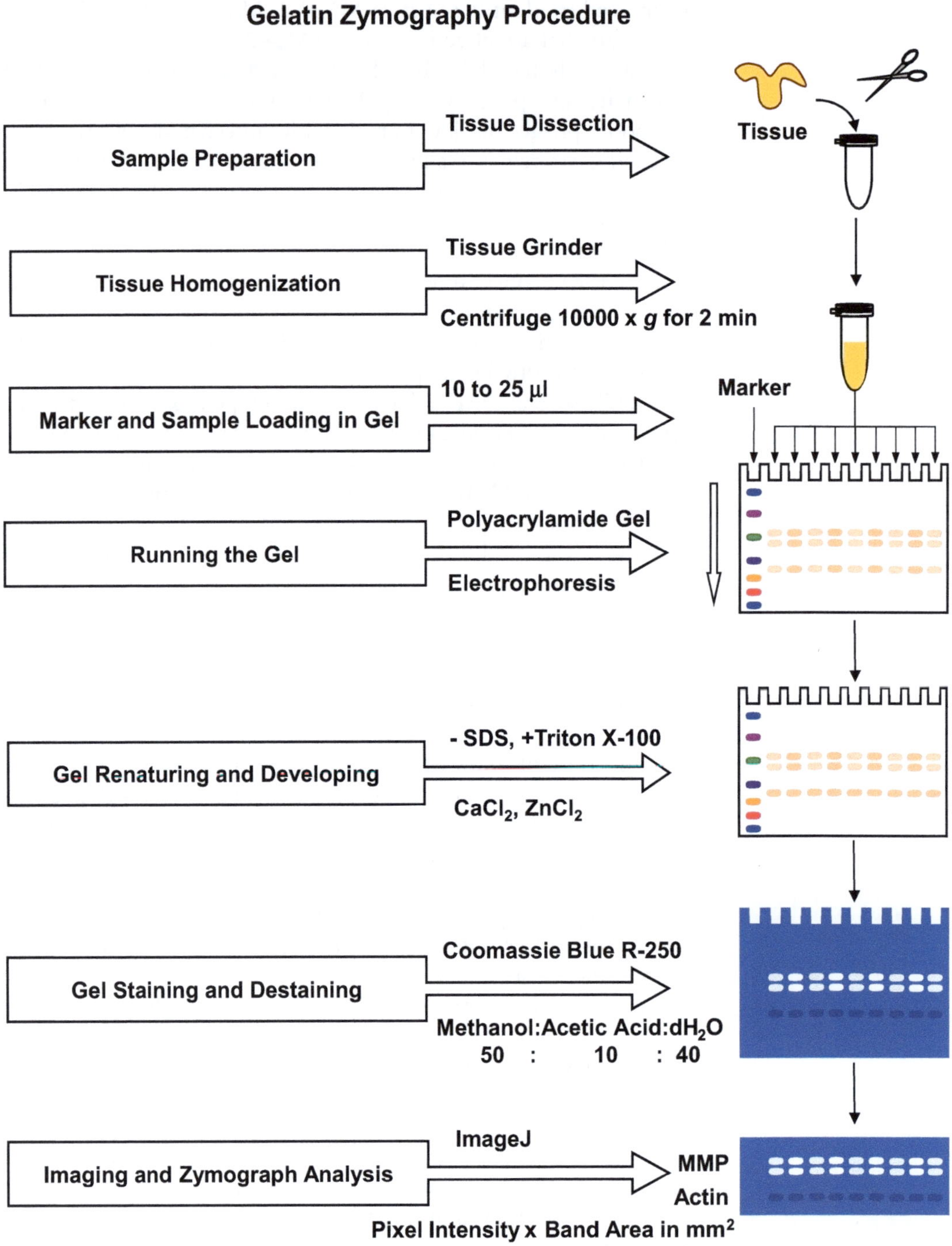

Fig. 2 Flow chart of gelatin zymography procedure

and interpretation. We will also provide helpful notes to troubleshoot problems and some tips to enhance the quality of the zymograms and assess the efficacy of MMPIs.

2 Materials

2.1 Materials for Zymography Experiment

The following materials should be prepared before starting a zymography experiment. All materials should be clean and stock solutions should be clear (*see* **Note 1**).

1. Glass plates with 0.75-mm-thick spacers and ten well combs.
2. 30% w/v Acrylamide/bis-acrylamide (A/C).
3. Separating gel buffer stock: 1.5 M Tris–HCl, pH 8.8. Use 27.23 g Tris-base in 150 mL dH_2O, adjust pH to 8.8 with 38% HCl.
4. Stacking gel buffer stock: 0.5 M Tris–HCl, pH 6.8. Use 6 g Tris-base in 100 mL dH_2O, adjust pH to 6.8 with 38% HCl.
5. 1% w/v gelatin. Use 100 mg gelatin in 10 mL dH_2O. Warm the solution to 60 °C in a water bath and vortex repeatedly until it becomes almost translucent (~20 min). Cool down the gelatin solution to room temperature before use.
6. 10% w/v SDS. Use 1 g SDS in 10 mL dH_2O.
7. 10% w/v ammonium persulfate (APS). Use 1 g APS in 10 mL dH_2O.
8. *N*,*N*,*N*′,*N*′-Tetramethylethylenediamine (TEMED).
9. Running buffer stock: For 1× running buffer, use 0.024 M or 2.9 g Tris-base, 0.192 M or 14.4 g glycine, and 0.1% w/v or 1 g SDS in 1 L dH_2O, pH 8.3. For 10× running buffer, use 0.24 M or 29 g Tris-base, 1.92 M or 144 g glycine and 1% w/v or 10 g SDS in 1 L dH_2O, pH 8.3. Remember to dilute the 10× running buffer 1:10 in dH_2O before use.
10. Sample buffer (2×): Use 2.5 mL of 0.5 M Tris/HCl pH 6.8, 2 mL glycerol, 4 mL of 10% w/v SDS (1 g SDS in 10 mL dH_2O), and 0.5 mL of 0.1% w/v bromophenol blue (0.001 g in 1 mL dH_2O). Add dH_2O to 10 mL. When ready to load the samples in the gel and depending on the protein concentration in the sample tissue homogenate, add one part of the sample to one part sample buffer.
11. Renaturing buffer. For 1× renaturing buffer (Triton X-100 2.5% v/v), add 5 mL Triton X-100 in 195 mL deionized dH_2O. For 10× solution stock (Triton X-100 25% v/v), add 25 mL Triton X-100 in 75 mL deionized dH_2O.
12. Developing buffer (1×): 50 mM or 6.06 g Tris-base, 0.2 M or 11.688 g NaCl, 5 mM or 0.555 g $CaCl_2$, 0.02% w/v or 0.2 g Brij35, and 1 μM or 0.136 mg $ZnCl_2$ in 1 L dH_2O, pH 7.6.
13. Staining solution: 0.5% w/v or 2 g Coomassie blue R-250, 25% v/v or 100 mL isopropanol, and 10% v/v or 40 mL acetic acid to 260 mL dH_2O. Filter through filter paper before use.

14. Destaining solution: Add methanol:acetic acid:dH_2O in the following proportion, 50:10:40. Final concentration is 50% (v/v) methanol and 10% (v/v) acetic acid in dH_2O.
15. Homogenization buffer: 0.02 M or 0.418 g 3-[*N*-Morpholino] propanesulfonicacid (MOPS), 4% w/v or 4 g SDS, 10% v/v or 10 mL glycerol, and 90 mL dH_2O. Immediately before use, add the following solutions: 1.5 mL homogenization buffer, 35 μL of 50 mM ethylenediaminetetraacetic acid (EDTA), and 75 μL of 20× anti-protease cocktail.
16. Anti-protease cocktail (20×): To make 1.5 mL stock, use 0.4% w/v bovine serum albumin (BSA) (take 0.3 mL of 2% w/v or 2 g BSA in 100 mL dH_2O), 0.11 mM or 165 μL of 1 mM leupeptin, 0.11 mM or 165 μL of 1 mM pepstatin, 0.15 TIU or 60 μL of 7.6 TIU/mL aprotinin, 0.4 mM or 60 μL of 10 mM *4*-(*2*-aminoethyl) benzenesulfonyl fluoride hydrochloride (AEBSF), and 0.1 M or 750 μL of 200 mM ethylene glycol tetraacetic acid (EGTA).

2.2 Materials for Studying MMPIs Using Zymography

1. Prepare all materials as for regular zymography.
2. Prepare stock solution for the MMPI of interest. For example, stock solution (10^{-2} M) of SB-3CT (MMP-2/MMP-9 inhibitor IV), Ro-28-2653 (5-biphenyl-4-yl-5-[4-(-nitro-phenyl)-piperazin-1-yl]-pyrimidinc-2,4,6-trione), and Batimastat (BB-94) is prepared in dimethylsulfoxide (DMSO) [54].

2.3 Materials for Reverse Zymography

1. Prepare all materials as for regular zymography.
2. Prepare human recombinant MMP-2 or MMP-9 (0.13 μg/mL).

3 Methods

3.1 Gel Preparation

1. Prepare 8% A/C separating gel (for two 0.75 mm gels) (*see* **Note 2**): 4.6 mL dH_2O, 1 mL 1% w/v gelatin (100 mg/10 mL) (*see* **Note 3**), 3.0 mL 1.5 M Tris/HCl (pH 8.8), 3.2 mL 30% A/C stock, 120 μL 10% w/v SDS (1 g/10 mL), 120 μL 10% w/v APS (1 g/10 mL).
2. Prepare 5% A/C stacking gel: 3.6 mL dH_2O, 1.56 mL 0.5 M Tris/HCl (pH 6.8), 0.6 mL 30% A/C stock, 20 μL 10% w/v SDS (1 g/10 mL), 140 μl 10% w/v APS (1 g/10 mL).
3. TEMED will polymerize the gel and should be added last to the freshly prepared separating and stacking gel, immediately before pouring into the cassette. Add 5 μL of TEMED to the separating gel solution to initiate the polymerization process. Swirl the solution rapidly and avoid bubble formation. Pipette 3 mL of the separating gel solution into each cassette, avoiding

the formation of bubbles. Carefully, overlay the separating gel with dH_2O, filling up to the top of the cassette. Do not disturb the surface of the separating gel solution. Allow the gel to polymerize for at least 30 min at room temperature. The warmer the condition the gel is in, the less the polymerization time. Do not touch the gels until polymerization is complete as indicated by a clear straight line of separation between the gel phase and the water phase.

4. Decant the water from the top of the separating gel and keep the gel upside down on bench top for a few minutes to drain away all the water. Use filter paper to absorb residual water.
5. Add 10 μL of TEMED to the stacking gel solution, swirl rapidly, and immediately pipette the solution on top of the already polymerized separating gel until it reaches the top of the front glass plate. Rapidly insert the appropriate comb (ten wells or lanes) into the liquid stacking gel and ensure that no bubbles are trapped under the comb. Allow the stacking gel to polymerize at room temperature (about 30 min). Place the gels (comb on top) in a sealed plastic bag or a container having 1× running buffer to keep them moist. The gels can be stored at 4 °C for up to 2–3 weeks before running.

3.2 Sample Preparation

1. Preparation of the biological sample is critical for a successful zymography. It is relatively easy to detect MMP activity in culture media and cell lysates. However, analysis of MMPs in tissues such as the aorta or uterus could be difficult (*see* **Note 4**).
2. Weigh approximately 50 mg of the tissue of interest. Cut the tissue into small pieces in a small weighing boat on ice. Transfer the tissue to a tissue grinder or mortar (Kimble) on ice. Add cold homogenization buffer with anti-protease cocktail (for 50 mg tissue add 300 μL homogenization buffer). The volume of homogenization buffer may vary depending on the tissue used. Homogenize the tissue completely using the tissue grinder or mortar and pestle (Kimble) on ice.
3. Transfer homogenate to labeled microcentrifuge tubes. Centrifuge the homogenate at 10,000 × *g* for 2 min at 4 °C. If the tissue homogenate contains floating lipids, repeat centrifugation at least two times to obtain a clear supernatant. Save the supernatant, discard the pellet, and measure the protein concentration in the supernatant using Bradford protein assay [55]. Before loading, adjust the protein concentration in the samples using sample buffer. Ideally, mix one part sample with one part sample buffer and let stand at room temperature for at least 30 min to allow coating of the protein with SDS. DO NOT HEAT.

3.3 Running the Gel and Electrophoresis

1. Gently pull the comb out of the stacking gel and peel off the rubber band at the bottom of the cassette.
2. Gently place the cassette in the gel protean II apparatus (Bio-Rad). Fill the buffer chambers with 1× running buffer. Reducing agents such as dithiothreitol (DDT) are omitted because of possible interference with subsequent refolding of gelatinases. Load the samples in the gel lanes. Typically, 10–25 μL of each sample which contains 0.1–10 μg of the enzyme is loaded to each of the gel lanes (*see* **Note 5**). Molecular weight markers are included on each gel preferably in lanes 1 and 6.
3. Run the gel in the protean II apparatus (Bio-Rad) using gel electrophoresis at the standard running conditions (125 V, constant voltage) and until the bromophenol blue tracking dye reaches the bottom of the gel (*see* **Notes 6** and 7). Running time could range between 60–120 min depending on the gel A/C percentage, running buffer concentration, and pH. The proteins will be separated according to their molecular weight; whereby the low molecular weight proteins will run faster and farther than the high molecular weight proteins which will be lagging behind (Fig. 2).

3.4 Renaturing, Developing, Staining, and Destaining the Gel

1. Dilute the renaturing buffer (10×) 1:10 with deionized dH_2O to obtain 1× renaturing buffer solution. Carefully remove the gel from the cassette and place it in a plastic tray containing 1× renaturing solution (50 mL for one mini-gel). Incubate the gel for 30 min at room temperature with gentle shaking in order to remove SDS which causes MMPs to denature and become inactive (*see* **Notes 8** and **9**).
2. Decant the zymogram 1× renaturing buffer and replace with fresh 1× developing buffer (50 mL for one mini-gel). Equilibrate the gel at room temperature for an additional 30 min in the developing buffer with gentle shaking. Decant the developing buffer and replace with 50 mL of fresh 1× developing buffer and incubate the gel at 37 °C overnight for ~18 h for maximum sensitivity (*see* **Note 10**). Optimal results are determined empirically by varying the sample load and incubation time (Fig. 3).
3. Decant the developing buffer and stain the gel with Commassie blue R-250 staining solution for at least 30 min until the gel is uniformly dark blue. Destain the gel with destaining solution until areas of gelatinolytic activity appear as clear sharp bands against dark blue background.

3.5 Zymogram Analysis

The goal of gelatin zymography is to obtain clear and sharp bands of the digested substrate against a blue background of the undegraded substrate (Fig. 3). Comparison of the location of the gelatinolytic with molecular weight standards run simultaneously on the

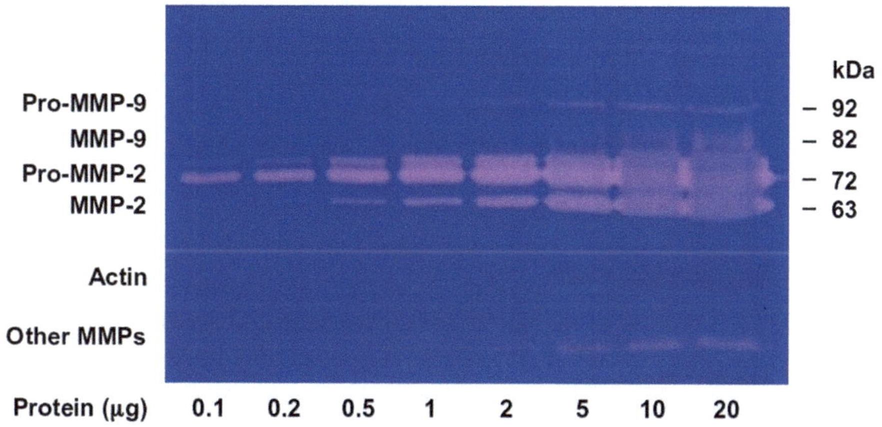

Fig. 3 Concentration-dependent MMP-2 and MMP-9 gelatinase activity in uterus of pregnant rats. Uterine tissue strips from normal pregnant rats were homogenized and prepared for gelatin zymography analysis using different concentrations of loaded protein (0.1–20 μg). Pro-MMP-2, MMP-2, pro-MMP-9, and MMP-9 showed concentration-dependent gelatinolytic bands because of the presence of their preferred substrate gelatin. Other MMPs are only detected at higher protein concentration and are less clear because gelatin is not their preferred substrate

same gel should help identify the specific MMP involved (*see* **Note 11**). The bands in the gel are quantified using ImageJ 1.38X (NIH). The gel image is scaled using a grey scale such that the intensity of each pixel would range from 0 to 255. The integrated intensity of the band of interest is calculated by first outlining and measuring the band area in pixels, then transferring it into mm^2 using a calibration bar. The total pixel intensity is measured by summing the pixel values within the band area, and the average pixel intensity is measured by dividing the total pixel intensity by the number of pixels. The average pixel density of the background is then subtracted from the average band intensity. The integrated intensity of the selected band is then measured as average pixel intensity × band area in mm^2 [54, 56]. The integrated intensity of the band can also be normalized to the housekeeping protein actin to correct for loading. Comparison of the integrated intensity of the bands would help determine the changes in MMP activity in specimens in different physiological and pathological conditions (*see* **Notes 12** and **13**).

3.6 Study of MMPI

1. Prepare all the materials and follow all the methods as described above for gelatin zymography.
2. For sample preparation, incubate the sample overnight in physiological buffer solution in the presence of appropriate concentration of MMPI. For example, SB-3CT (MMP-2/MMP-9 inhibitor IV, 10^{-6} M), BB-94 (10^{-6} M), or Ro-28-2653 (10^{-6} M) (*see* **Note 14**).

3. Compare the integrated intensity of the gelatinolytic bands in control samples in the absence of MMPI with that in the presence of MMPI. The integrated intensity of the gelatinolytic bands should be less in the presence of MMPI compared to the control samples in the absence of MMPI.

3.7 Reverse Zymography

1. Prepare all the materials and follow the methods as described above for regular gelatin zymography.
2. When preparing the separating gel, make the 8% separating gel as in regular gelatin zymography, but add human recombinant MMP-2 or MMP-9 (0.13 μg/mL) (*see* **Note 15**).
3. For the staining step, decant the developing buffer and stain the gel with Coomassie Blue R-250 staining solution for at least 30 min until the areas representing the undigested gelatin due to the presence of endogenous TIMPs appear as clear dark blue bands against faint blue background of digested gelatin caused by the added MMP-2 or MMP-9.

4 Notes

Although zymography could be a sensitive and quantifiable assay to analyze MMP activity, problems related to the nature, source, and preparation of samples, the substrate in the gel, and the distinction between inactive and active forms of MMPs could compromise the validity of the technique and complicate interpretation of the results. In order to obtain consistent and reliable zymograms, it is important to pay attention to all the different steps including the gel preparation, sample preparation, running and developing the gels, and analysis of the zymograms.

1. For gel electrophoresis, all stock solutions should be prepared using electrophoresis-grade reagents and kept fresh. If any precipitations are observed, the quality of the solution and the concentration of the different ingredients are likely altered and should not be used. Also, avoid bacterial contamination of the buffers and solutions as bacterial proteases may result in the appearance of nonspecific bands. This can be minimized by filter sterilization of the buffers and stock solutions and storage at 4 °C.
2. An 8% polyacrylamide gel is generally used for separating gelatinases. However, the percentage of acrylamide and the thickness of the separating gel may vary depending on the MMP type, form, and molecular mass. For instance, to better visualize the dimeric form of MMP-9 (~200 kDa) or to obtain a better resolution of bands with close molecular weights (latent and active forms), a lower 4–6% polyacrylamide solution

should be used. However, by taking this approach the gelatinolytic bands may become less sharp. On the other hand, to better visualize the lower molecular weight MMPs such as MMP-1 and -7, a higher 10–12% polyacrylamide solution can be used. The gels can be prepared in advance and stored at 4 °C for 2 or 3 weeks without significant effects on the resolution.

3. Gelatin is commonly used as the protein substrate because it is inexpensive, easily hydrolyzed by several peptidases, and does not tend to migrate out of the resolving gel in electrophoretic tests performed at 4 °C [57]. Gelatin is the substrate of choice to detect the gelatinases MMP-2 and MMP-9. Other MMPs such as MMP-1, MMP-8, and MMP-13 can degrade gelatin; however, the gelatinolytic bands will be weak because gelatin is not their preferred substrate [58, 59]. For improved detection of these MMPs, modified zymography has been developed by incorporating a more suitable substrate such as casein or collagen into the gel, or by enhancing the gelatinolytic signal with the addition of heparin to the samples [60–62].

4. Fresh tissues should be used for measuring MMP activity, and the whole procedure of tissue preparation and homogenization should be performed on ice at 4 °C. Because of inherent differences in tissue structure, protein extractability may vary between and within specimens [63]. For example, tissues such as the liver and placenta are easy to cut and homogenize due to their fragile nature, while the aorta and uterus are more difficult to cut into small pieces and to homogenize because of their large content of collagen and elastin fibers. The homogenized samples should not be boiled as high temperature causes protein precipitation of the enzymes. Also, reducing agents such as DTT or 2-mercaptoethanol should not be added as reducing agents break the disulfide bond and thereby prevent some forms of tertiary protein folding, inhibit MMP refolding after electrophoresis [53], and break up quaternary protein structure in oligomeric subunits. The tissue extraction procedure itself may activate MMPs or cause inhibition of active enzymes by interacting with some of the components of the homogenization buffer. Some studies suggest that EDTA, other Zn^{2+} chelators, and protease inhibitors should not be added to the tissue extract [64]. While EDTA may prevent MMP activation by binding with the Ca^{2+} and Mg^{2+} ion in the homogenization buffer, MMPs are reactivated after incubation in the developing buffer which contains Ca^{2+} and Zn^{2+} ions.

5. The amount of sample loaded in the gel is critical for successful zymography as large amounts of tissue extracts may produce saturated and distorted bands. The sensitivity of zymography is much greater than that of Western blots, which depends on the

antibody affinity for MMPs. Because zymography is a sensitive technique, gelatinolytic bands can be detected with MMP levels as low as 10 pg [56, 65]. However, these low MMP levels are not often detected because the ratio of MMPs to total protein in crude samples is extremely low. This may make it necessary to load larger amounts of the tissue extract. Overloading of total protein extracts into the wells or lanes may lead to saturated gelatinolytic bands in the zymogram. For instance, in pregnant rat uterus extracts, gelatin zymography using different protein amounts showed dose-dependent increases in MMP-2 and MMP-9 proteolytic activity at 0.1, 0.2, and 0.5 μg, clearly discernible bands at 1–2 μg, but saturated bands at 5, 10, and 20 μg (Fig. 3).

6. For the proteins to move from the cathode to the anode through the gel, the gel system and running buffer should have the proper pH as it may affect the mobility of the different components of the gel system relative to the proteins. For example, depending on the pH, glycine can exist in three different charge states; positive, neutral, or negative. Control of the charge of glycine in the different buffers is key to the mobility of proteins in the gel. All of the proteins in the gel have an electrophoretic mobility that is intermediate between the mobility of glycine and Cl^-, so that as the glycine and Cl^- fronts sweep through the sample well, the proteins are concentrated in the narrow zone between the two fronts. This process continues through the stacking gel until the proteins hit the separating gel, where the pH switches to 8.8. At this pH, the glycine becomes mostly negatively charged and migrates much faster than the proteins. As the glycine front accelerates past the proteins, the proteins become concentrated in a very narrow region at the interface of the stacking and separating gels. Because the separating gel has a greater acrylamide concentration, it slows the mobility of the proteins according to their molecular weight and the protein separation begins [66]. If any of the buffers or gel system pH is altered, the protein mobility will be affected in areas of the gel with "improper" pH. This explains the odd protein migration behavior and the distorted shape of the protein bands in some gels.
7. Careful attention to the temperature could provide the best conditions for running the gel. It is important to keep the gels cold while running, otherwise the lower part of the gels may become distorted and show wavy bands. Putting the gel apparatus on ice during running could minimize overheating of the apparatus and distortion of the gel.
8. SDS is a strongly denaturing anionic detergent which unfolds and fully denatures all proteins including MMPs, essentially disregarding secondary structures or hydrophobic domains,

and generates SDS-protein complexes that are mostly characterized by a uniform charge-to-mass ratio. This makes SDS-PAGE in general a very simple and reliable technique for protein separation and molecular mass determination. The ratio of 1.4 g of SDS bound per gram of protein is often quoted as a typical stoichiometric value [67]. When the proteins are saturated with micelles of SDS (SDS-protein), this amount of highly charged surfactant molecules is sufficient to overwhelm the intrinsic charges on the protein chains so that the net charge per unit of mass becomes almost constant, thus allowing the protein chains to separate through SDS-PAGE, mostly according to their molecular weight [68].

9. Pro-MMPs are secreted as inactive zymogens with an inhibitory propeptide domain. The pro-MMP architecture in which Cys^{73} is located in the vicinity of the Zn^{2+} ion makes the Cys^{73}-Zn^{2+} complex vulnerable to disruption by multiple stimuli. Dissociation of the Cys^{73} residue from the Zn^{2+} ion "switches" it from a non-catalytic to catalytic Zn^{2+}. Because the sequences surrounding Cys^{73} in the propeptide and the Zn^{2+}-binding site in the catalytic domain of the MMPs are highly homologous, the "cysteine switch" mechanism applies to all MMPs. For example, during the electrophoresis step of the gelatin zymography experiment, the propeptide is unfolded in the denaturing conditions induced by SDS. After electrophoresis, other non-ionic detergents such as Triton X-100 are used in the renaturing buffer to replace SDS and remove it from the SDS-complex. This allows the pro-MMPs in the sample to renature, become partially refolded, and autoactivate, resulting in the appearance of a partially catalytically active pro-MMP portion of the originally inactive pro-MMPs. Only about 35% of the MMP catalytic activity is recovered during the protein refolding [63], which may not represent the true biological activity of the pro-MMP when activated in vivo. Because these propeptides are covalently anchored to the enzyme pro-forms, the pro-MMPs are detected at higher molecular weights than the activated MMPs from which the propeptides are cleaved off. Furthermore, non-covalently bound complexes, such as TIMP-MMP complexes, are dissociated by SDS during the electrophoresis step [56, 64, 69]. Hence, the gelatinolytic bands of the zymography may not be a measure of the net MMPs activity present in the sample, but should rather be seen as a measure of potential activity of MMPs [56, 69, 70].

10. The incubation time of the gel in the developing buffer is critical for proper renaturation and proteolysis. Since the appearance of the gelatinolytic bands depends on enzymatic activity, changing the incubation time will affect the intensity of the bands. Incubating the gel in the developing buffer at 37 °C for

4 h in a closed tray may be sufficient to detect the gelatinolytic bands. However, in most cases, overnight incubation (16–18 h) may be needed to obtain better resolution and reproducible results. If the bands remain barely visible, it may be necessary to develop the gels for a longer period of time, even up to 72 h.

11. The identity of the MMP type in the gelatinolytic band is usually determined by comparing the band location with known molecular weight standards run simultaneously in the same gel. This could also help in discerning the latent inactive from the active forms of MMPs [51]. Also, pro-MMPs are usually activated in a process involving the generation of an inactive intermediate forms which are then processed to generate the fully mature active forms [71, 72]. Some commercially available molecular weight standards contain a reducing agent, and when they are used under nonreducing conditions, they may indicate different molecular weights [51]. Also, detection of small differences in molecular weight between intermediate and fully active MMP species would require further optimization of the conditions to enhance the sensitivity of the zymography assay. Hence, for an exact identification of MMPs, Western blot analysis using specific antibodies should be performed. While Western blot analysis is more specific than zymography, antibodies may not be sensitive enough to detect low levels of MMPs. On the other hand, while zymography is more sensitive than Western blots in detecting small amounts of MMPs, some of the limitations regarding the resolution of gelatin zymography may make it difficult to analyze the data. Both gelatin zymography and Western blots techniques could complement each other in studying MMPs.

12. Gelatin zymography can been used to detect the changes in specific MMP activity in vascular remodeling and angiogenesis [1], and in the uterine and vascular changes associated with pregnancy [8]. We have recently used gelatin zymography to measure the changes in MMP activity during uterine wall stretch and in response to sex hormones during pregnancy in rats [54]. We found that oxytocin-induced contraction of uterine strips was reduced in pregnant compared with virgin rats. Gelatin zymography showed increased activity of MMP-2 and -9 in uterus of pregnant versus virgin rats. Prolonged stretch of uterine strips of virgin rats was associated with reduced contraction and enhanced activity of MMP-2 and -9. Treatment of stretched uterus of virgin rats with 17β-estradiol (E2) or progesterone (P4) or E2 + P4 caused further reduction in contraction and increases in MMP activity. MMP-2 and -9 decreased oxytocin-induced

contraction in uterus of virgin rat. These data suggested that during pregnancy uterine stretch and increased sex hormone levels cause increases in the activity of MMP-2 and -9, which in turn reduce uterine contraction and enhance uterine relaxation [54].

13. Gelatin zymography can be used to measure the changes in MMP activity in vascular disease. We have used gelatin zymography to measure the changes in uteroplacental and vascular MMPs in an animal model of hypertension in pregnancy produced by reduction in uterine perfusion pressure (RUPP) [73]. We observed a decrease in gelatinase activity of MMP-2 and -9 in uterus, placenta, and aorta of RUPP compared with normal pregnant rats. Also, collagen was more abundant in uterus, placenta, and aorta of RUPP than in those of Norm-Preg rats. The antiangiogenic factor-soluble fms-like tyrosine kinase-1 (sFlt-1) decreased MMP activity in uterus, placenta, and aorta of normal pregnant rats, and vascular endothelial growth factor (VEGF) reversed the decreases in MMPs in tissues of RUPP rats. These observations suggested that placental ischemia and antiangiogenic sFlt-1 decrease uterine, placental, and vascular MMP-2 and -9, leading to increased uteroplacental and vascular collagen and growth restriction in hypertensive pregnancy, and that angiogenic factors may reverse the decrease in MMP activity and enhance uteroplacental and vascular growth in preeclampsia [73].
14. Gelatin zymography can be used to test the specificity and efficacy of MMPIs. Selective MMPIs can be added to a tissue sample ex vivo to determine if they reduce MMP activity and alter tissue function. MMPIs can also be added to the gel (or part of a gel cut in half) during the incubation in the developing buffer to determine if they decrease the activity of the MMP of interest. The optimal dose and reaction time of MMPIs with MMPs can be determined by performing a concentration-response curve and time course studies. We have recently tested the effects of MMPIs on uterine function and MMP activity. We found that pretreatment of uterine strips with the MMPIs SB-3CT, BB-94, or Ro-28-2653 (10^{-6} M) did not change oxytocin contraction in virgin uterus, but enhanced contraction in uterine strips of pregnant rats. Also, gelatin zymography revealed that the intensity of the pro-MMP-2-, MMP-2-, and MMP-9-digested gelatin bands was reduced in uterine strips treated with the MMP-2/MMP-9 inhibitor IV (SB-3CT, 10^{-6} M) compared with control non-treated uterine strips of virgin or pregnant rats [54]. These observations support the contention that MMPIs could be used to test the role of changes in MMP activity in modulating the function of different systems.

15. There is mounting evidence that MMPs regulate tissue remodeling under physiological and pathological conditions, and reliable detection and quantitative methods such as zymography are needed to determine the role of MMPs in various processes. This classical technique for measuring MMP activity is a highly sensitive, cost-effective, and relatively simple to perform. Modification of the substrate in the zymography assay could enhance the detection spectrum to include MMPs with different substrate preferences. Improvement in the sensitivity and accuracy of gelatin zymorgraphy will further enhance its value in assessing the changes in MMP activity in biological samples and in disease conditions. Tissue remodeling is controlled by a balance between endogenous TIMPs and MMPs such that an increase in TIMP/MMP activity ratio would decrease ECM protein degradation and vice versa. Reverse zymography could be utilized to test the activity of endogenous TIMPs (Fig. 4). Also, synthetic MMPIs have been evaluated as diagnostic and therapeutic tools in cancer, autoimmune, and vascular disease [15]. Zymography could be a valuable technique to test the specificity and efficacy of MMPIs and could help in the development of more potent and selective inhibitors for specific MMPs.

Fig. 4 TIMP activity in uterus of pregnant rats. Uterine tissue strips from normal pregnant rats were homogenized and prepared for reverse zymography. Separating gel was prepared as in regular gelatin zymography experiment except that MMP-9 (0.13 μg/mL) was added. TIMP-1 appears as a darker blue band of the undigested substrate against a faint blue background of the degraded gelatin substrate

Acknowledgments

This work was supported by grants from National Heart, Lung, and Blood Institute (HL-65998, HL-98724, HL-111775). Dr. Zongli Ren was a visiting scholar from the Department of Cardiovascular Surgery, Renmin Hospital of Wuhan University, Wuhan 430060, Hubei, PR China, and a recipient of scholarship from the China Scholarship Council. Dr. Juanjuan Chen was a visiting scholar from the Department of Obstetrics & Gynecology, The Third Affiliated Hospital of Guangzhou Medical University, Guangzhou, China 510150.

References

1. Benjamin MM, Khalil RA (2012) Matrix metalloproteinase inhibitors as investigative tools in the pathogenesis and management of vascular disease. EXS 103:209–279
2. Visse R, Nagase H (2003) Matrix metalloproteinases and tissue inhibitors of metalloproteinases: structure, function, and biochemistry. Circ Res 92(8):827–839
3. Galis ZS, Khatri JJ (2002) Matrix metalloproteinases in vascular remodeling and atherogenesis: the good, the bad, and the ugly. Circ Res 90(3):251–262
4. English WR, Holtz B, Vogt G, Knauper V, Murphy G (2001) Characterization of the role of the "MT-loop": an eight-amino acid insertion specific to progelatinase A (MMP2) activating membrane-type matrix metalloproteinases. J Biol Chem 276(45):42018–42026
5. Kucukguven A, Khalil RA (2013) Matrix metalloproteinases as potential targets in the venous dilation associated with varicose veins. Curr Drug Targets 14(3):287–324
6. Pei D, Kang T, Qi H (2000) Cysteine array matrix metalloproteinase (CA-MMP)/MMP-23 is a type II transmembrane matrix metalloproteinase regulated by a single cleavage for both secretion and activation. J Biol Chem 275(43):33988–33997
7. Ellerbroek SM, Wu YI, Overall CM, Stack MS (2001) Functional interplay between type I collagen and cell surface matrix metalloproteinase activity. J Biol Chem 276(27): 24833–24842
8. Palei AC, Granger JP, Tanus-Santos JE (2013) Matrix metalloproteinases as drug targets in preeclampsia. Curr Drug Targets 14(3):325–334
9. Bode W, Fernandez-Catalan C, Grams F, Gomis-Ruth FX, Nagase H, Tschesche H, Maskos K (1999) Insights into MMP-TIMP interactions. Ann N Y Acad Sci 878:73–91
10. Nagase H, Visse R, Murphy G (2006) Structure and function of matrix metalloproteinases and TIMPs. Cardiovasc Res 69(3):562–573
11. Brew K, Dinakarpandian D, Nagase H (2000) Tissue inhibitors of metalloproteinases: evolution, structure and function. Biochim Biophys Acta 1477(1–2):267–283
12. Baker AH, Edwards DR, Murphy G (2002) Metalloproteinase inhibitors: biological actions and therapeutic opportunities. J Cell Sci 115(Pt 19):3719–3727
13. Murphy G (2011) Tissue inhibitors of metalloproteinases. Genome Biol 12(11):233
14. Strickland DK, Ashcom JD, Williams S, Burgess WH, Migliorini M, Argraves WS (1990) Sequence identity between the alpha 2-macroglobulin receptor and low density lipoprotein receptor-related protein suggests that this molecule is a multifunctional receptor. J Biol Chem 265(29):17401–17404
15. Bode W, Maskos K (2003) Structural basis of the matrix metalloproteinases and their physiological inhibitors, the tissue inhibitors of metalloproteinases. Biol Chem 384(6):863–872
16. Jacobsen JA, Major Jourden JL, Miller MT, Cohen SM (2010) To bind zinc or not to bind zinc: an examination of innovative approaches to improved metalloproteinase inhibition. Biochim Biophys Acta 1803(1):72–94
17. Whittaker M, Floyd CD, Brown P, Gearing AJ (1999) Design and therapeutic application of matrix metalloproteinase inhibitors. Chem Rev 99(9):2735–2776
18. Macaulay VM, O'Byrne KJ, Saunders MP, Braybrooke JP, Long L, Gleeson F, Mason CS, Harris AL, Brown P, Talbot DC (1999) Phase I study of intrapleural batimastat (BB-94), a matrix metalloproteinase inhibitor, in the treatment of malignant pleural effusions. Clin Cancer Res 5(3):513–520

19. Fingleton B (2006) Matrix metalloproteinases: roles in cancer and metastasis. Front Biosci 11:479–491
20. Miller KD, Saphner TJ, Waterhouse DM, Chen TT, Rush-Taylor A, Sparano JA, Wolff AC, Cobleigh MA, Galbraith S, Sledge GW (2004) A randomized phase II feasibility trial of BMS-275291 in patients with early stage breast cancer. Clin Cancer Res 10(6):1971–1975
21. Leighl NB, Paz-Ares L, Douillard JY, Peschel C, Arnold A, Depierre A, Santoro A, Betticher DC, Gatzemeier U, Jassem J, Crawford J, Tu D, Bezjak A, Humphrey JS, Voi M, Galbraith S, Hann K, Seymour L, Shepherd FA (2005) Randomized phase III study of matrix metalloproteinase inhibitor BMS-275291 in combination with paclitaxel and carboplatin in advanced non-small-cell lung cancer: National Cancer Institute of Canada-Clinical Trials Group Study BR.18. J Clin Oncol 23(12):2831–2839
22. Lutz J, Yao Y, Song E, Antus B, Hamar P, Liu S, Heemann U (2005) Inhibition of matrix metalloproteinases during chronic allograft nephropathy in rats. Transplantation 79(6):655–661
23. Lee M, Bernardo MM, Meroueh SO, Brown S, Fridman R, Mobashery S (2005) Synthesis of chiral 2-(4-phenoxyphenylsulfonylmethyl)thiiranes as selective gelatinase inhibitors. Org Lett 7(20):4463–4465
24. Olson MW, Gervasi DC, Mobashery S, Fridman R (1997) Kinetic analysis of the binding of human matrix metalloproteinase-2 and -9 to tissue inhibitor of metalloproteinase (TIMP)-1 and TIMP-2. J Biol Chem 272(47):29975–29983
25. Kleifeld O, Kotra LP, Gervasi DC, Brown S, Bernardo MM, Fridman R, Mobashery S, Sagi I (2001) X-ray absorption studies of human matrix metalloproteinase-2 (MMP-2) bound to a highly selective mechanism-based inhibitor. Comparison with the latent and active forms of the enzyme. J Biol Chem 276(20):17125–17131
26. Grams F, Brandstetter H, D'Alo S, Geppert D, Krell HW, Leinert H, Livi V, Menta E, Oliva A, Zimmermann G, Gram F, Livi VE (2001) Pyrimidine-2,4,6-triones: a new effective and selective class of matrix metalloproteinase inhibitors. Biol Chem 382(8):1277–1285
27. Fisher JF, Mobashery S (2006) Recent advances in MMP inhibitor design. Cancer Metastasis Rev 25(1):115–136
28. Reiter LA, Mitchell PG, Martinelli GJ, Lopresti-Morrow LL, Yocum SA, Eskra JD (2003) Phosphinic acid-based MMP-13 inhibitors that spare MMP-1 and MMP-3. Bioorg Med Chem Lett 13(14):2331–2336
29. Dive V, Georgiadis D, Matziari M, Makaritis A, Beau F, Cuniasse P, Yiotakis A (2004) Phosphinic peptides as zinc metalloproteinase inhibitors. Cell Mol Life Sci 61(16):2010–2019
30. Sorsa T, Tjaderhane L, Konttinen YT, Lauhio A, Salo T, Lee HM, Golub LM, Brown DL, Mantyla P (2006) Matrix metalloproteinases: contribution to pathogenesis, diagnosis and treatment of periodontal inflammation. Ann Med 38(5):306–321
31. Rudek MA, Figg WD, Dyer V, Dahut W, Turner ML, Steinberg SM, Liewehr DJ, Kohler DR, Pluda JM, Reed E (2001) Phase I clinical trial of oral COL-3, a matrix metalloproteinase inhibitor, in patients with refractory metastatic cancer. J Clin Oncol 19(2):584–592
32. Schechter I, Berger A (1968) On the active site of proteases. 3. Mapping the active site of papain; specific peptide inhibitors of papain. Biochem Biophys Res Commun 32(5):898–902
33. MacColl E, Khalil RA (2015) Matrix metalloproteinases as regulators of vein structure and function: implications in chronic venous disease. J Pharmacol Exp Ther 355(3):410–428
34. Park HI, Jin Y, Hurst DR, Monroe CA, Lee S, Schwartz MA, Sang QX (2003) The intermediate S1′ pocket of the endometase/matrilysin-2 active site revealed by enzyme inhibition kinetic studies, protein sequence analyses, and homology modeling. J Biol Chem 278(51):51646–51653
35. Aureli L, Gioia M, Cerbara I, Monaco S, Fasciglione GF, Marini S, Ascenzi P, Topai A, Coletta M (2008) Structural bases for substrate and inhibitor recognition by matrix metalloproteinases. Curr Med Chem 15(22):2192–2222
36. Fabre B, Ramos A, de Pascual-Teresa B (2014) Targeting matrix metalloproteinases: exploring the dynamics of the s1′ pocket in the design of selective, small molecule inhibitors. J Med Chem 57(24):10205–10219
37. Hu Y, Xiang JS, DiGrandi MJ, Du X, Ipek M, Laakso LM, Li J, Li W, Rush TS, Schmid J, Skotnicki JS, Tam S, Thomason JR, Wang Q, Levin JI (2005) Potent, selective, and orally bioavailable matrix metalloproteinase-13 inhibitors for the treatment of osteoarthritis. Bioorg Med Chem 13(24):6629–6644
38. Udi Y, Fragai M, Grossman M, Mitternacht S, Arad-Yellin R, Calderone V, Melikian M, Toccafondi M, Berezovsky IN, Luchinat C, Sagi I (2013) Unraveling hidden regulatory sites in structurally homologous metalloproteases. J Mol Biol 425(13):2330–2346
39. Overall CM, Kleifeld O (2006) Towards third generation matrix metalloproteinase inhibitors for cancer therapy. Br J Cancer 94(7):941–946
40. Ndinguri MW, Bhowmick M, Tokmina-Roszyk D, Robichaud TK, Fields GB (2012) Peptide-based selective inhibitors of matrix

metalloproteinase-mediated activities. Molecules 17(12):14230–14248

41. Koivunen E, Arap W, Valtanen H, Rainisalo A, Medina OP, Heikkila P, Kantor C, Gahmberg CG, Salo T, Konttinen YT, Sorsa T, Ruoslahti E, Pasqualini R (1999) Tumor targeting with a selective gelatinase inhibitor. Nat Biotechnol 17(8):768–774
42. Suojanen J, Salo T, Koivunen E, Sorsa T, Pirila E (2009) A novel and selective membrane type-1 matrix metalloproteinase (MT1-MMP) inhibitor reduces cancer cell motility and tumor growth. Cancer Biol Ther 8(24):2362–2370
43. Devy L, Dransfield DT (2011) New strategies for the next generation of matrix-metalloproteinase inhibitors: selectively targeting membrane-anchored MMPs with therapeutic antibodies. Biochem Res Int 2011:191670
44. Devy L, Huang L, Naa L, Yanamandra N, Pieters H, Frans N, Chang E, Tao Q, Vanhove M, Lejeune A, van Gool R, Sexton DJ, Kuang G, Rank D, Hogan S, Pazmany C, Ma YL, Schoonbroodt S, Nixon AE, Ladner RC, Hoet R, Henderikx P, Tenhoor C, Rabbani SA, Valentino ML, Wood CR, Dransfield DT (2009) Selective inhibition of matrix metalloproteinase-14 blocks tumor growth, invasion, and angiogenesis. Cancer Res 69(4):1517–1526
45. Hoet RM, Cohen EH, Kent RB, Rookey K, Schoonbroodt S, Hogan S, Rem L, Frans N, Daukandt M, Pieters H, van Hegelsom R, Neer NC, Nastri HG, Rondon IJ, Leeds JA, Hufton SE, Huang L, Kashin I, Devlin M, Kuang G, Steukers M, Viswanathan M, Nixon AE, Sexton DJ, Hoogenboom HR, Ladner RC (2005) Generation of high-affinity human antibodies by combining donor-derived and synthetic complementarity-determining-region diversity. Nat Biotechnol 23(3):344–348
46. Hu J, Van den Steen PE, Houde M, Ilenchuk TT, Opdenakker G (2004) Inhibitors of gelatinase B/matrix metalloproteinase-9 activity comparison of a peptidomimetic and polyhistidine with single-chain derivatives of a neutralizing monoclonal antibody. Biochem Pharmacol 67(5):1001–1009
47. Martens E, Leyssen A, Van Aelst I, Fiten P, Piccard H, Hu J, Descamps FJ, Van den Steen PE, Proost P, Van Damme J, Liuzzi GM, Riccio P, Polverini E, Opdenakker G (2007) A monoclonal antibody inhibits gelatinase B/MMP-9 by selective binding to part of the catalytic domain and not to the fibronectin or zinc binding domains. Biochim Biophys Acta 1770(2):178–186
48. Sela-Passwell N, Kikkeri R, Dym O, Rozenberg H, Margalit R, Arad-Yellin R, Eisenstein M, Brenner O, Shoham T, Danon T, Shanzer A, Sagi I (2011) Antibodies targeting the catalytic zinc complex of activated matrix metalloproteinases show therapeutic potential. Nat Med 18(1):143–147
49. Heussen C, Dowdle EB (1980) Electrophoretic analysis of plasminogen activators in polyacrylamide gels containing sodium dodecyl sulfate and copolymerized substrates. Anal Biochem 102(1):196–202
50. Murphy G, Crabbe T (1995) Gelatinases A and B. Methods Enzymol 248:470–484
51. Hawkes SP, Li H, Taniguchi GT (2010) Zymography and reverse zymography for detecting MMPs and TIMPs. Methods Mol Biol 622:257–269
52. Lombard C, Saulnier J, Wallach J (2005) Assays of matrix metalloproteinases (MMPs) activities: a review. Biochimie 87(3–4):265–272
53. Snoek-van Beurden PA, Von den Hoff JW (2005) Zymographic techniques for the analysis of matrix metalloproteinases and their inhibitors. Biotechniques 38(1):73–83
54. Yin Z, Sada AA, Reslan OM, Narula N, Khalil RA (2012) Increased MMPs expression and decreased contraction in the rat myometrium during pregnancy and in response to prolonged stretch and sex hormones. Am J Physiol Endocrinol Metab 303(1):E55–E70
55. Bradford MM (1976) A rapid and sensitive method for the quantitation of microgram quantities of protein utilizing the principle of protein-dye binding. Anal Biochem 72: 248–254
56. Kleiner DE, Stetler-Stevenson WG (1994) Quantitative zymography: detection of picogram quantities of gelatinases. Anal Biochem 218(2):325–329
57. Michaud D, Cantin L, Raworth DA, Vrain TC (1996) Assessing the stability of cystatin/cysteine proteinase complexes using mildly-denaturing gelatin-polyacrylamide gel electrophoresis. Electrophoresis 17(1):74–79
58. Knauper V, Lopez-Otin C, Smith B, Knight G, Murphy G (1996) Biochemical characterization of human collagenase-3. J Biol Chem 271(3):1544–1550
59. Bjornland K, Winberg JO, Odegaard OT, Hovig E, Loennechen T, Aasen AO, Fodstad O, Maelandsmo GM (1999) S100A4 involvement in metastasis: deregulation of matrix metalloproteinases and tissue inhibitors of matrix metalloproteinases in osteosarcoma cells transfected with an anti-S100A4 ribozyme. Cancer Res 59(18):4702–4708
60. Fernandez-Resa P, Mira E, Quesada AR (1995) Enhanced detection of casein zymography of

matrix metalloproteinases. Anal Biochem 224(1):434–435

61. Gogly B, Groult N, Hornebeck W, Godeau G, Pellat B (1998) Collagen zymography as a sensitive and specific technique for the determination of subpicogram levels of interstitial collagenase. Anal Biochem 255(2):211–216
62. Yu WH, Woessner JF Jr (2001) Heparin-enhanced zymographic detection of matrilysin and collagenases. Anal Biochem 293(1):38–42
63. Woessner JF Jr (1995) Quantification of matrix metalloproteinases in tissue samples. Methods Enzymol 248:510–528
64. Kupai K, Szucs G, Cseh S, Hajdu I, Csonka C, Csont T, Ferdinandy P (2010) Matrix metalloproteinase activity assays: importance of zymography. J Pharmacol Toxicol Methods 61(2):205–209
65. Masure S, Proost P, Van Damme J, Opdenakker G (1991) Purification and identification of 91-kDa neutrophil gelatinase. Release by the activating peptide interleukin-8. Eur J Biochem 198(2):391–398
66. Garfin DE (2009) One-dimensional gel electrophoresis. Methods Enzymol 463:497–513
67. Pitt-Rivers R, Impiombato FS (1968) The binding of sodium dodecyl sulphate to various proteins. Biochem J 109(5):825–830
68. Shapiro AL, Vinuela E, Maizel JV Jr (1967) Molecular weight estimation of polypeptide chains by electrophoresis in SDS-polyacrylamide gels. Biochem Biophys Res Commun 28(5): 815–820
69. Ikeda M, Maekawa R, Tanaka H, Matsumoto M, Takeda Y, Tamura Y, Nemori R, Yoshioka T (2000) Inhibition of gelatinolytic activity in tumor tissues by synthetic matrix metalloproteinase inhibitor: application of film in situ zymography. Clin Cancer Res 6(8):3290–3296
70. Van den Steen PE, Dubois B, Nelissen I, Rudd PM, Dwek RA, Opdenakker G (2002) Biochemistry and molecular biology of gelatinase B or matrix metalloproteinase-9 (MMP-9). Crit Rev Biochem Mol Biol 37(6):375–536
71. Springman EB, Angleton EL, Birkedal-Hansen H, Van Wart HE (1990) Multiple modes of activation of latent human fibroblast collagenase: evidence for the role of a Cys73 active-site zinc complex in latency and a "cysteine switch" mechanism for activation. Proc Natl Acad Sci U S A 87(1):364–368
72. Van Wart HE, Birkedal-Hansen H (1990) The cysteine switch: a principle of regulation of metalloproteinase activity with potential applicability to the entire matrix metalloproteinase gene family. Proc Natl Acad Sci U S A 87(14):5578–5582
73. Li W, Mata KM, Mazzuca MQ, Khalil RA (2014) Altered matrix metalloproteinase-2 and -9 expression/activity links placental ischemia and anti-angiogenic sFlt-1 to uteroplacental and vascular remodeling and collagen deposition in hypertensive pregnancy. Biochem Pharmacol 89(3):370–385

Chapter 9

Detection and Characterization of Bacterial Proteinases Using Zymography

Madathiparambil G. Madanan and Ambili Mechoor

Abstract

Proteinases play a crucial role in invasion and pathogenesis of bacteria, especially the extracellular and membrane-bound forms. Analysis of these proteinases demands the isolation by retaining the enzymatic activity. The isolation procedures maintaining the native structure of the enzyme in its soluble form are also of extreme importance. The qualitative analyses of these proteinases are carried out by electrophoresis and zymography. Enzymatic characterization based on the effect of inhibitors and activators on gelatinase activity also can be assessed using this zymography. The membrane-bound proteinases can be isolated in their native and soluble form, still retaining the activity using 6-aminocaproic acid and sodium deoxycholate; the procedure of which is explained in this chapter.

Key words Bacteria, Gelatinase, Proteinase, Zymography, Leptospira, Extracellular matrix

1 Introduction

Degradation and remodeling of the extracellular matrix (ECM) components play a vital role in maintaining the physiological tissue functions. Abnormal degradation of the ECM components also transpires in several pathological conditions like cancer, hepatitis, and in several bacterial infections as well. One of the major ECM targets that undergo drastic degradation is collagen which is executed primarily by a group of proteinases, which includes collagenases and gelatinases [1–3]. Several of these enzymes are produced by bacteria as well [4]. Bacterial proteinases are mainly seen in two forms, viz.: the secretory forms and the membrane-bound forms [5–8]. The usual extraction procedures for the total proteins including membrane proteins involve the use of strong detergents and chaotropic ions which can affect the enzymatic activity. Hence, these procedures implemented for enzyme extraction cannot be implemented for the isolation of the membrane-bound bacterial gelatinases and collagenases as it may affect the enzyme activity. An extraction procedure which preserves the enzyme activity and

Jeff Wilkesman and Liliana Kurz (eds.), *Zymography: Methods and Protocols*, Methods in Molecular Biology, vol. 1626,
DOI 10.1007/978-1-4939-7111-4_9, © Springer Science+Business Media LLC 2017

solubility only can enable the electrophoretic separation and facilitate zymography development for detection of proteinase activity.

Described below is an isolation procedure of the total proteins including the membrane-bound bacterial gelatinases using sodium deoxycholate and 6-aminocaproic acid. The procedure isolates the enzymes in its intact native soluble form, which enables electrophoretic separation and zymographic detection. Sodium deoxycholate, a bile salt-related anionic detergent, can solubilize the lipid bilayer of the bacterial membrane releasing the bound proteinases [9–11]. The protein extracted by this procedure may not be readily soluble. The routine method of protein solubilization is using sodium chloride, which cannot be used here as it interferes with electrophoretic separation. Hence, as an alternate safe solution, 0.75 M 6-aminocaproic acid is used to solubilize the extracted proteins of outer membrane [12, 13]. The 6-aminocaproic acid is not found to be interfering with the electrophoresis and can be loaded mixing with the sample buffer.

1.1 Zymography

Zymography is based on the following principles: (1) during electrophoresis, the substrate is retained in the gel; (2) proteinase/gel atinase/collagenase activity is reversibly inhibited by SDS during electrophoresis. An additional advantage of zymography is that both the proenzymes and the active forms of enzymes can be distinguished on the basis of their molecular weight [14].

Zymography is a powerful technique for analyzing hydrolytic enzymes based on substrate degradation. During zymography, the proteinases are separated by electrophoresis under denaturing, nonreducing conditions. The separation occurs in a polyacrylamide gel containing a specific substrate that is copolymerized with the acrylamide [15]. During electrophoresis, sodium dodecyl sulfate (SDS) denatures the proteinases into an inactive form. On completion of the electrophoretic run, the gel is washed with Triton X-100, which causes the exchange of the SDS with Triton X-100. This helps in partial renaturation and recovery of the enzyme activity [16]. Subsequently, the gel is incubated in the appropriate activation buffer. During incubation, the concentrated, renatured proteinase in the gel will digest/degrade the impregnated substrate. After incubation, the gel is stained with Coomassie blue, and the proteinase activity is detected as clear bands against a blue background of undegraded substrate. The clear bands in the gel can be quantified by densitometry if needed.

1.2 Bacterial Proteinase Detection Using Zymography

This method is used to find all expressed gelatinases of the laboratory grown *Leptospira*—a spirochete pathogen causing Leptospirosis. The results indicated the presence of 12 gelatinases in the molecular mass range of 240–32 kDa, representing different

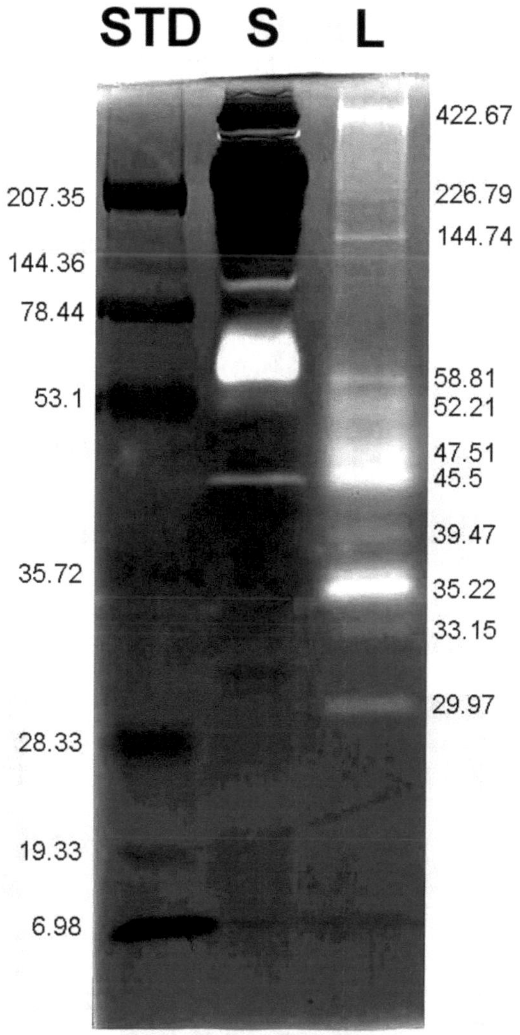

Fig. 1 Zymography of total protein extracted from *Leptospira* using 50 mM Tris–HCl, pH 7.2, containing 0.5% deoxycholate and 0.75 M 6-aminocaproicacid. Molecular weight standard (STD), 1 μL (~70 μg protein) Human serum (S), and 50 μg *Leptospira* protein extract (L), loaded on 10% zymography gel

types of proteinases (Fig. 1). The protein extract can be pretreated with aminophenyl mercuric acetate (APMA) to bring the proenzymes into active form. Treatment with different activators and inhibitors of proteinases was also carried out in the same buffer system used for the extraction (Fig. 2). This enabled further characterization of the proteinases avoiding further buffer exchange and loss of solubility of the membrane-bound forms. Furthermore, this helped in the direct application of the test onto the zymography to assess the change in activity.

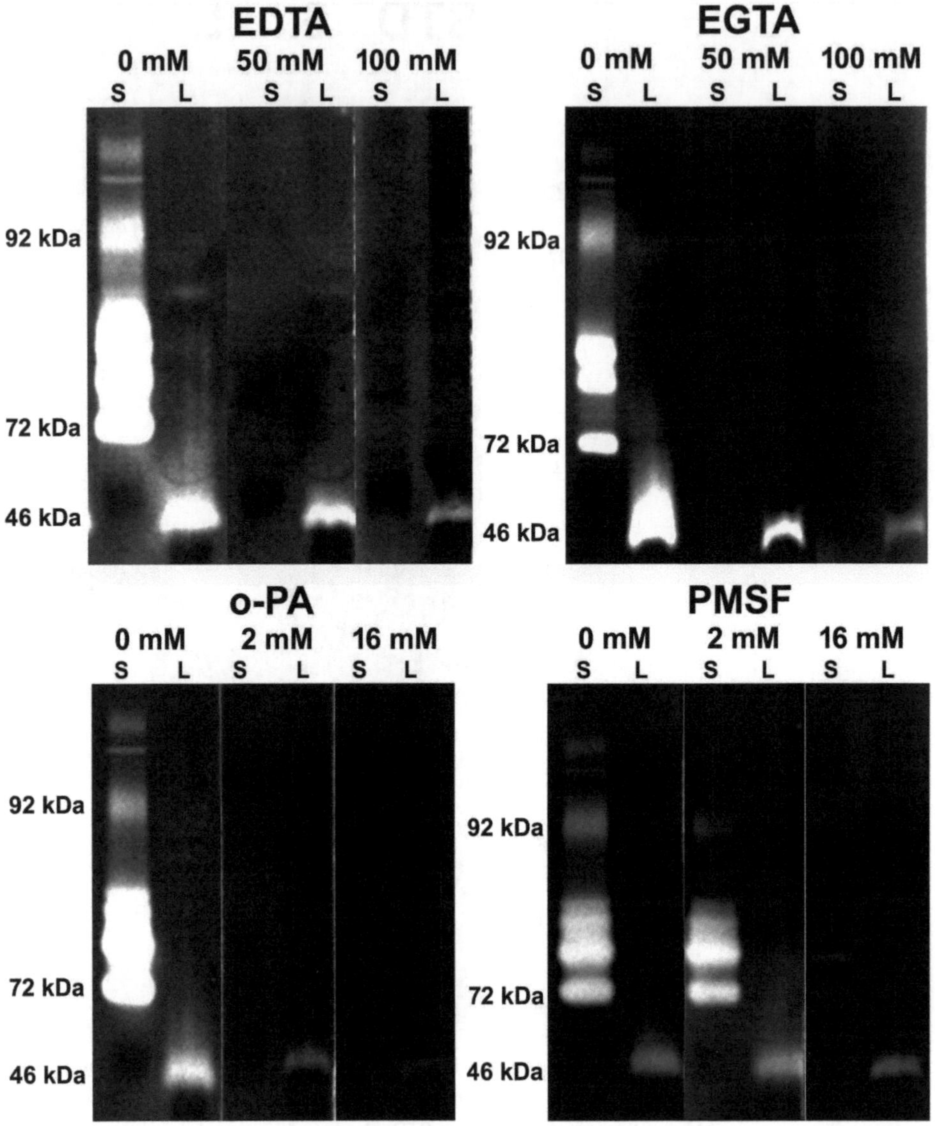

Fig. 2 Effect of inhibitors on leptospiral proteinase activity: Zymographic analysis on 10% zymography gel. *Leptospira* proteinase activity was tested in the presence of EDTA, EGTA, *o*-PA, or PMSF. The inhibitors were added reaching the indicated concentration and incubated for 18 h at 37 °C. Each well contains 1 μL (~70 μg protein) Human serum (S), and 5 μg protein from 46 kDa region of gel filtration fraction of *Leptospira* total protein extract (L)

2 Materials

2.1 Chemicals

1. Gelatin from bovine skin, Triton X-100, sodium deoxycholate, 6-aminocaproic acid, 1,10-phenanthroline monohydrate (*o*-PA), phenylmethylsulfonyl fluoride (PMSF), ethylenediaminetetraacetic acid (EDTA), ethylene glycol tetraacetic acid

(EGTA), deionized water to attain a sensitivity of 18 MΩ cm at 25 °C.

2. All other reagents/chemicals are of analytical grade (*see* **Note 1**).

2.2 Instruments

1. Electrophoresis apparatus: Any brand/model that can cast 10 × 7 cm^2 or more sized gel of thickness 1 mm.
2. Gel trays: Gel trays must be of at least 5% larger in dimensions and having smooth and flat surface with lid.
3. Incubator: For incubating the gel trays for 18 h at 37 °C.
4. Shaker: Rocking or rotary shaker with low speed for staining the gel.

2.3 Buffers and Stock Solutions

2.3.1 Stock Solutions for Electrophoresis

1. *Acrylamide-bis acrylamide solution (30% T, 3% C):* Dissolve acrylamide 29.1 g and bis-acrylamide 0.9 g in 40 mL water. Make up to 100 mL with water and filter through a Whatman grade 1 filter paper. Store the stock solution at 4 °C in an amber-colored bottle.
2. Resolving gel buffer (1.5 M Tris–HCl, pH 8.8): Dissolve 18.18 g Tris base with 80 mL water. Adjust the pH to 8.8 by titrating with HCl. Make up the volume to 100 mL with water and store at 4 °C.
3. Stacking gel buffer (0.5 M Tris–HCl, pH 6.8): Dissolve 6.06 g Tris base with 80 mL water. Adjust the pH by titrating with HCl to 6.8. Make up the volume to 100 mL with water and store at 4 °C.
4. Sodium dodecyl sulfate 10%: Dissolve 10 g SDS in 80 mL of water and make up the volume to 100 mL and store at room temperature.
5. Ammonium persulfate (APS) 10%: Dissolve 100 mg ammonium persulfate in 1 mL of water and store at 4 °C. This stock is stable up to 2 weeks.
6. Gelatin 10%: Dissolve 100 mg gelatin in 1 mL of water by warming at 40 °C. This can be stored in aliquots of 100 μL at −20 °C. Thaw to dissolve completely before use. This stock is stable up to 2 weeks.
7. 10× Electrode buffer (0.025 M Tris–HCl, 0.192 M glycine, 0.1% SDS pH 8.3): Dissolve 30.28 g Tris base, 144 g glycine, and 10 g SDS in 800 mL of water. Make up the solution to 1000 mL. The pH of the solution thus prepared would be approximately 8.3. The buffer required for running the electrophoresis may be diluted to 1× and the stock buffer stored at 4 °C.
8. Sample Buffer (2×): Take 5 mL of water and 2.5 mL of 0.5 M Tris–Cl, pH 6.8 (Resolving Gel Buffer as above) in a 10 mL graduated tube and dissolve 200 mg SDS slowly without clogging and

foaming. Add 2 mL of glycerol mix completely and dissolve 20 μg of bromophenol blue. Make up the volume to 10 mL and store at room temperature.

2.3.2 Reagents for Protein Extraction

1. Wash Buffer: Phosphate buffered saline containing 5 mM $MgCl_2$: Take 100 mL of PBS and dissolve 47.6 mg of $MgCl_2$.
2. Extraction buffer (50 mM Tris–HCl, pH 7.2, containing 0.5% deoxycholate and 0.75 M 6-aminocaproicacid): 500 mM Tris–Cl pH 7.2 was prepared by dissolving 6.06 g Tris base in 80 mL of water and pH adjusted to 7.2 with HCl. The volume made up to 100 mL with water and stored at 4 °C.

   Take 1 mL of 500 mM Tris–HCl pH 7.2 in a 10 mL graduated tube and add 5 mL of water. Dissolve 50 mg of deoxycholate and 984 mg of 6-aminocaproicacid (*see* **Note 2**). Make up the volume to 10 mL while dissolving the reagents completely. Store at 4 °C (*see* **Note 3**).

2.3.3 Reagents for Zymography

1. Renaturing Buffer (2.5% Triton X-100 solution): Dissolve 2.5 mL of Triton X-100 in 97.5 mL of water (*see* **Note 4**).
2. Stock 500 mM Tris–Cl: 500 mM Tris–Cl pH 7.5 is prepared by dissolving 6.06 g Tris base in 80 mL of water and pH brought to 7.5 with HCl. Volume made up to 100 mL with water and store at 4 °C.
3. Stock 250 mM $CaCl_2$: Dissolve 2.77 g $CaCl_2$ in 100 mL water and store at room temperature.
4. Activation buffer (50 mM Tris–HCl pH 7.5, 1% Triton X-100 and 25 mM $CaCl_2$): Dissolve 1 mL of Triton X-100 in 79 mL of water. Add 10 mL of 500 mM Tris–Cl pH 7.5 and 10 mL of 250 mM $CaCl_2$ (*see* **Note 5**).
5. Stock 500 mM EDTA: Dissolve 18.61 g EDTA in 80 mL of water, add NaOH pellets (approximately 2 g) to bring the pH 8, adjust the volume to 100 mL.
6. Stock 500 mM EGTA: Dissolve 19.02 g EGTA in 80 mL of water, add NaOH pellets (approximately 6.5 g) to bring the pH 8, adjust the volume to 100 mL.
7. Stock 50 mM *o*-PA: Dissolve 99.11 mg *o*-PA in 10 mL of ethanol.
8. Stock 100 mM PMSF: Dissolve 17.14 mg PMSF in 10 mL of isopropanol and store at −20 °C.

2.3.4 Reagents for Staining

1. Coomassie blue stain: To make 100 mL Coomassie blue stain mix 40 mL methanol (or ethanol), 10 mL acetic acid, 50 mL of water, and dissolve 10 mg Coomassie brilliant blue R 250.
2. Destaining solution: Destaining solution is a mixture of 40 mL methanol (or ethanol), 10 mL acetic acid, and 50 mL water (total volume will be 100 mL).

3 Methods

3.1 Electrophoresis

1. Assembly of gel cassette: The glass plates should be thoroughly cleaned using detergent without leaving any greasy areas left on the surface. Fingerprints on the glass surface also should be avoided. Align the plates perfectly without any leak. The leak can be checked by pouring water into the casting setup and wait for few minutes to find the leak based on decrease in the water level. The water should be then drained and the inner side of the casting setup should be thoroughly cleaned with long strips of filter paper.
2. Resolving gel: The upper level of resolving gel may be decided by placing the comb in position and marking the bottom of each well using a marking pen. Make the resolving gel mix by adding the components at required volume, with reference to the Table 1, to cast the gel. The system mentioned (8 × 10 cm^2) required 10 mL of resolving gel mix. Add ammonium persulfate just before casting. Pour the resolving gel mix into the casting chamber through the middle portion of the setup to about 5 mm below the marking of bottom line of the wells. Care should be taken to avoid air bubbles. Slowly layer some water about 1 mm level on the top of the gel to make an even layer and removing bubbles. Allow to set the gel for 1 h (*see* **Note 6**).
3. Stacking gel: Decant the water from the top of the gel by tilting the gel setup to one side so that the water gets collected to the corner which can be removed using a narrow strip of filter paper. Make the stacking gel mix by adding the components at required volume, with reference to the Table 2, to cast the gel. Pour the staking gel onto the resolving gel and place the comb in position to form the wells with about 5 mm stalking gel between the resolving gel and the bottom of wells. Wait for at least 3 h for complete polymerization. This is required for

Table 1
Composition of resolving gel

Resolving gel 10% T	10 mL	20 mL	50 mL
H_2O	3.86	7.71	19.27
Tris–HCl pH 8.8	2.5	6.68	16.7
ABA 30%	3.34	5	12.5
10% Gelatin	0.1	0.2	0.5
10% SDS	0.1	0.2	0.5
TEMED	0.006	0.012	0.03
10% APS	0.1	0.2	0.5

Table 2
Composition of stacking gel

Stacking gel 3.5% T	3 mL	6 mL	10 mL
H_2O	1.84	3.68	6.13
ABA 30%	0.35	0.7	1.17
Tris–HCl pH 6.6	0.75	1.5	2.5
10% SDS	0.03	0.06	0.1
TEMED	0.006	0.009	0.012
10% APS	0.03	0.06	0.1

good cross-linking of the gelatin. Remove the lower spacer and fix the gel on the electrophoresis apparatus. Slowly keep the gels in the tank containing 1× electrode buffer without having trapped air in the bottom of the gel.

4. Preparation of Sample: Bacterial culture (10 mL culture of *Leptospira*) is centrifuged at 5000 × *g* for 10 min and the pellet shall be washed two times with 1 mL of wash buffer and centrifuged at 5000 × *g* for 10 min at room temperature. To the pellet obtained after second wash, add 100 μL of extraction buffer and vortex for 5 min. The supernatant of the extract will be centrifuged 12,000 × *g* for 10 min at room temperature and the supernatant can be used as the sample. Required amount (about 50–100 μg protein per well) of this sample should be mixed with equal amount of 2× sample buffer and incubate at 37 °C for 30 min before loading onto the gel.
5. Loading the sample: Load the sample into the wells carefully with a micropipette and gel loading tips to avoid cross-contamination of samples. Carefully top-up the remaining portion of the well and the entire top of the gel with electrode buffer without disturbing the loaded samples. Fill the upper tank of the electrophoresis unit also using the electrode buffer.
6. Running the gel: Electrophoresis must be carried out at 4 °C in a cold room. Alternately, electrophoresis can be performed inside a refrigerator (*see* **Note 7**). Electrophoresis unit is switched on initially at 6 mA per gel until the dye front enters the resolving gel. Increase the current to 12 mA per gel for remaining run until the dye front reaches 1 cm above from the bottom level of the gel (*see* **Note 8**). In a mini gel format (8 × 10 cm), the entire run takes about 2 h.
7. Marking of bands of the standard: On completing the electrophoresis, carefully open the glass plates using a plastic spatula or wonder wedge-plate separation tool and mark the position of dye front and prestained marker by punching holes or scanning the gel.

3.2 Zymography

1. Place the gel submerged in renaturing solution in a gel tray. Incubate at room temperature for 30 min without shaking (*see* **Note 9**).
2. Remove the renaturing solution and add activation buffer in order to completely submerge the gel.
3. Incubate in an incubator at 37 °C for 18 h.

3.3 Characterization of Proteinases

For protein characterization, the gel will be subjected to incubation with different reagents and concentrations to be tested. To achieve this, the reagent to be tested must be added to the activation buffer. This can be made from stock solution of the reagents which is made in appropriate solvents. This stock must be added, in order to reach the required final concentration of the reagent, while making the activation buffer by replacing equivalent amount of water from the buffer composition.

1. After renaturation the gel should be kept on a clean glass plate.
2. Lanes to be tested in different reagents must be cut through the inter-lane space recognizable through the space between the wells on top and space between the color of dye front at the bottom.
3. Mark the orientation of gel on each lane.
4. Place the gel in a small tray sufficient to occupy the gel. Small trays used for incubation of Western blots or 15 mL screw-caped centrifuge tubes (for single lane from 8 × 10 cm gel and careful to avoid folding and partial immersion in the buffer) may be used. For each lane from 8 × 10 cm^2 format gel 10 mL of the activation buffer will be sufficient.
5. Incubate the gels in an incubator at 37 °C for 18 h for activity (*see* **Note 10**).

3.4 Staining

1. After the incubation, remove the activation buffer and briefly rinse the gel with water for 1 min. This removes the extra Triton X-100 on the gel which may cause unnecessary foaming while shaking.
2. Add staining solution into the tray submerging the gel completely.
3. Keep the gel shaking at lower speed to avoid damage to the gel.
4. Staining may be continued about 1 h or until the gel becomes completely blue.
5. Remove the stain and rinse the gel for 1 min with water to remove extra stain from the surface of the gel.
6. Add destaining solution into the tray and keep shaking.
7. Change the destaining solution several times until the desired band contrast is observed with respect to the blue background.

8. Scan the gel using a gel documentation system or document scanner.
9. Determine the molecular mass of protein bands with respect to the protein standards.

4 Notes

1. The reagents can be stored at room temperature (unless indicated otherwise). The waste disposal regulations should be strictly followed while disposing waste materials.
2. Crystallization of the reagents may occur at 4 °C which can be redissolved at room temperature before extraction procedure
3. The 6-aminocaproic acid is also known as EACA and is reported to inhibit chymotrypsin, Factor VIIa, lysine carboxypeptidase, plasmin, and plasminogen activator [17–21]; but the inhibitory activity was not tested in our experiment, as the samples were of bacterial origin.
4. This is done by taking Triton X-100 with a wide bore pipette and pouring into water while continuously stirring on a magnetic stirrer.
5. The volume required may vary in order to completely submerge the gel in the buffer with respect to the size of the gel tray used.
6. We found that the 1 mm thick gel is ideal for good resolution and contrast of bands. The thinner gels make lighter background and the thicker ones make darker background with less visibility of bands. This may be due to more destaining and less amount of gelatin in the thinner gels in contrast to high amount of gelatin and difficulty in destaining in the thicker gels. Also, the gelatin in the thicker gels may not have completely digested by the proteinase due to requirement of more time; the enzyme may get degraded during the course of time or may not be having proper renaturation and buffering due to thickness of the gel.
7. In our electrophoresis system, about three fourth lower portions of the plates are submerged in the electrode buffer (anode) tank. This system exchanges heat generated during electrophoresis and prevents loss of proteinase activity. We have used prechilled electrode buffer (stored at 4 °C) which had an advantage of avoiding requirement of cold room or refrigerator for running the gel.
8. In case of other gels the voltage and run time can be calculated as per the standard SDS–PAGE.

9. In many publications and according to other procedures, shaking enhances SDS elimination, as well as multiple washes (3 × 15 min each wash) are suggested. Sometimes with shaking conditions we have noticed that the background become pale and bands not sharp. We assume that this is because the gel is not fixed and, while incubation, the protein leaches out and the bands spread more. Even without shaking, we were able to get good bands.
10. In the experiment we have tested the inhibition of proteinases with different inhibitors such as EDTA (0, 50 and 100 mM), EGTA (0, 50 and 100 mM), *o*-PA (0, 2 and 16 mM), PMSF (0, 2 and 16 mM) by developing zymogram gels in activation buffer containing these reagents incubated for 18 h at 37 °C as shown in Table 3. The results show differences in band intensity caused by the presence of the inhibitors (Fig. 2). While EDTA, EGTA, and *o*-PA inhibit metalloproteases by chelation, PMSF inhibits serine or cysteine proteases. Similarly, activation by different metal ions, effect of pH, and temperature on the proteinase activity can be demonstrated [4, 7] in order to characterize the proteinases.

Table 3
Modification of activation buffer with inhibitors

Final concentration (mM) of inhibitors in activation buffer	EDTA			EGTA			*o*-PA			PMSF		
	0	50	100	0	50	100	0	2	16	0	2	16
Water (mL)	7.9	6.9	5.9	7.9	6.9	5.9	7.9	7.5	4.7	7.9	7.7	6.3
500 mM EDTA (mL)	0	1	2	–	–	–	–	–	–	–	–	–
500 mM EGTA (mL)	–	–	–	0	1	2	–	–	–	–	–	–
50 mM *o*-PA (mL)	–	–	–	–	–	–	0	0.4	3.2	–	–	–
100 mM PMSF (mL)	–	–	–	–	–	–	–	–	–	0	0.2	1.6
500 mM Tris-HCl, pH 7.5 (mL)	1	1	1	1	1	1	1	1	1	1	1	1
250 mM $CaCl_2$ (mL)	1	1	1	1	1	1	1	1	1	1	1	1
Triton X-100 (mL)	0.1	0.1	0.1	0.1	0.1	0.1	0.1	0.1	0.1	0.1	0.1	0.1
Total volume (mL)	10	10	10	10	10	10	10	10	10	10	10	10

Acknowledgments

All the demonstrated experiments were carried out by Mr. Sandhanakrishnan Cattavarayane and the work was supported by grant Nos. SR/SO/HS-29/2005 and SR/S0/HS/0108/2012, SERB Division, Department of Science and Technology, New Delhi, India to M.G.M.

References

1. Ambili M, Jayasree K, Sudhakaran PR (1998) 60K gelatinase involved in mammary gland involution is regulated by beta-oestradiol. Biochim Biophys Acta 1403:219–231
2. Birkedal-Hansen H, Moore WG, Bodden MK et al (1993) Matrix metalloproteinases: a review. Crit Rev Oral Biol Med 4:197–250
3. Sato H, Takino T, Okada Y et al (1994) A matrix metalloproteinase expressed on the surface of invasive tumour cells. Nature 370: 61–65
4. Madathiparambil MG, Cattavarayane S, Manickam GD et al (2011) A zymography analysis of proteinase activity present in Leptospira. Curr Microbiol 62:917–922
5. Haiko J, Suomalainen M, Ojala T et al (2009) Invited review: breaking barriers—attack on innate immune defences by omptin surface proteases of enterobacterial pathogens. Innate Immun 15:67–80
6. Le Sage V, Zhu L, Lepage C et al (2009) An outer membrane protease of the omptin family prevents activation of the Citrobacter rodentium PhoPQ two-component system by antimicrobial peptides. Mol Microbiol 74:98–111
7. Madathiparambil MG, Cattavarayane S, Perumana SR et al (2011) Presence of 46 kDa gelatinase on the outer membrane of Leptospira. Curr Microbiol 62:1478–1482
8. Sole M, Scheibner F, Hoffmeister AK et al (2015) Xanthomonas campestris pv. vesicatoria secretes proteases and xylanases via the Xps type II secretion system and outer membrane vesicles. J Bacteriol 197:2879–2893
9. Levy D, Gulik A, Seigneuret M et al (1990) Phospholipid vesicle solubilization and reconstitution by detergents. Symmetrical analysis of the two processes using octaethylene glycol mono-n-dodecyl ether. Biochemistry 29: 9480–9488
10. Lichtenberg D, Ahyayauch H, Alonso A et al (2013) Detergent solubilization of lipid bilayers: a balance of driving forces. Trends Biochem Sci 38:85–93
11. Seddon AM, Curnow P, Booth PJ (2004) Membrane proteins, lipids and detergents: not just a soap opera. Biochim Biophys Acta 1666:105–117
12. Schagger H, Cramer WA, Von Jagow G (1994) Analysis of molecular masses and oligomeric states of protein complexes by blue native electrophoresis and isolation of membrane protein complexes by two-dimensional native electrophoresis. Anal Biochem 217:220–230
13. Schagger H, Von Jagow G (1991) Blue native electrophoresis for isolation of membrane protein complexes in enzymatically active form. Anal Biochem 199:223–231
14. Woessner JF Jr (1995) Quantification of matrix metalloproteinases in tissue samples. Methods Enzymol 248:510–528
15. Leber TM, Balkwill FR (1997) Zymography: a single-step staining method for quantitation of proteolytic activity on substrate gels. Anal Biochem 249:24–28
16. Snoek-Van Beurden PA, Von Den Hoff JW (2005) Zymographic techniques for the analysis of matrix metalloproteinases and their inhibitors. Biotechniques 38:73–83
17. Adelman B, Rizk A, Hanners E (1988) Plasminogen interactions with platelets in plasma. Blood 72:1530–1535
18. Dessaint JP, Katz SP, Waksman BH (1979) Catheptic carboxypeptidase B as a major component in "T-cell activating factor" of macrophages. J Immunopharmacol 1:399–414
19. Humphries JE, Vasudevan J, Gonias SL (1993) Fibrinogenolytic and fibrinolytic activity of cell-associated plasmin. Arterioscler Thromb 13:48–55
20. Krishnamurti C, Vukelja SJ, Alving BM (1994) Inhibitory effects of lysine analogues on t-PA induced whole blood clot lysis. Thromb Res 73:419–430
21. Soter NA, Austen KF, Gigli I (1975) Inhibition by epsilon-aminocaproic acid of the activation of the first component of the complement system. J Immunol 114:928–932

Chapter 10

A Sensitive, Rapid, and Specific Technique for the Detection of Collagenase Using Zymography

Shivcharan Prasad and Ipsita Roy

Abstract

In-gel zymography is a commonly employed tool to identify active enzymes in a quantitative and qualitative manner. In this work, apart from the incorporation of substrate which is traditionally employed in zymography, the identification of collagenase by incubation of the enzyme resolved on a polyacrylamide gel with substrate solution is described. The two methods are quite fast and result in specific detection of bacterial collagenase.

Key words Collagen, Collagenase, Matrix metalloproteinase, Zymography

1 Introduction

Matrix metalloproteinases (MMPs) are a group of structurally and functionally related calcium-dependent endopeptidases which act on cellular matrices. These are synthesized as inactive precursors and are converted into the functionally active form as per physiological need. An imbalance in the ratio of MMPs and their inhibitors is a characteristic of many disease conditions like cancer metastasis, cardiovascular diseases, rheumatoid arthritis, etc. [1]. Hence, accurate measurement of active MMPs is important. Different methods have been reported in the literature to assay the activity of matrix metalloproteinases [2]. Although native proteins have been employed as substrates, most of the enzymatic assays for matrix metalloproteinases rely on synthetic substrates as the natural substrates are difficult to dissolve in aqueous buffers. These reactions are also quite slow. The synthetic substrates are often labeled. Triple helical peptides derived from collagen have also been employed [2]. Immunological methods like ELISA have limited application as they are unable to distinguish between the inactive precursor, isoforms, the degraded protein, and the active enzyme. Zymography (visualization of an enzyme on a gel by substrate conversion) is a commonly used tool

Jeff Wilkesman and Liliana Kurz (eds.), *Zymography: Methods and Protocols*, Methods in Molecular Biology, vol. 1626,
DOI 10.1007/978-1-4939-7111-4_10, © Springer Science+Business Media LLC 2017

for measuring the activity of MMPs, mainly gelatinases and collagenases [3–6]. In fact, the initial identification of an active collagenase involved in collagen degradation employed zymography [7]. Bacterial collagenases are defined by their ability to cleave helical structures in fibrillar collagen in a physiological milieu [8]. The collagenase ColG from *Clostridium histolyticum* is classified as a Class I collagenase [9] while ColH from the same organism is placed in Class II [10], based on the presence and arrangement of their domains and subdomains. Bacterial collagenases are employed in a wide range of applications including debridement of wounds and burns and other areas of medical interest, food technology, tannery, pharmaceutical and cosmetic industries [11]. In this work, we have described two protocols for specific detection of very low amounts (nanogram level) of collagenases [5]. In the first case, collagen is incorporated into the polyacrylamide gel matrix while in the second method, the polyacrylamide gel-immobilized collagenase is incubated with the substrate solution post electrophoretic run. Although our method is developed using purified collagenase, we show its applicability in a crude sample containing a mixture of enzymes including collagenase.

2 Materials

All solutions were prepared in ultrapure deionized water of 18 MΩ cm at 25 °C.

2.1 Material Description

The following materials are used:

1. Crude collagenase containing collagenase Type I (125 U/mg).
2. Collagenase (EC 3.4.24.3, Type VII from *Clostridium histolyticum*).
3. Collagen (Type I).

2.2 Preparation of Collagen Solution

1. Add collagen to 0.013 M HCl (4 mg/mL). Incubate at 60 °C overnight [3].
2. Filter the collagen solution with Whatman grade 1 filter paper (*see* **Note 1**).
3. Adjust the pH of collagen solution to assay condition (*see* **Note 2**). For this, add one-fourth volume of a buffer concentrate prepared by mixing equal volumes of 0.2 M Na_2HPO_4 and 0.5 M NaCl. Adjust pH to 7.4 with 0.1 M NaOH [3]. The final concentration of collagen is adjusted to 3.2 mg/mL.

2.3 Non-denaturing Polyacrylamide Gel Electrophoresis

1. Prepare 30% (w/v) acrylamide-bisacrylamide solution (30% T, 2.6% C) (29.2:0.8 acrylamide:bisacrylamide). For this, take 40 mL distilled water in an amber bottle of 100 mL capacity. Weigh 29.2 g of acrylamide crystal and 0.8 g *N*,*N*′-methylene

bisacrylamide and transfer to the bottle. Stir on a magnetic stirrer for 30 min (*see* **Note 3**). Make up the volume to 100 mL with distilled water. Filter through Whatman grade 1 filter paper. Store the solution at 4 °C till use.

2. For 4× resolving gel buffer, weigh 18.15 g of Tris and dissolve in 80 mL of distilled water in a glass beaker. Adjust the pH to 8.8 with 3 M HCl and make up the final volume to 100 mL. Store the solution at 4 °C till use.
3. For 4× stacking gel buffer, weigh 6.0 g of Tris and dissolve in 80 mL of distilled water in a glass beaker. Adjust the pH to 6.8 with 3 M HCl and make the final volume up to 100 mL. Store the solution at 4 °C till use.
4. Prepare 10% (w/v) ammonium persulfate (APS) in distilled water (*see* **Note 4**).
5. Prepare electrophoresis running buffer by stirring 6.05 g Tris buffer and 28.80 g glycine in 900 mL distilled water in a 1 L graduated cylinder for 30 min. Adjust the pH to 8.3 with 3 M HCl and make up the final volume to 1 L with distilled water.
6. Prepare sample loading buffer by mixing 100 mM Tris–HCl, pH 6.8, 0.2% (w/v) bromophenol blue and 20% (v/v) glycerol.
7. Prepare staining solution by dissolving 250 mg Coomassie Brilliant Blue R-250 in a solution of 45 mL each of methanol and distilled water and 10 mL glacial acetic acid.
8. Prepare destaining solution by mixing 375 mL methanol (37.5% v/v) and 100 mL glacial acetic acid (10% v/v) and making up the volume to 1 L with distilled water.

3 Method

3.1 Zymography of Collagenase (Protocol 1)

1. Thaw all solutions at room temperature for at least 15 min.
2. Prepare the resolving gel by adding 3.2 mL distilled water, 2 mL resolving gel buffer, 2.64 mL acrylamide-bisacrylamide solution and 963 μL collagen in a 15 mL falcon tube.
3. Add 80 μL ammonium persulfate solution and 6 μL TEMED to the above. Mix gently and add the solution in-between two glass plates with the help of a pipette. Stand the gel on a horizontal table for 30–45 min for complete polymerization.
4. Add 1:1 isopropanol solution on the top of the gel.
5. Once the gel has polymerized, decant off the solvent and wash the top of the gel once with distilled water.
6. Prepare the stacking gel by adding 1.83 mL distilled water, 0.75 mL stacking gel buffer and 0.39 mL acrylamide-bisacrylamide

solution in a 15 mL falcon tube. Add 15 μL ammonium persulfate solution and 3 μL TEMED to this solution.

7. Mix gently and add the solution on top of the resolving gel in-between two glass plates with the help of a pipette.
8. Insert the comb gently between the gel plates and avoid trapping of air bubble below the comb. Allow the gel to stand for at least 30 min to polymerize.
9. Remove the combs and wash the wells with electrophoresis running buffer.
10. Mix the collagenase samples with loading buffer (1:1, v/v) and load in the wells.
11. Fill the upper and lower chambers (mini VE vertical electrophoresis system, GE Healthcare) with electrophoresis running buffer.
12. Apply a constant voltage (150 V, 25 mA) through an external power supply.
13. After the completion of the run, soak the polyacrylamide gel in 0.01 M sodium phosphate buffer, pH 7.4 containing 0.001 M Ca^{2+} for 4 h (*see* **Note 5**) (Fig. 1).
14. Stain the gel with staining solution for about 1 h (*see* **Note 6**).
15. Destain the gel with destaining solution (*see* **Note 6**).
16. Scan the gel on any image scanner with an illuminator.

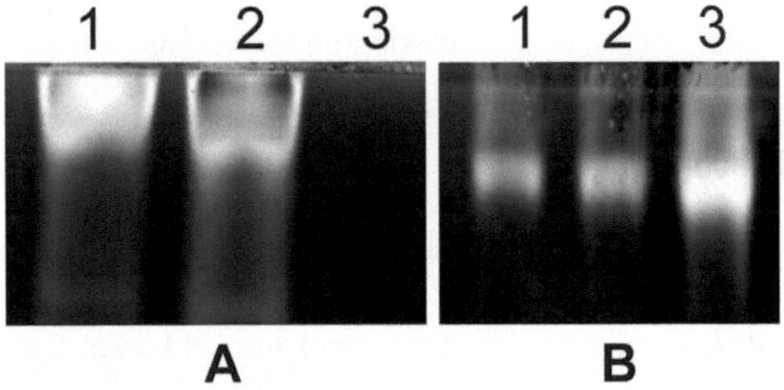

Fig. 1 Zymography of collagenase by protocol 1. (**a**) After the completion of the electrophoretic run, the polyacrylamide gel was soaked in 0.01 M sodium phosphate buffer, pH 7.4 containing 0.001 M Ca^{2+} for 3 h. *Lane 1*: 2 U collagenase, *Lane 2*: 4 U collagenase. The specificity of the method was checked by running 4 μg of BSA (*Lane 3*) as the control. (**b**) Electrophoresis of collagenase (*Lane 1*: 32 mU, *Lane 2*: 40 mU, *Lane 3*: 50 mU) was carried out at 4 °C, as described before. After the completion of the run, the polyacrylamide gel was incubated in 0.01 M sodium phosphate buffer, pH 7.4 containing 0.001 M Ca^{2+} for 4 h. One unit of collagenase activity is defined as the amount of enzyme required to produce 1 μmol of leucine in 5 h, as determined by the ninhydrin method [12], under standard assay conditions [13]. Reproduced with permission from [5].

3.2 Zymography of Collagenase (Protocol 2)

1. Prepare 10% crosslinked polyacrylamide gel as described before (*see* Subheading 3.1, **steps 1–4**) but without adding collagen. Insert a comb gently between the gel plates and avoid trapping of air bubble below the comb.
2. Load collagenase samples in the wells (along with the sample loading buffer) by mixing equal volume of each (*see* **Note** 7) (Fig. 2).
3. Run the gel at room temperature under constant voltage condition (150 V, 25 mA) in a discontinuous buffer system [14].
4. After the completion of the run, wash the gel twice with distilled water and soak in a solution containing 20 mL collagen solution, 8 mL 0.05 M sodium phosphate buffer, pH 7.4 (final concentration 0.01 M) and 28 μL 1 M Ca^{2+} (final concentration 0.001 M).
5. Repeat **steps 14–16** from Subheading 3.1 (*see* **Note 6**).

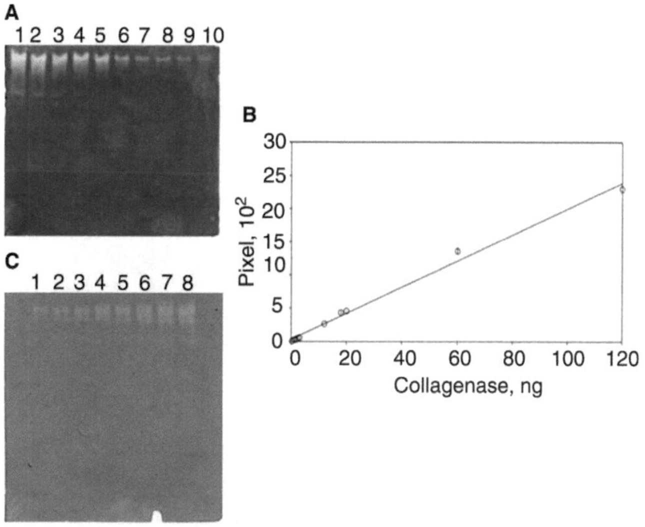

Fig. 2 Zymography of collagenase by protocol 2. Electrophoresis of collagenase (*Lane 1*: 1.2 mU, *Lane 2*: 0.6 mU, *Lane 3*: 0.2 mU, *Lane 4*: 0.18 mU, *Lane 5*: 0.12 mU, *Lane 6*: 0.03 mU, *Lane 7*: 0.024 mU, *Lane 8*: 0.018 mU, *Lane 9*: 0.012 mU, *Lane 10*: 0.003 mU) was carried out at 4 °C. (**b**) Densitometric analysis of the bands was carried out using the Image QuantTL software (GE Healthcare). Values shown are mean of three independent zymograms. (**c**) Electrophoresis of a crude extract of *Clostridium histolyticum*, known to contain elastase, clostripain and sulfhydryl protease (Worthington website), apart from collagenase, was carried out at 4 °C (*Lane 1*: 7.5 μg protein, *Lane 2*: 15 μg, *Lane 3*: 22.5 μg, *Lane 4*: 30 μg, *Lane 5*: 37.5 μg, *Lane 6*: 45 μg, *Lane 7*: 56.2 μg, *Lane 8*: 75 μg). Reproduced with permission from [5]

4 Notes

1. Filtration is required to remove undissolved collagen, so that a homogenous solution can be prepared.
2. The optimal pH for collagenase activity is 7.4.
3. Wear face mask and gloves when weighing and handling acrylamide crystals. Residual unpolymerized acrylamide and bisacrylamide are neurotoxic.
4. It is better to prepare fresh solution every time and store on ice. Store APS salt in a desiccator at 4 °C.
5. Ca^{2+} is required for collagenase activity.
6. Both overstaining and overdestaining lead to loss of sensitivity (Fig. 3). Combination of staining and destaining protocols is reported to increase sensitivity [15].
7. Samples should not be boiled before loading in the wells.

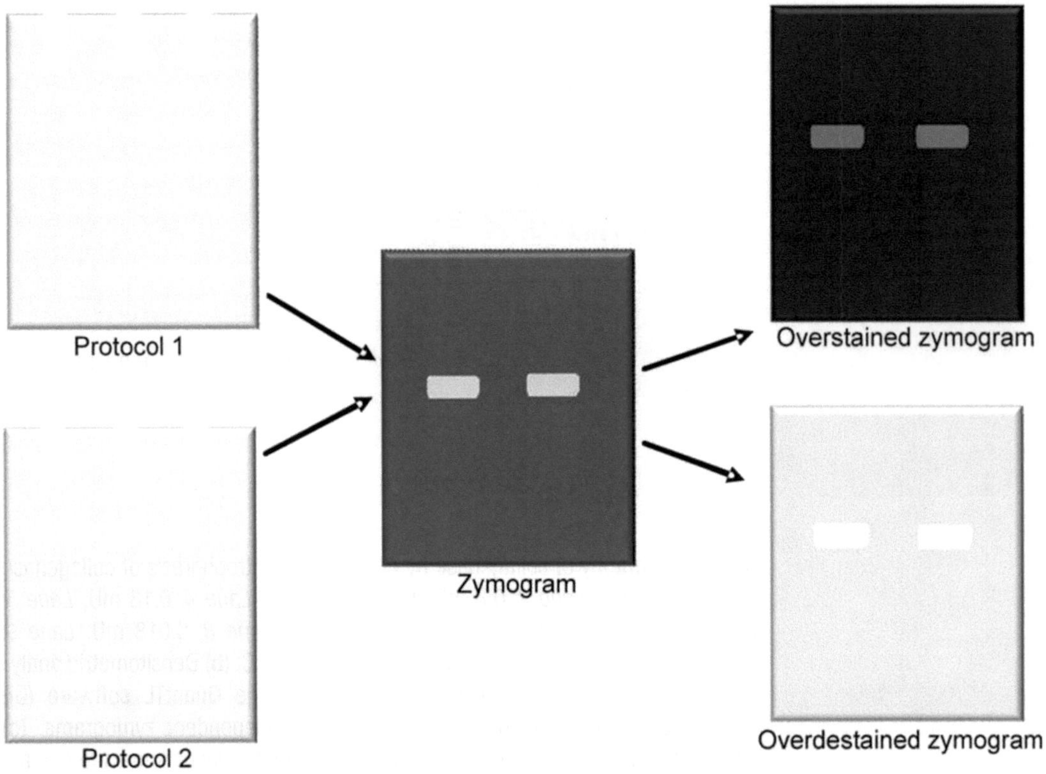

Fig. 3 Schematic representation of the consequences of overstaining or overdestaining of the zymogram

Acknowledgements

Partial financial support from Department of Science and Technology is acknowledged. The authors thank K. Thanzami for her contribution to the work described here.

References

1. Vandenbroucke RE, Libert C (2014) Is there new hope for therapeutic matrix metalloproteinase inhibition? Nat Rev Drug Discov 13:904–927
2. Lombard C, Saulnier J, Wallach J (2005) Assays of matrix metalloproteinases (MMPs) activities: a review. Biochimie 87:265–272
3. Birkedal-Hansen H, Taylor RE (1982) Detergent-activation of latent collagenase and resolution of its component molecules. Biochem Biophys Res Commun 107: 1173–1178
4. Snoek-van Beurden PA, Von den Hoff JW (2005) Zymographic techniques for the analysis of matrix metalloproteinases and their inhibitors. Biotechniques 38:73–83
5. Thanzami K, Roy I (2008) A sensitive, rapid and specific technique for the detection of collagenase using zymography. Electrophoresis 29:1585–1588
6. Vandooren J, Geurts N, Martens E, Van den Steen PE, Opdenakker G (2013) Zymography methods for visualizing hydrolytic enzymes. Nat Methods 10:211–220
7. Gross J, Lapière CM (1962) Collagenolytic activity in amphibian tissues: a tissue culture assay. Proc Natl Acad Sci U S A 48: 1014–1022
8. Harrington DJ (1996) Bacterial collagenases and collagen-degrading enzymes and their potential role in human disease. Infect Immun 64:1885–1891
9. Matsushita O, Koide T, Kobayashi R, Nagata K, Okabe A (2001) Substrate recognition by the collagen-binding domain of *Clostridium histolyticum* class I collagenase. J Biol Chem 276:8761–8770
10. Yoshihara K, Matsushita O, Minami J, Okabe A (1994) Cloning and nucleotide sequence analysis of the colH gene from *Clostridium histolyticum* encoding a collagenase and a gelatinase. J Bacteriol 176:6489–6496
11. Duarte AS, Correia A, Esteves AC (2014) Bacterial collagenases – a review. Crit Rev Microbiol. doi:10.3109/1040841X.2014.904270
12. Moore S, Stein WH (1948) Photometric ninhydrin method for use in the chromatography of amino acids. J Biol Chem 176:367–388
13. Mandl I, MacLennan J, Howes E (1953) Isolation and characterization of proteinase and collagenase from *Cl. histolyticum.* J Clin Investig 32:1323–1329
14. Walker JM (2009) SDS polyacrylamide gel electrophoresis of proteins. In: Walker JM (ed) The protein protocols handbook. Humana Press, Totowa, NJ, pp 177–185
15. Leber TM, Balkwill FR (1997) Zymography: a single-step staining method for quantitation of proteolytic activity on substrate gels. Anal Biochem 249:24–28

Acknowledgements

[illegible]

[illegible]

Part III

Reverse Zymography and In Situ Zymography

Chapter 11

Reverse Zymography: Overview and Pitfalls

Kanika Sharma and Debasish Bhattacharyya

Abstract

Reverse zymography is a technique by which protease inhibitor(s) in a sample could be electrophoretically separated in a substrate-impregnated acrylamide gel and their relative abundance could be semi-quantified. The gel after electrophoresis is incubated with a protease when the impregnated substrate and all other proteins of the sample are degraded into small peptides except the inhibitor(s) that show clear bands against a white background. Since reverse zymography cannot distinguish between a protease inhibitor and a protein that is resistant against proteolysis, the results should be confirmed from inhibition of protease activity by solution state assay.

Key words Substrate electrophoresis, Inhibitor, Protease digestion, Protease-resistant proteins

1 Introduction

Proteolytic enzymes form the core of human pathophysiology by executing innumerable degradative as well as developmental functions [1]. Regulation of proteolytic activity by their specific or nonspecific inhibitors helps in maintaining a pivotal balance between the synthesis and turnover of proteins [2]. Matrix metalloproteinases (MMPs) are calcium-dependent zinc-containing endopeptidases belonging to metzincin superfamily. Apart from Extra Cellular Matrix (ECM) degradation, they are involved in cellular proliferation, migration, apoptosis etc. Tissue inhibitors of metalloproteinases (TIMPs) are the family of four inhibitors: TIMP-1, TIMP-2, TIMP-3, and TIMP-4. They regulate MMP activity by chelating the zinc group present at the active site. Maintenance of balance between MMPs and TIMPs is crucial to avoid chronic pathological conditions [3]. Several methods have been developed to analyze MMP/TIMP activity in biological samples. Zymography and its modification, reverse zymography, are two such widely used methods.

Zymography as a substrate electrophoretic technique is widely used to identify proteolytic activity under non-reducing conditions.

Jeff Wilkesman and Liliana Kurz (eds.), *Zymography: Methods and Protocols*, Methods in Molecular Biology, vol. 1626,
DOI 10.1007/978-1-4939-7111-4_11, © Springer Science+Business Media LLC 2017

Post electrophoresis, replacement of SDS by a non-ionic detergent of low critical micelle concentration (CMC) allows the proteases to refold partially resulting in a catalytically active enzyme. It is the method of choice to detect the activity of different matrix metalloproteinase (MMP) isoenzymes from a wide range of biological samples [4]. It also discriminates between functional monomers, multimers, complexes and degraded products, thus providing an activity-based evaluation of the biological sample [5, 6].

Reverse zymography is a modified zymographic technique which enables the detection of protease inhibitors in biological samples. It facilitates the characterization of inhibitors, their relative abundance and stability against the protease. This method was initially developed to identify TIMPs in biological extracts by copolymerizing gelatin, MMP-2/MMP-9 with acrylamide whereby TIMP forms a 1:1 complex with MMP which appears as bands against white background [7]. Figure 1 shows the visual difference between zymography (white band against blue background) and reverse zymography (blue bands against white background).

Basic protocol of reverse zymography involves electrophoretic separation of sample on a substrate-copolymerized SDS-PAGE. Post run, the gel is washed with Triton-X 100 to reduce SDS present in the gel. This is followed by incubation of the gel in developing buffer incorporated with specific protease, against which inhibitor is being searched, for 12–24 h at 37 °C. The protease digests the background substrate and results in appearance of bands where the inhibitor is present. Staining and destaining provides blue bands of protease inhibitor(s) against white background.

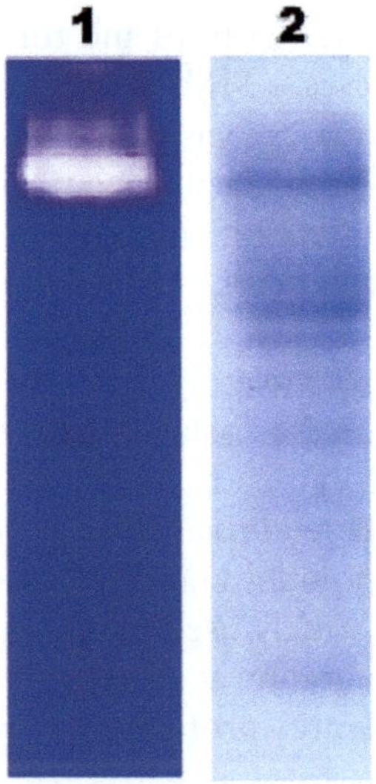

Fig. 1 (*1*) Gelatin zymography of leaf bromelain. (*2*) Reverse zymography of Soyabean Trypsin Inhibitors (STI) (25 μg) digested with trypsin (0.5 mg)

On the basis of the protease being used, one can deduce information regarding the protease family to which the inhibitor belongs. Similar to zymography, reverse zymography offers advantages like detection of inhibitors in impure samples, non-interference of optically absorbing molecules, electrophoretic separation of inhibitors, their relative abundances and molecular weights.

2 Materials

All reagents were prepared using ultrapure distilled water and stored at 4 °C until otherwise stated.

1. *Resolving gel buffer:* 3 M Tris–HCl, pH 8.8. Weigh 363.1 g of Tris and dissolve in approximately 250 mL distilled water. Adjust the pH with 6 M HCl and complete to 1 L. Store at 4 °C (*see* **Note 1**).
2. *Stacking gel buffer:* 0.5 M Tris–HCl, pH 6.8. Weigh 60.5 g of Tris and prepare 1 L buffer solution as described in **step 1**. Store at 4 °C.
3. *30% acrylamide/bisacrylamide solution:* weigh 29.2 g of acrylamide (monomer) and 0.8 g of bisacrylamide (cross-linker) and transfer to 100 mL graduated cylinder containing 50 mL water. Mix the solution for approximately 30 min on magnetic stirrer to get a clear solution. Make up the volume to 100 mL using water, centrifuge and pass through a 0.45 μm filter. Store at 4 °C in an amber-colored glass bottle covered with aluminum foil to protect from light (*see* **Note 2**).
4. *Ammonium persulfate:* 10% solution in water. Always use a freshly prepared solution.
5. *N,N,N′,N′-Tetramethyl-ethylenediamine (TEMED):* use as purchased. Store at 4 °C.
6. *Zymography running buffer:* 0.025 M Tris–HCl, pH 8.3, 0.192 M glycine, 0.1% SDS. Store at room temperature.
7. *Developing buffer:* 10 mM Tris–HCl, pH 7.6, 200 mM NaCl, 10 mM $CaCl_2$, 0.02% Brij-35 and variable types of protease against which the inhibitor is searched for.
8. *Staining solution:* dissolve 0.025 g of Coomassie Brilliant Blue in 40 mL methanol, 10 mL acetic acid and make up the solution to 100 mL.
9. *Destaining solution:* methanol:acetic acid:water (40:10:50) (AR/GR grade).
10. *Native sample loading buffer* (2×): 0.5 mL of 0.5 M Tris-HCl, pH 6.8, 0.002% bromophenol blue, 0.22 mL glycerol added to 1.28 mL water.

3 Methods

1. Electrophoresis is performed using a vertical electrophoresis apparatus with a 10 × 8 × 0.2 cm^3 gel cassette.
2. Prepare the 10% resolving gel (% with respect to acrylamide concentration) by mixing 0.1% substrate (BSA, gelatin, collagen etc as par choice) with 3.32 mL of 30% acrylamide solution, 1.25 mL of resolving buffer, 0.1 mL 10% SDS, 5.21 mL distilled water, 0.02 mL of TEMED, and 0.1 mL of 10% ammonium persulfate solution. Fill the gel cassette covering 3/4th of the available space. Carefully cover the resolving gel layer with water so that the gel surface remains horizontal. Follow polymerization of the gel by tilting the cassette. Once polymerized, the water layer is removed by soaking with a tissue paper (*see* **Notes 3** and **4**).
3. Prepare the stacking gel by mixing 0.625 mL of acrylamide solution, 1.25 mL of stacking buffer, 0.05 mL of 10% SDS solution, 3.045 mL of water, 0.01 mL of TEMED and 0.05 mL of 10% ammonium persulfate solution. Insert a 7 or 10-well gel comb immediately without introducing air bubbles.
4. Add native sample loading buffer to the samples which are to be electrophoresed. Then load them in consecutive wells while avoiding any lateral diffusion. Electrophorese at 15–25 mA until the dye front has reached the bottom of the gel. Maintaining the dye front within the gel ensures retention of all protein molecules in it (*see* **Notes 5–7**).
5. After electrophoresis, open the gel plates with a spatula and gently transfer to a container.
6. Wash the gel with 2.5% Triton-X 100 solution two times for 30 min each followed by water several times. Under these conditions, SDS from the gel is partly removed, that allows embedded proteins to restore majority of their native structures which were lost because of strong denaturing potency of SDS (*see* **Notes 8** and **9**).
7. Incubate the gel in developing buffer for 12–24 h at 37 °C. At this step, extraneously added proteolytic enzymes degrade the embedded substrate and all protein components of the test sample except the protease inhibitor(s) (*see* **Note 10**).
8. Stain the gel using 0.025% Coomassie Brilliant Blue R-250 in 40% methanol and 10% acetic acid for 3 h at 37 °C (*see* **Notes 11–13**).
9. Wash the gels with destaining solution resulting in appearance of bands.

4 Notes

1. While making the buffer solutions, first the salt is dissolved in approximately 1/4th of the final volume. Then, the pH is adjusted by adding 6 M HCl in small increments. Finally the volume is made up using water and the pH is rechecked.
2. It is recommended to wear a mask and gloves while weighing dry acrylamide. Inhalation of acrylamide dust can be neurotoxic while dermal contact can cause skin irritation and redness. Acrylamide in solution is less hazardous and upon polymerization, becomes nontoxic.
3. After destaining, the body of the gel that originally contained substrate should be colorless. This is possible only when the substrate is degraded to very small peptides that are difficult to be stained. Therefore, the protein used for copolymerization of the gel should be a good substrate of the protease which is used for digestion of the gel, e.g. collagen or gelatin (denatured form of collagen) for collagenase; BSA or casein for trypsin, etc. The protease must degrade the gel bound substrate efficiently and extensively. Otherwise, the background of the gel will appear as light to dark blue. This is the reason why inhibitors of a protease of narrow specificity cannot be detected in reverse zymography. In other words, inhibitors of proteases of broad specificity like chymotrypsin, papain, bromelain, Proteinase K, etc. and proteases like trypsin that can easily cleave the substrates could be better detected by reverse zymography.
4. The amount of substrate being incorporated into the gel should be optimized. Excess substrate might lead to a residual blue color in the background which might require excess protease for digestion and may interfere with correct interpretation of the results.
5. The duration of electrophoresis is a vital parameter which determines adequate separation of multiple bands within the sample. If a sample contains both high and low molecular mass inhibitors, then visualization of both categories of bands in a single gel needs to be carefully standardized. Low molecular mass inhibitors might run out of the gel if the gel is run for a longer duration. While high molecular mass inhibitors might not sufficiently enter into the gel if the gel is run for a shorter duration. Thus for an unknown sample, two electrophoretic runs of different durations might be necessary for complete evaluation of the inhibitor profile.
6. Unlike zymography, in principle the rate of migration of the sample should not affect the electrogram. However, too fast migration may cause trailing of protein bands. Also, a high

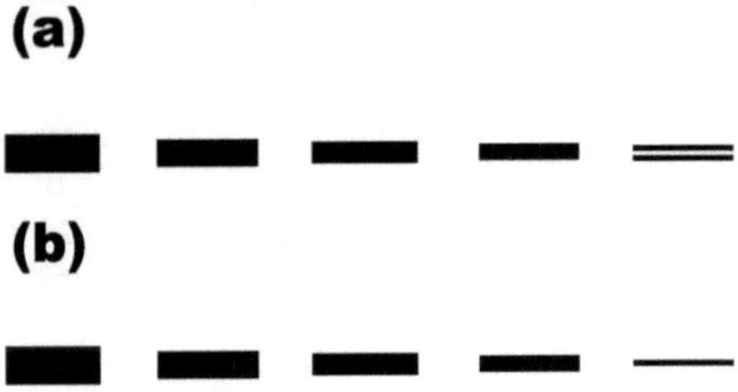

Fig. 2 Loading of samples. Both (**a**) and (**b**) show that overloading and underloading are necessary to ascertain the purity of sample

voltage can adversely affect sharpening of protein bands due to thermal diffusion.

7. Overloading of samples might often lead to the appearance of a broad band. Under such conditions, absence of bands of high or low molecular weight beyond detection limit ensures homogeneity of the preparation. However, this can be confusing with respect to the number of bands actually present within the broad band. Figure 2 shows that gradually reducing the concentration of the sample would ultimately reveal if more than one band is present or not. This aspect of resolution is true for any type of gel electrophoresis.
8. In the analysis of reverse zymogram, it is assumed that the inhibitor molecule retains its inhibitory property after electrophoresis in presence of SDS followed by washing with Brij-35 or Triton-X 100 to reduce the load of SDS in the gel. If the inhibitor is structurally unstable and is unable to restore reversibly its inhibitory potency, there remains a possibility that the inhibitor is to be digested by the protease. This is a matter of concern when the inhibitor is structurally unstable and its native state conformation is a requirement to bind and inactive the protease. On the other hand, if an inhibitor resists denaturation in the presence of SDS, this concern does not arise.
9. In continuation to the previous consideration, it is possible that a protein in the test sample which is not an inhibitor of the protease but is resistant to protease digestion due to either structural stability or absence of proteolytically sensitive bonds or inaccessibility of such bonds to the protease should appear as a band in reverse zymography. Therefore, reverse zymography cannot distinguish between an inhibitor and a protein resistant to proteolysis. In absence of background knowledge, an inhibitor detected by reverse zymography should be verified by independent means like solution state enzyme assay. This fact is illustrated in Fig. 3 where lysozyme is used as the test sample. Lysozyme is resistant to proteolytic digestion [8]. Therefore, it appears as an inhibitor of trypsin. In reality, lysozyme is not an

Fig. 3 Lysozyme (25 μg) digested with trypsin (0.5 mg) appears as an inhibitor in reverse zymography

inhibitor of trypsin. This limitation of reverse zymography has been described in details in reference [9]. This indicates that a solution state enzyme assay should be necessarily performed along with reverse zymography to confirm the inhibitory nature of a band.

10. In reverse zymography, it is assumed that the externally applied proteases should digest the impregnated substrate along with all proteins that are present in the test sample except the inhibitor/s. If it happens that the relative abundance of a protein in the sample is very large and the protease is unable to digest the protein completely under the conditions employed, a false band may appear along with the inhibitor(s), if any, leading to wrong interpretation of the results. While optimizing the experimental conditions, one should ensure that digestion of susceptible proteins must be complete.

11. Ideally, the substrate copolymerized with acrylamide should not migrate under electrophoretic conditions. However, a portion of the substrate remains free and may migrate under electrophoretic field according to its isoelectric point. Usually substrates of low molecular weight migrate more, e.g., casein (*Mr.* 22–23 kDa, p*I* 4.2–4.7) is more likely to migrate than BSA (*Mr.* 67 kDa, p*I* 4.8). Electrophoresis of a gel copolymerized with casein after staining shows two distinct regions of different color intensities—one lighter region due to migration of unbound substrate and the other one containing both

Fig. 4 Blank run of casein-incorporated zymogram. The gel is divided into two sections. *Region 1* is lighter than *region 2* showing migration of substrate

bound and unbound substrates (Fig. 4). Thus, the extent of migration of the substrate should be evaluated using a blank run in absence of the test sample. After staining, if the gel shows two distinct regions—then the concentration of the impregnated substrate should be decreased along with modification of the gel polymerization protocol.

12. As stated above, after staining the gel of a reverse zymography, two regions of darker and lighter shades may appear. In that case, the free substrate may be electrophoretically removed before application of the test sample. Otherwise, it will be difficult to generate a good-quality photographic presentation of the gel.
13. The Coomassie Blue solution used for staining should not contain any precipitate or semisolid lumps. These particles form intense spots on the gel surface that are difficult to be destained. For uniform staining, the dye should be passed through Whattman filter paper 1 prior usage.

Acknowledgement

K.S. is supported by fellowship from University Grants Commission, New Delhi (UGC/NET-SRF).

References

1. López-Otín C, Matrisian LM (2007) Emerging roles of proteases in tumour suppression. Nat Rev Cancer 7:800–808
2. Rawlings ND, Tolle DP, Barrett AJ (2004) Evolutionary families of peptidase inhibitors. Biochem J 378:705–716
3. Visse R, Nagase H (2003) Matrix metalloproteinases and tissue inhibitors of metalloproteinases: structure, function, and biochemistry. Circ Res 92:827–839
4. Wilkesman J, Kurz L (2009) Protease analysis by zymography: a review on techniques and patents. Recent Pat Biotechnol 3:175–184
5. Vandooren J, Geurts N, Martens E, Van den Steen PE, Opdenakker G (2013) Zymography methods for visualizing hydrolytic enzymes. Nat Methods 10:211–220
6. Snoek-van Beurden PAM, Von den Hoff JW (2005) Zymographic techniques for the analysis of matrix metalloproteinases and their inhibitors. Biotechniques 38:73–83
7. Hawkes SP, Li H, Taniguchi GT (2010) Zymography and reverse zymography for detecting MMPs and TIMPs. Methods Mol Biol 622:257–269
8. Nonaka Y, Akieda D, Aizawa T, Watanabe N, Kamiya M, Kumaki Y, Mizuguchi M, Kikukawa T, Demura M, Kawano K (2009) X-ray crystallography and structural stability of digestive lysozyme from cow stomach. FEBS J 276: 2192–2200
9. Dutta S, Bhattacharyya D (2013) Reverse zymography alone does not confirm the presence of an inhibitor. Protein J 32:155–162

Chapter 12

Cell In Situ Zymography: Imaging Enzyme–Substrate Interactions

Aastha Chhabra and Vibha Rani

Abstract

Zymography has long been used for the detection of substrate-specific enzyme activity. In situ zymography (ISZ), an adaptation from the conventional substrate zymography, is a widely employed technique useful for the detection, localization, and estimation of enzyme–substrate interactions in tissues. Here, we describe a protocol to detect 'in position' matrix metalloproteinase (MMP) activity in cells utilizing H9c2 cardiomyoblasts as a model. This technique is primarily adopted from the method used for histological sections and is termed as 'Cell in situ Zymography'. It is a simple, sensitive, and quantifiable methodology to assess the functional activity of an enzyme 'on site/in position' in cell culture.

Key words Cytochemistry, Dye-quenched casein, Enzymes, Fluorescent assay, Image analysis/quantification, In situ zymography, Matrix metalloproteinases (MMPs), Proteolytic activity

1 Introduction

Zymography is a vividly used technique that complements other molecular techniques for the assessment of enzyme–substrate interactions through its ability to detect functional activity of an enzyme. One of the well-established zymography techniques, known as 'Substrate Zymography', is an electrophoresis-based method that involves detection of substrate-specific enzyme activity as distinct bands indicating digestion of substrate in a polyacrylamide gel wherein the substrate is either present in the incubation buffer (as in Superoxide Dismutase Zymography) or is incorporated in gel (as in Gelatin Zymography) [1]. However, one of the major drawbacks of this technique concerns the loss of enzyme localization in the cells/tissue during sample preparation and has compelled researchers to devise an alternate method [2–4].

Thus, with a primary objective of detection and localization of the functional activity of an enzyme, in situ Zymography (ISZ), a variant of substrate zymography, was developed for histological sections [5–7]. 'In situ' is a Latin phrase that refers to 'in position'

Jeff Wilkesman and Liliana Kurz (eds.), *Zymography: Methods and Protocols*, Methods in Molecular Biology, vol. 1626,
DOI 10.1007/978-1-4939-7111-4_12, © Springer Science+Business Media LLC 2017

and this technique offers the advantage to observe and estimate the substrate-specific enzyme activity 'on site'. The basic principle involves digestion of substrate by the functionally active form of enzyme present in its native location, followed by detection of this interaction through liberation of signal from the photophore/fluorophore conjugated to the substrate by light or fluorescent microscopy. Based on the nature of substrate, ISZ has been classified into two methodologies, wherein the first method utilizes a photographic emulsion overlay containing a substrate like gelatin while the second uses a fluorophore labeled substrate [3, 8, 9]. The modified version of the latter recommends usage of substrate conjugated to quenched fluorophore where the enzyme activity is a direct function of the fluorescent signal obtained on cleavage of substrate and liberation of the fluorophore [10]. Although the scope of this technique ranges for a broad set of enzymes, it has essentially been used in studies pertaining to Matrix Metalloproteinases (MMPs), a class of calcium- and zinc-dependent endopeptidases that play a key role in extracellular matrix (ECM) remodeling under various physiological and pathological conditions [11].

Cell culture has been a valuable material of study with its application extending to various areas of biological research. The introduction of three-dimensional culture systems has advanced the 'in vitro' approach further [12]. To our knowledge, very few studies cite the use of ISZ in cell lines. Cha and colleagues have reported assessment of gelatinolytic activity in mouse skeletal myoblasts (C2C12) and rat cardiac myoblasts (H9c2) utilizing DQ-gelatin in a matrix of low-melting agarose [13]. In this chapter, we explain a simple protocol and help extend the use of in situ Zymography in cell culture [14].

2 Materials

Make all solutions in ultrapure water by purifying de-ionized water to attain a sensitivity of 18 MΩ cm at 25 °C. Prepare and store all reagents at room temperature (unless indicated otherwise). Ensure that light-sensitive steps are carried out in dark. Follow Good Laboratory Practices (GLP) guidelines while performing the experiment.

2.1 Cell Culture Components

1. Cell line: Rat heart-derived H9c2 cardiomyoblasts obtained from National Centre for Cell Science (NCCS), Pune, India (*see* **Note 1**).
2. Growth media: Dulbecco's Modified Eagle Medium (pH 7.4) containing 4.5 g/L glucose, 0.584 g/L L-glutamine, 0.11 g/L sodium pyruvate, 3.7 g/L sodium bicarbonate and 2.6 g/L HEPES, supplemented with antibiotic (100 units/mL

Penicillin and 100 µg/mL Streptomycin) and 10% Fetal Bovine Serum (FBS). Prepare and filter sterilize the media with a 0.22 µm filter. Store at 4 °C (*see* **Note 2**).

3. Cover slips: Microcoverglass No. 1, 18 × 18 mm^2, sterile.
4. Multi-well plate: 6-well, total well area—9.6 cm^2, well volume—15.5 mL, flat-bottom, with lid, sterile.
5. Pointed forceps—sterile.
6. Humidified CO_2 incubator: 5% CO_2, 37 °C.
7. Methanol: HPLC grade, purity ≥99.9%.
8. 1× Phosphate-buffered saline: 0.2 g/L KCl, 0.2 g/L KH_2PO_4, 1.14 g/L $Na_2HPO_4{\bullet}2H_2O$ and 8 g/L NaCl. Weigh the components and dissolve in 500 mL of ultrapure water. Make up the volume to 1 L once all the salts dissolve completely. Autoclave and keep at 4 °C for long-term storage. Pre-warm to 37 °C before use.

2.2 Cell In Situ Zymography Components

1. Digestion buffer: 50 mM Tris–Cl (pH 7.4), 150 mM NaCl, 5 mM $CaCl_2$ and 0.1% Brij-35. Weigh the components and dissolve in ultrapure water. Make up the volume of buffer once all salts dissolve completely. Add 0.1% Brij-35 (non-ionic detergent) with a cut tip in the final solution. Use freshly prepared buffer.
2. FITC-Casein: Casein fluorescein isothiocyanate type III from bovine milk, lyophilized and salt free (*see* **Notes 3** and **4**).
3. 0.5% Agarose solution: Agarose for routine use (Molecular Biology Reagent grade). Prepare in digestion buffer as described in methods section (*see* Subheading 3.2, **step 3** and **Note 5**).
4. Microscope glass slide: Glass slides for microscopes, 76 × 26 × 1.35 mm^3, dust-free, pre-warmed to 45–50 °C.
5. Pointed forceps—Sterile.
6. Shallow trough—Amber colored/covered with foil, 0.5–1 cm in depth, with lid, dust-free (*see* **Note 6**).
7. Laboratory incubator: Humidified, maintained at 37 °C
8. Standard fluorescent microscope: With FITC filter (Excitation—485 nm; Emission—530 nm), with DAPI filter (Excitation—364 nm; Emission—454 nm), with camera attached to the microscope (*see* **Note 7**).

3 Methods

Carry out all procedures at room temperature unless otherwise specified.

3.1 Cell culture

1. Follow good laboratory practices for cell culture techniques (*see* **Note 8**). Place a sterile cover slip in a 6-well plate using forceps and seed the suspension of H9c2 cells covering the entire surface of cover slip. Carefully add 2 mL of growth media per well from one corner of the well (*see* **Note 9**).
2. Gently swirl the plate to allow the cell suspension and media to spread uniformly in the well.
3. Allow the cells to adhere to the cover slip and grow to a confluency of about 50–60% in a humidified CO_2 incubator maintained at 37 °C and 5% CO_2 (*see* **Notes 10** and **11**).
4. Give suitable treatment to the cells as per the plan of experiment to be conducted.
5. On the day of assessing MMP activity, aspirate the media and add 1 mL of 1× phosphate buffered saline (PBS, pH 7.4) per well to wash the cells. Gently rock the plate on a gel-rocker at 30 rpm for 15 s. Remove PBS and repeat the process twice.
6. Add 600 μL methanol to each well and fix the cells by keeping the plate at −20 °C for 15 min (*see* **Note 12**).
7. Wash the cells with 1× PBS as described in **step 5** (however, at 50 rpm for 15 s) to remove traces of methanol completely.

3.2 Cell In Situ Zymography

All steps concerning fluorogenic substrate must be carried in dark. A schematic flowchart summarizes the technique briefly (Fig. 1). It must be ensured that the conduct of the experiment is well-controlled as suggested in Table 1.

1. Prepare fresh digestion buffer containing 50 mM Tris–Cl, pH 7.4, 150 mM NaCl, 5 mM $CaCl_2$ and 0.1% Brij-35 (*see* **Note 13**).

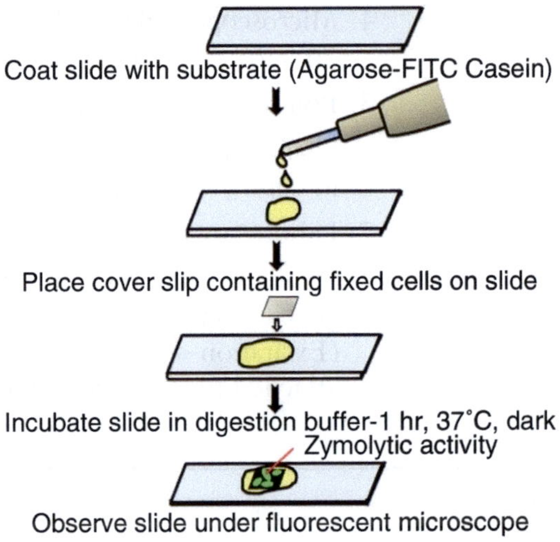

Fig. 1 Graphical overview: Fluorescently labeled substrate-based cell in situ zymography

Table 1
Appropriate controls for cell in situ zymography

Compound	Class	Remarks
1. Non-Metalloprotease-specific inhibitors		
PMSF	Serine protease	No inhibition of MMP-specific activity
Pepstatin	Aspartic protease	–
Leupeptin	Threonine protease	–
2. Metalloprotease-specific inhibitors		
o-Phenanthroline	Zinc chelator	Inhibits Zn^{2+}-dependent MMP activity
EDTA	Divalent cation chelator	Inhibits Ca^{2+}- and Zn^{2+}-dependent MMP activity
Tetracycline and its derivatives	Broad spectrum antibiotic; Cycline derivative	Nonspecific MMP inhibitor
3. MMP activator		
Amino-phenyl mercuric acetate (APMA)	Organomercurial	Activates latent form of MMPs
4. Enzyme activity control		
Trypsin (Serine protease)	Positive control	Digests substrate including casein and gelatin
5. Substrate control		
FITC-Albumin	Negative control	Nonspecific substrate for MMP activity

2. Make 0.5 μg/mL FITC-casein solution in digestion buffer from the stock solution of FITC-casein substrate prepared as per Product Datasheet. Add 1 μL of FITC-casein solution from a stock of 0.5 mg/mL to 500 μL of digestion buffer and mix well. This step must be carried in dark and in an amber tube (*see* **Note 14**).
3. Also, prepare 0.5% agarose solution in digestion buffer. Add 0.1 g agarose to 10 mL digestion buffer. Boil the solution to dissolve agarose completely as indicated by a clear solution. Allow the solution to cool down to a temperature of 45–50 °C (*see* **Note 15**).
4. Make 1:1 solution of FITC-casein with agarose from the solutions prepared in **steps 2** and **3**. Add 500 μL of agarose solution (prepared in **step 3**) to an equal volume of FITC-casein solution (prepared in **step 2**) just before it begins to solidify. Pipette up and down to thoroughly mix the solution.

5. Immediately drop 150 μL of agarose solution containing the fluorescently labeled substrate on the center of a clean microscope glass slide, pre-warmed to a temperature of 40–50 °C (*see* **Notes 16** and **17**).
6. Pick the cover slip with fixed cells (described in Subheading 3.1) from one corner using pointed forceps and quickly place it on the microscope glass slide with the cell side facing the gel solution.
7. Allow the agarose solution to spread evenly and solidify such that the cover slip embeds in the gel (*see* **Note 18**).
8. Place the microscope glass slide in a shallow bottom trough and keep the setup in an incubator set at 37 °C. Pour digestion buffer from one corner of the trough so as to just reach the height of the slide (*see* **Note 19**).
9. Cover the trough with a lid on top and keep the slide undisturbed for 1 h in dark.
10. Remove the slide from the trough and dry it with a KimWipe (Kimberly Clark). Observe the slide under fluorescent microscope in the FITC filter. Spot regions of zymolytic activity observed as green fluorescence signal of FITC molecule liberated from casein with its intensity as a direct function of caseinolytic activity (*see* **Notes 20** and **21**).
11. Fix the image capture settings such as exposure, gamma, magnification using the Control (non-treated) set and capture images of all experimental groups with the same specifications through a camera attached to the microscope (Fig. 2a–f) (*see* **Notes 22–24**).
12. Follow the same protocol for assessing MMP activity in presence of inhibitor, *o*-phenanthroline (*see* **Note 25**).

3.3 Image Analysis and Representation

1. Total fluorescence per cell can be quantified using Image J, a widely available image processing software (http://rsbweb.nih.gov/ij/), and be represented as a bar or box and whisker plot (*see* **Note 26**). Open the image to be quantified in Image J and go to 'Analyze' in the tool bar. Open 'Set Measurements' and select parameters—'Area', 'Mean Gray Value' and 'Integrated Density'. Redirect the specifications to the selected image and click 'OK'.
2. Draw the boundary of the cell to be quantified in the image using free hand selection tool and click 'Measure' in the 'Analyze' drop down list (Shortcut key—Ctrl + M).
3. Also, draw a region in the background close to the cell selected in **step 2** using free hand selection tool and click 'Measure' in the 'Analyze' drop down list to measure the background fluorescence. Repeat this step for 4–5 background selections per cell.

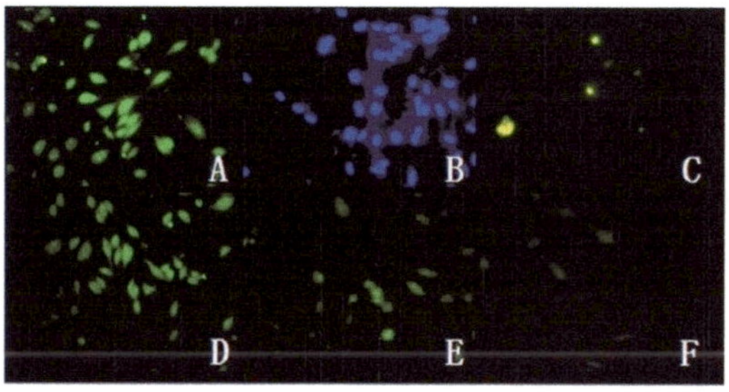

Fig. 2 Cell in situ zymography at a glance: (**a**) Digestion of FITC-conjugated casein present in agarose gel observed as green fluorescent signal indicates MMP activity in H9c2 cells. (**b**) Negative control shows DAPI-stained nuclei of cells fixed on cover slip embedded in agarose gel without FITC-Casein. (**c**) Negative control shows no zymolytic activity in absence of cells on the cover slip embedded in FITC-Casein containing agarose gel. (**d**) Digestion of DQ Gelatin (alternate substrate) present in agarose gel observed as a function of fluorescent signal indicates substrate-specific gelatinolytic activity of MMPs in H9c2 cells. (**e**) Fluorescent signal liberated from FITC-Casein denotes caseinolytic activity of MMPs in H9c2 cells fixed on a cover slip embedded in a matrix of agar (alternate for agarose). (**f**) Zymolytic activity in presence of MMP inhibitor, *o*-phenanthroline. A major drop in fluorescent signal, in presence of a zinc chelator, signifies the specificity of enzyme–substrate interaction

4. Now repeat **steps 2** and **3** for at least 8–10 cells per image and quantify a minimum of four images captured from different fields for each experimental group (accounts to minimum 40 values for each group).
5. Export the data from the 'Output' window to a Microsoft Excel sheet for calculations.
6. Compute the average 'Mean Gray Value' obtained as a measure of background signal for each selected cell (known as 'Mean fluorescence of background'). Calculate the Corrected Total Cell Fluorescence (CTCF), which accounts for any noise due to background, for each cell using the formula, ***CTCF = Integrated density of selected cell – (Area of selected cell × Mean fluorescence of background)***.
7. A number of CTCF values will be obtained for each experimental group that can be plotted as a bar or box and whisker plot. However, any other representation method may also be used.

4 Notes

1. This protocol is optimized utilizing H9c2 cardiomyoblasts (rat-derived cardiac cell line) as the model for assessment of MMP activity. However, any other adherent cell type may also be freely used.

2. DMEM growth medium with the formulation as described in Subheading 2 was utilized for maintenance of H9c2 cells. The general information, characteristics, and culture method for each cell line are detailed by the concerned cell culture repository. Hence, the selection of the culture conditions may therefore be carried out in consideration with the same (e.g. Use RPMI-1640 medium to culture Human leukemic monocytes (U937) and T cells (Jurkat)).
3. FITC-Casein substrate is light sensitive and water soluble. It should be stored at −20 °C to remain stable for at least 2 years. As per manufacturer's instructions, it is recommended to resuspend and aliquot FITC-Casein to smaller volumes upon arrival. Repeated freeze–thaw cycles can lead to slight increase in the background signal and lower the sensitivity of substrate. Details of product used in the present study—Catalog number C0528 (Sigma Aldrich).
4. Alternate substrate—DQ Gelatin (suggested product—DQ™ Gelatin, D12054, From Pig Skin, Fluorescein Conjugate, Thermo Fischer Scientific). It is a highly quenched, fluorescein labeled gelatin substrate with great degree of sensitivity that gives bright green fluorescence upon digestion by gelatinolytic MMPs (Excitation—485 nm; Emission—530 nm) (Fig. 2d). Interestingly, FITC-conjugated substrates for different types of collagen are also available now.
5. Alternate gelling agent—1% agar prepared in digestion buffer may also be used (Fig. 2e).
6. The cover of 96-well microplate or a 100 mm culture dish may be used for this purpose.
7. It is important to view the sample using a fluorescence microscope with appropriate filters. The selection of an optimal filter requires the user to match the optical filter specifications to the spectral characteristics of the fluorophore being employed. Details of fluorescent microscope and image capture software used in the current work—Olympus CKX31 (Japan) and ProgRes Capture Pro 2.7, respectively.
8. The key for good laboratory practice in cell culture is to ensure that all procedures are carried out to a standard that precludes contamination by bacteria, fungi and mycoplasma as well as cross-contamination with other cell lines. Some of the basic precautions are suggested as follows: (a) Surface sterilize the work space, materials and equipments with 70% ethanol before commencement of work. (b) Ensure that the growth media and other reagents are sterile and opened inside the hood only. Screw the cap of reagent bottle tightly and seal the neck before taking it out from the hood. (c) Disinfect all materials before removing them from the hood. Also, surface sterilize the work

area with 70% ethanol and UV, after use. (d) Routinely examine the cultures and growth media for evidence of any gross microbial contamination. (e) Clean the incubator, cabinet, centrifuge and microscope regularly. (f) Follow appropriate guidelines for disposal of biological waste (infectious and non-infectious) that includes sharps contaminated with biological waste, liquid as well as solid waste.

9. Place only one cover slip per well in the 6-well plate. It is convenient to use pointed forceps with a good grip for holding and placing the cover slip. Seed the cell suspension on cover slip in a drop by drop fashion and gently fill the well with media from one corner rather than directly on the cover slip to prevent dislodging of cells.
10. It is possible that the cells do not adhere to the cover slip—user must confirm the cell type (suspension/adherent) and check for cell viability since dead cells tend to round-off and detach.
11. Seed the cells at a density such that they achieve ~60% confluency at the time of experiment. A very high cell density can lead to over-digestion and an excessively bright fluorescent signal in which cells are not distinct. This affects the image quality and makes the quantification troublesome. Hence, it is recommended to have cells at a density such that they are discrete.
12. We prefer to use chilled absolute methanol to fix the cells before proceeding for zymography. Fixation helps to immobilize the antigen while retaining the cellular and subcellular structures. The fixative and protocol for cell fixation may be modified as per the properties of the cell line being used.
13. Brij-35 is a non-ionic light detergent that ensures enzyme stabilization and minimizes the risk of nonspecific interactions.
14. FITC-casein is diluted in only half the volume of digestion buffer (1 μL FITC-casein from stock of 5 mg/mL in 500 μL digestion buffer) such that its effective concentration in the substrate–agarose (1:1) mixture remains 0.5 μg/mL.
15. Agarose is dissolved in only half the volume of digestion buffer (0.1 g agarose in 10 mL buffer) such that its effective concentration in FITC-casein containing gel (1:1) solution remains 0.5%. Do not add the fluorophore-conjugated substrate to the agarose solution until it cools down to a temperature of ~45 °C (i.e. temperature below which it begins to solidify).
16. Cut the end of the tip or use a wide-orifice tip for a smooth flow while dropping the gel solution on glass slide. This will also help avoid air bubbles on the slide.
17. Pre-warming of microscope glass slide allows just enough time to drop the gelling agent on the slide and place the cover slip over it before the solution solidifies. A little delay prevents the

cover slip from embedding in the gel and thus, increases the chance of its displacement when incubated in the digestion buffer.

18. The agarose solution spreads evenly by capillary action and covers the entire surface area of the cover slip as it is placed over the gel mixture. Care must be taken while placing the cover slip to ensure that there is no air bubble between the cover slip and gel solution present on the glass slide.
19. Care must be taken to avoid displacement of the cover slip from its position during this step. It is thus suggested, to first place the trough containing slides in the incubator and then carefully add the digestion buffer in the trough. Adding a sufficient volume of digestion buffer in the trough ensures that the setup does not dry during the incubation period.
20. Take out the slide carefully from the trough such that the cover slip does not displace from its position during handling. Dab the excess liquid over a piece of KimWipe/tissue paper before setting up the slide for imaging.
21. The fluorescein label on FITC-Casein substrate is highly quenched. Upon digestion by MMPs present in the cells, the substrate is cleaved into smaller peptides, thus, terminating the quenching of the fluorescent label.
22. Capture the same field in case of dual staining (i.e. simultaneous execution of cell in situ zymography (Green—Fluorescein-conjugated substrate) with nuclear staining (Blue—DAPI) so that well-defined overlays may be created. Additionally, the cells must also be viewed in the bright field.
23. Nuclei of cells may be stained with a commonly used dye, DAPI (4′,6-diamidino-2-phenylindole), which binds to nucleic acids. It is a light-sensitive, water-soluble stain with an excitation/emission maxima at 358/461 nm. Stock solution of DAPI must preferably be aliquoted and stored at −20 °C. However, the solution may be stored at 2–8 °C for a short term. *Preparation of DAPI solution:* Resuspend DAPI in storage solution (Stock concentration 1 mg/mL; Working concentration: 50 ng/mL). The storage solution may also be utilized for all serial dilutions. Storage solution (Tris Buffer Solution): 10 mM Tris–HCl pH 7.4, 10 mM EDTA pH 8, 100 mM NaCl. *Caution:* DAPI is a known mutagen and hence, precaution must be taken while handling it. Safe disposal of dye in compliance with safety regulations is recommended.
24. In case of very weak/strong fluorescent signal, the user must troubleshoot based on parameters including substrate concentration, MMP activation, incubation period and temperature.

25. *o*-Phenanthroline (1,10-phenanthroline monohydrate) is a metalloproteinase inhibitor that chelates divalent metals like zinc and iron. It is soluble in methanol and can be prepared at a stock concentration of 200 mM. The stock remains stable at −20 °C for months. Effective concentration: 1–10 mM. The inhibitor is added to the digestion buffer used for making the gel solution as well as that utilized for incubation (Fig. 2f).
26. Image J is one of the most commonly used image processing software. However, other softwares like Image-Pro Plus (Media Cybernetics) may also be used. Additionally, we suggest one useful link to make a box and whisker plot (http://boxplot.tyerslab.com).

Acknowledgement

This work was supported by the research grant awarded to Dr. Vibha Rani by the Department of Science and Technology (DST), Government of India (SR/FT/LS-006/2009) and Department of Biotechnology (DBT), Government of India (BT/PR3978/17/766/2011).

References

1. Hawkes SP, Li H, Taniguchi GT (2001) Zymography and reverse zymography for detecting MMPs, and TIMPs. In: Clark I (ed) Matrix metalloproteinases protocols. Humana Press, Totowa, NJ, pp 399–410
2. Troeberg L, Nagase H (2003) Zymography of metalloproteinases. Curr Protoc Protein Sci 21:1–12
3. Snoek-van Beurden PAM, Von den Hoff JW (2005) Zymographic techniques for the analysis of matrix metalloproteinases and their inhibitors. Biotechniques 38:73–83
4. Vandooren J, Geurts N, Martens E, Van den Steen PE, Opdenakker G (2013) Zymography methods for visualizing hydrolytic enzymes. Nat Methods 10:211–220
5. Nemori R, Tachikawa T (1999) A review for *in situ* zymography: method for localization of protease activities in a tissue. Tissue Cult Eng 25:29–32
6. Yan SJ, Blomme EA (2003) *In situ* zymography: a molecular pathology technique to localize endogenous protease activity in tissue sections. Vet Pathol 40:227–236
7. George SJ, Johnson JL (2010) *In situ* zymography. Methods Mol Biol 622:271–277
8. Galis ZS, Sukhova GK, Libby P (1995) Microscopic localization of active proteases by *in situ* zymography: detection of matrix metalloproteinase activity in vascular tissue. FASEB J 9:974–980
9. Mungall BA, Pollitt CC, Collins R (1998) Localisation of gelatinase activity in epidermal hoof lamellae by *in situ* zymography. Histochem Cell Biol 110:535–540
10. Frederiks WM, Mook ORF (2004) Metabolic mapping of proteinase activity with emphasis on *in situ* zymography of gelatinases: review and protocols. J Histochem Cytochem 52:711–722
11. Löffek S, Schilling O, Franzke CW (2011) Biological role of matrix metalloproteinases: a critical balance. Eur Respir J 38:191–208
12. Pampaloni F, Reynaud EG, Stelzer EHK (2007) The third dimension bridges the gap between cell culture and live tissue. Nat Rev Mol Cell Biol 8:839–845
13. Cha MC, Purslow PP (2010) Matrix metalloproteinases are less essential for the in-situ gelatinolytic activity in heart muscle than in skeletal muscle. Comp Biochem Physiol A Mol Integr Physiol 156:518–522
14. Chhabra A, Jaiswal A, Malhotra U, Kohli S, Rani V (2012) Cell *in situ* zymography: an *in vitro* cytotechnology for localization of enzyme activity in cell culture. In Vitro Cell Dev Biol Anim 48:463–468

Part IV

2D Zymography

Part IV

Chapter 13

Examination of Gelatinase Isoforms in Rodent Models of Acute Neurodegenerative Diseases Using Two-Dimensional Zymography

Shanyan Chen, Fanjun Meng, Zhenzhou Chen, Zhe Qu, Jiankun Cui, and Zezong Gu

Abstract

Pathological activation of gelatinases (matrix metalloproteinase-2 and -9; MMP-2/-9) has been shown to cause a number of detrimental outcomes in neurodegenerative diseases. In gel gelatin zymography is a highly sensitive methodology commonly used in revealing levels of gelatinase activity and in separating the proform and active form of gelatinases, based on their different molecular weights. However, this methodology is inadequate in resolving complex enzyme isoforms, because gelatinase expression and activity can be regulated at transcriptional and/or post-translational levels under in vivo conditions resulting in alternation of their isoelectric focusing (IEF) points. In this chapter, we describe an advanced methodology, termed two-dimensional zymography, combining IEF with zymographic electrophoresis under non-reducing conditions to achieve significant improvement in separation of the gelatinase isoforms in both cell-based and in vivo models for acute brain injuries and neuroinflammation.

Key words Matrix metalloproteinase-2 and -9, Gelatinases, Two-dimensional zymography, Post-translational modification isoforms, Neurodegeneration

1 Introduction

Matrix metalloproteinases (MMPs) are members of a family of 26 zinc-dependent endopeptidases that contain structurally similar hemopexin, propeptide, and catalytic domains. They are so-called metalloproteinases due to their dependence on metal ions, particularly Zn^{2+} and Ca^{2+}, as cofactors in the enzymatic active site [1, 2]. MMPs regulate homeostasis of the extracellular matrix (ECM) by proteolysis of its components, such as collagen, laminin, fibronectin and collagens, as well as many non-matrix bioactive molecules

Fanjun Meng (Deceased)

Jeff Wilkesman and Liliana Kurz (eds.), *Zymography: Methods and Protocols*, Methods in Molecular Biology, vol. 1626,
DOI 10.1007/978-1-4939-7111-4_13, © Springer Science+Business Media LLC 2017

[3]. MMPs are needed for modulating interactions between cells during development and for repair, including tissue remodeling, cell migration, wound healing, angiogenesis, and embryogenesis [4, 5]. However, dysregulation of MMPs are also known to cause progression of damage in various diseases. In neurological diseases, excessive activation of MMPs is implicated in multiple sclerosis, HIV-associated dementia, spinal cord injury, traumatic brain injury (TBI), amyotrophic lateral sclerosis (ALS), and stroke [6–10].

In the MMP family, MMP-2 and MMP-9 are also called gelatinases due to their ability to degrade gelatin, the denatured form of collagen. Reportedly, pathological activation of gelatinases has been shown to cause a number of detrimental outcomes, including disruption of the blood–brain barrier (BBB), hemorrhage, neuronal apoptosis [11], and brain damage in ischemic stroke [8] and traumatic brain injury (TBI) [10]. MMP-9 (Gelatinase B) activity, in particular, is significantly elevated in humans after stroke [12–14]. High plasma MMP-9 concentration in the acute phase of cerebral infarct is considered as an independent surrogate of hemorrhagic transformation in all stroke subtypes [15]. Our group and others have shown that aberrant MMP-9 proteolytic activity is associated with an increase in BBB permeability and neuronal injury after acute cerebral ischemia [8, 16–18].

MMPs are expressed in the cells and secreted into the surrounding ECM as inactive zymogens. The Zn^{2+} ion in the active site of the enzyme is blocked by a cysteine residue located in the pro-domain of the MMP peptide. MMP enzyme becomes proteolytically active only when the Zn^{2+} ion in active site is exposed by dissociation of the binding between the cysteinyl sulfhydryl in the propeptide domain and the catalytic Zn^{2+} ion. This mechanism of activation is referred to as the "cysteine switch" [19]. The removal of the cysteine residue from the MMP active site can be achieved by cleavage of the pro-domain, resulting in a lower molecular weight active MMP [20]. The activity of MMPs can also be regulated by the levels of post-translational modifications, including *S*-nitrosylation, *S*-glutathiolation, glycosylation, sialylation, and phosphorylation [8, 21, 22]. It has been reported that MMPs could be activated by the pro-oxidant species peroxynitrite ($ONOO^-$) without removal of the inhibitory propeptide domain. $ONOO^-$ may react with the cysteine thiol within the PRCGVPD sequence of the propeptide domain by *S*-glutathiolation, resulting in an increase in proteolytic activity. It suggests that full-length form of the MMP may also be proteolytically active [23–25]. Phosphorylation is another post-translational modification playing a significant role in regulating MMPs. It has been demonstrated that phosphorylation of MMP-2 (Gelatinase A) by inflammatory factor protein kinase C increases its activity [26].

In-gel zymography (IGZ) was first introduced in 1978 to study plasminogen activators [27]. In an IGZ experiment, hydrolytic enzymes are separated by their molecular weights and detected by their ability to degrade the substrate copolymerized in the separating gel of SDS-PAGE. Briefly, after separation by non-reducing SDS-PAGE gel, the sodium dodecyl sulfate (SDS) is replaced by a non-ionic detergent with a lower critical micelle concentration. This step allows the enzymes to be partially refolded to their active forms; these enzymes are previously unfolded by SDS ion. Next, the gel is incubated in a buffer with essential cofactors in certain temperature, which allows the reactivated enzymes to degrade the copolymerized substrate. Proteolytic activity is detected by light zones on a dark background following staining procedures [28]. With IGZ, information that can be obtained includes: proteolytic activity of specific enzyme, molecular weights, and the presence of covalent complexes or fragments in complex biological samples. However, accumulated evidences show that complex post-translational modification is involved in the activation of MMPs without changing molecular weights by the removal of the inhibitory pro-peptide domain [24, 26, 29]. Therefore, identification of different isoforms of gelatinases based on molecular weights after electrophoresis is not sufficient in complex biological samples.

We complemented conventional IGZ with an additional dimension of protein separation by in-gel IEF to achieve significant improvement in separating the enzymatic isoforms due to their various p*I* values [30]. In this so-called 2D zymography method, gelatinase isoforms with charge differences can be separated in the first IEF dimension, followed by the conventional zymography. We have found in LPS-stimulated microglial cells and ischemic/TBI mouse brain tissue, this extra IEF dimension is particularly useful for identifying different MMP-9 isoforms [30]. We also found the MMP-9 p*I* values were shifted in 2D zymography after treatment with alkaline phosphatase, suggesting the significance of phosphorylation in MMP-9 activation [30]. Therefore, 2D zymography is demonstrated as an effective method to separate isoforms of gelatinases with different post-translational modifications in acute brain injuries.

2 Materials

2.1 Brain Extraction and Gelatinase Purification

1. Brain Extraction Buffer: 50 mM Tris, pH 7.6, 150 mM NaCl, 5 mM $CaCl_2$, 0.05% Brij35, 1% Triton X 100.
2. Gelatin Sepharose 4B.
3. Elution Buffer: 8 M urea, 4% CHAPS.

2.2 Isoelectric Focusing

1. Rehydration Buffer: 8 M urea, 4% CHAPS, 0.5% IPG buffer (*see* **Note 1**).
2. SDS Equilibration Buffer: 50 mM Tris–HCl, pH 8.8, 6 M urea, 30% glycerol, 2% SDS, 0.002% bromophenol blue.

2.3 Gelatin SDS-PAGE Gel

1. Separating gels: 0.375 M Tris–HCl, pH 8.8, 10% acrylamide/bis-acrylamide (29:1), 0.1% SDS, 0.05% ammonium persulfate (APS), 0.005% TEMED, 10% glycerol, 0.1% gelatin.
2. Stacking gels: 0.125 M Tris–HCl pH 6.8, 4% acrylamide/bis-acrylamide (29:1), 0.1% SDS, 0.05% APS, 0.001% TEMED.
3. Agarose Sealing Mixture: 0.5% agarose, 0.002% bromophenol blue in 1× electrophoresis running buffer.
4. Electrophoresis Running Buffer: 2.9 g Tris-base, 14.4 g glycine, 1 g SDS, doubly distilled H_2O to 1 L.
5. Renature Buffer: 2.5% Triton X-100.
6. Developing Buffer: 50 mM Tris–HCl pH 7.6, 0.2 M NaCl, 5 mM $CaCl_2$, 0.2% Brij35.
7. Coomassie Blue Staining Solution: 0.2% Coomassie brilliant blue R-250, 40% ethanol, 10% acetic acid.
8. Destaining solution: 40% ethanol, 10% acetic acid.

3 Methods

In this section, we describe how to examine isoforms of MMP-2 and MMP-9, both in lysates prepared from brain samples, and from conditioned media of BV-2 microglial cultures.

3.1 Protein Extraction and Gelatinase Purification

The filament-induced transient middle cerebral artery occlusion (MCAo) or TBI is performed as described previously [8, 10, 30–32]. Mice are sacrificed with an overdose of isoflurane and transcardially perfused with saline to remove intravascular blood. Mouse brains are rapidly collected.

Immortalized mouse BV-2 microglial cells are originally from Dr. R. Donato (University of Perugia, Italy) [33]. BV-2 cells are maintained in DMEM (Dulbecco's Modified Eagle Medium) containing 5% heat-inactivated FBS (fetal bovine serum) at 37 °C in a saturated humidity atmosphere containing 95% air and 5% CO_2. At 70–80% confluence, BV-2 cells are starved with no serum medium for 4 h and treated with 100 or 500 ng/mL LPS in the conditioned medium for 16 h as described previously [30].

1. Weigh the brain tissues.
2. Add extraction buffer with proteinase inhibitors. Volume of extraction buffer (μL) = weight of sample (mg) × 6.
3. Homogenize the brain tissues on ice with a pestle in a microcentrifuge tube. Keep the homogenate on ice for 20 min.
4. Centrifuge (17,000 × *g*) the homogenate for 20 min at 4 °C. Collect the supernatant.

5. To measure protein concentration, 10 μL of each sample is added into 190 μL of ddH_2O, making a 20-fold dilution. Perform BCA assay following manufacturer's manual.
6. Make all samples from **step 4** to the same concentration according to the results of BCA assay.
7. Wash the Gelatin Sepharose 4B with extraction buffer for three times.
8. After the last wash, carefully aspirate the supernatant. Add protein samples from **step 6** to the Gelatin Sepharose 4B. For cell-based studies, add conditioned media to the Gelatin Sepharose 4B.
9. Rotate the samples in tubes slowly at 4 °C overnight to purify gelatinases.

3.2 Gelatinase Elution

1. Wash Gelatin Sepharose 4B with extraction buffer three times and then discard supernatant.
2. Add 75 μL of elution buffer to each tube and rotate the tubes for 40 min at room temperature.
3. Centrifuge (17,000 × *g*) the samples for 30 min at room temperature. Collect the supernatant (75 μL) for the following steps.

3.3 First Dimensional Separation: Isoelectric Focusing

1. Mix the samples and rehydration buffer. Volume of this mixture depends on the length of IPG strips. For 7-cm strips, 125 μL mixtures are applied.
2. Distribute the mixture in strip holders evenly. Lay the IPG strips on the mixture.
3. Cover the IPG strips with mineral oil to minimize evaporation.
4. IPG strips are swelled with the protein samples for 12 h under 50 V (active rehydration).
5. Proteins are separated by IEF (PROTEAN IEF Cell, Bio-Rad) using the following conditions with: 250 V (rapid) for 250 voltage-hours (V·h); 500 V (rapid) for 500 V·h; 1000 V (rapid) for 1000 V·h; 5000 V (linear) for 10,000 V·h and 5000 V (rapid) for 20,000 V·h.
6. Equilibrate the IPG strips with the SDS equilibration buffer twice, each time 15 min with shaking.
7. Wash the IPG strips with running buffer before loading to the second-dimensional gels.

3.4 Second-Dimensional Separation: Gelatin SDS-PAGE

1. First, prepare separating gels (30 mL for four gels): mix 3 mL of ddH_2O; 3 mL of 1% gelatin; 6 mL of 50% (v/v) glycerol; 10 mL of 30% acrylamide–bis-acrylamide; 7.5 mL of separating gel buffer stock (1.5 M Tris buffer, pH 8.8); 300 μL of 10% (w/v) SDS; 150 μL of 10% (w/v) APS. Degas the solution for

approximately 5 min. To speed up the process, dissolve the gelatin with warm water.

2. Add 15 μL of TEMED to the separating gel solution to initiate polymerization. Immediately, pipette approximately 6.2 mL of the separating gel solution into the gel casting chamber (Mini-PROTEAN Tetra Cell, Bio-Rad). Avoid formation of bubbles.
3. Carefully overlay the separating gel solution in the cassette with ddH_2O using a syringe. Do not disturb the surface of the separating gel solution.
4. Let the assembly stand for at least 1 h at room temperature to polymerize. Polymerization is complete when discrete lines of separation can be noted between the gels and the water overlay.
5. Prepare stacking gels: decant the overlay water from the separating gels. Immediately, add 6 μL of TEMED to the stacking gel solution, swirl rapidly and pipet the solution onto the top of the polymerized separating gels until it reaches the top of the front plate.
6. Rapidly, insert the appropriate combs into the liquid stacking gels, making sure that no bubbles remain trapped under the combs. Let the stacking gel polymerize at room temperature for 30 min.

3.5 Running and Developing of Gelatin SDS-PAGE Gels

1. Apply the equilibrated IPG strips to the second-dimensional gels carefully.
2. Add 1 mL of agarose sealing mixture over the IPG strips to cover them completely.
3. Run the gels at constant voltage (120 V) until the 37 kDa band of the prestained maker runs out of the gels.
4. Carefully remove the gels from the cassette and place it in plastic tray containing 100 mL of renaturing buffer. Incubate the gels for 30 min at room temperature with gentle agitation for three times. Replace the renaturing buffer with 100 mL of developing buffer. Wash the gels for 30 min at room temperature.
5. Replace the developing buffer with 100 mL of fresh developing buffer. Incubate the gels at 37 °C for about 18 h in the incubator (*see* **Note 2**).
6. Stain the gels with Coomassie blue staining solution for 4 h at room temperature, agitating on a rotary shaker.
7. Destain the gels with destaining solution for 15–30 min at room temperature, agitating on a rotary shaker. Replace the solution with water when the clear lanes and spots are revealed (*see* Fig. 1) [30].

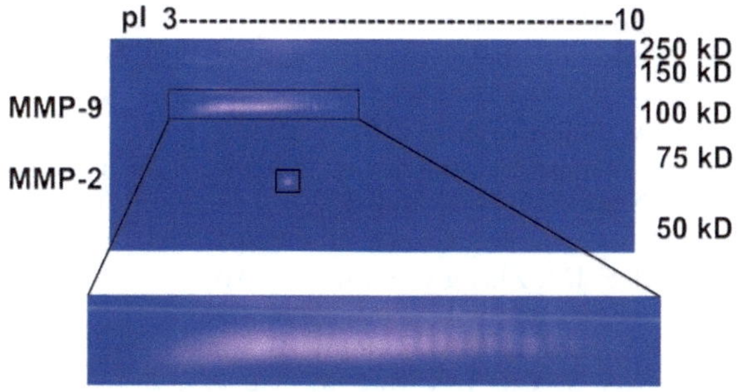

Fig. 1 2D zymography reveals gelatinase isoforms from conditioned medium of LPS-stimulated microglial BV-2 cells. MMP-9 isoforms were visualized as a serious transparent spots with p*I* values ranging at 3.5–7 and molecular weight at 100 kDa; MMP-2 was visualized as a single spot with p*I* value between 5 and 6, and molecular weight at 65 kDa

4 Notes

1. No reducing reagents (e.g. DTT) are allowed for 2-D zymography.
2. Developing time can be adjusted based on the amount and the activity of the gelatinases in different samples.

Acknowledgments

This work was supported in part by the American Heart Association National Scientist Development award (09SDG2260983), The Dana Foundation, The National Football Leagues (NFL) Charities Foundation, The University of Missouri Mizzou Advantage One Health One Medicine Program and the Department of Pathology Research funds to Z.G.

References

1. Kessenbrock K, Plaks V, Werb Z (2010) Matrix metalloproteinases: regulators of the tumor microenvironment. Cell 141(1):52–67
2. Parks WC, Wilson CL, Lopez-Boado YS (2004) Matrix metalloproteinases as modulators of inflammation and innate immunity. Nat Rev Immunol 4(8):617–629
3. Mott JD, Werb Z (2004) Regulation of matrix biology by matrix metalloproteinases. Curr Opin Cell Biol 16(5):558–564
4. Yong VW (2005) Metalloproteinases: mediators of pathology and regeneration in the CNS. Nat Rev Neurosci 6(12):931–944
5. Rosenberg GA (2009) Matrix metalloproteinases and their multiple roles in neurodegenerative diseases. Lancet Neurol 8(2):205–216
6. Kaplan A, Spiller KJ, Towne C, Kanning KC, Choe GT, Geber A, Akay T, Aebischer P, Henderson CE (2014) Neuronal matrix metalloproteinase-9 is a determinant of selective neurodegeneration. Neuron 81(2):333–348
7. Agrawal SM, Lau L, Yong VW (2008) MMPs in the central nervous system: where the good guys go bad. Semin Cell Dev Biol 19(1):42–51
8. Gu Z, Kaul M, Yan B, Kridel SJ, Cui J, Strongin A, Smith JW, Liddington RC, Lipton SA

(2002) S-nitrosylation of matrix metalloproteinases: signaling pathway to neuronal cell death. Science 297(5584):1186–1190
9. Chaturvedi M, Kaczmarek L (2014) Mmp-9 inhibition: a therapeutic strategy in ischemic stroke. Mol Neurobiol 49(1):563–573
10. Hadass O, Tomlinson BN, Gooyit M, Chen S, Purdy JJ, Walker JM, Zhang C, Giritharan AB, Purnell W, Robinson CR 2nd, Shin D, Schroeder VA, Suckow MA, Simonyi A, Sun GY, Mobashery S, Cui J, Chang M, Gu Z (2013) Selective inhibition of matrix metalloproteinase-9 attenuates secondary damage resulting from severe traumatic brain injury. PLoS One 8(10):e76904
11. Chen ZL, Strickland S (1997) Neuronal death in the hippocampus is promoted by plasmin-catalyzed degradation of laminin. Cell 91(7):917–925
12. Montaner J, Alvarez-Sabin J, Molina CA, Angles A, Abilleira S, Arenillas J, Monasterio J (2001) Matrix metalloproteinase expression is related to hemorrhagic transformation after cardioembolic stroke. Stroke 32(12):2762–2767
13. Horstmann S, Kalb P, Koziol J, Gardner H, Wagner S (2003) Profiles of matrix metalloproteinases, their inhibitors, and laminin in stroke patients: influence of different therapies. Stroke 34(9):2165–2170
14. Switzer JA, Hess DC, Ergul A, Waller JL, Machado LS, Portik-Dobos V, Pettigrew LC, Clark WM, Fagan SC (2011) Matrix metalloproteinase-9 in an exploratory trial of intravenous minocycline for acute ischemic stroke. Stroke 42(9):2633–2635
15. Castellanos M, Leira R, Serena J, Pumar JM, Lizasoain I, Castillo J, Davalos A (2003) Plasma metalloproteinase-9 concentration predicts hemorrhagic transformation in acute ischemic stroke. Stroke 34(1):40–46
16. Gasche Y, Fujimura M, Morita-Fujimura Y, Copin JC, Kawase M, Massengale J, Chan PH (1999) Early appearance of activated matrix metalloproteinase-9 after focal cerebral ischemia in mice: a possible role in blood-brain barrier dysfunction. J Cereb Blood Flow Metab 19(9):1020–1028
17. Lapchak PA, Chapman DF, Zivin JA (2000) Metalloproteinase inhibition reduces thrombolytic (tissue plasminogen activator)-induced hemorrhage after thromboembolic stroke. Stroke 31(12):3034–3040
18. Wang J, Tsirka SE (2005) Neuroprotection by inhibition of matrix metalloproteinases in a mouse model of intracerebral haemorrhage. Brain 128(Pt 7):1622–1633
19. Van Wart HE, Birkedal-Hansen H (1990) The cysteine switch: a principle of regulation of metalloproteinase activity with potential applicability to the entire matrix metalloproteinase gene family. Proc Natl Acad Sci U S A 87(14):5578–5582
20. Park AJ, Matrisian LM, Kells AF, Pearson R, Yuan ZY, Navre M (1991) Mutational analysis of the transin (rat stromelysin) autoinhibitor region demonstrates a role for residues surrounding the "cysteine switch". J Biol Chem 266(3):1584–1590
21. Vandooren J, Van den Steen PE, Opdenakker G (2013) Biochemistry and molecular biology of gelatinase B or matrix metalloproteinase-9 (MMP-9): the next decade. Crit Rev Biochem Mol Biol 48(3):222–272
22. Rossano R, Larocca M, Riviello L, Coniglio MG, Vandooren J, Liuzzi GM, Opdenakker G, Riccio P (2014) Heterogeneity of serum gelatinases MMP-2 and MMP-9 isoforms and charge variants. J Cell Mol Med 18(2):242–252
23. Okamoto T, Akaike T, Nagano T, Miyajima S, Suga M, Ando M, Ichimori K, Maeda H (1997) Activation of human neutrophil procollagenase by nitrogen dioxide and peroxynitrite: a novel mechanism for procollagenase activation involving nitric oxide. Arch Biochem Biophys 342(2):261–274
24. Okamoto T, Akaike T, Sawa T, Miyamoto Y, van der Vliet A, Maeda H (2001) Activation of matrix metalloproteinases by peroxynitrite-induced protein S-glutathiolation via disulfide S-oxide formation. J Biol Chem 276(31): 29596–29602
25. Viappiani S, Nicolescu AC, Holt A, Sawicki G, Crawford BD, Leon H, van Mulligen T, Schulz R (2009) Activation and modulation of 72 kDa matrix metalloproteinase-2 by peroxynitrite and glutathione. Biochem Pharmacol 77(5): 826–834
26. Sariahmetoglu M, Crawford BD, Leon H, Sawicka J, Li L, Ballermann BJ, Holmes C, Berthiaume LG, Holt A, Sawicki G, Schulz R (2007) Regulation of matrix metalloproteinase-2 (MMP-2) activity by phosphorylation. FASEB J 21(10):2486–2495
27. Granelli-Piperno A, Reich E (1978) A study of proteases and protease-inhibitor complexes in biological fluids. J Exp Med 148(1):223–234
28. Vandooren J, Geurts N, Martens E, Van den Steen PE, Opdenakker G (2013) Zymography methods for visualizing hydrolytic enzymes. Nat Methods 10(3):211–220
29. Schulz R (2007) Intracellular targets of matrix metalloproteinase-2 in cardiac disease: rationale and therapeutic approaches. Annu Rev Pharmacol Toxicol 47:211–242
30. Chen S, Meng F, Chen Z, Tomlinson BN, Wesley JM, Sun GY, Whaley-Connell AT,

Sowers JR, Cui J, Gu Z (2015) Two-dimensional zymography differentiates gelatinase isoforms in stimulated microglial cells and in brain tissues of acute brain injuries. PLoS One 10(4):e0123852

31. Gu Z, Cui J, Brown S, Fridman R, Mobashery S, Strongin AY, Lipton SA (2005) A highly specific inhibitor of matrix metalloproteinase-9 rescues laminin from proteolysis and neurons from apoptosis in transient focal cerebral ischemia. J Neurosci 25(27):6401–6408

32. Cui J, Chen S, Zhang C, Meng F, Wu W, Hu R, Hadass O, Lehmidi T, Blair GJ, Lee M, Chang M, Mobashery S, Sun GY, Gu Z (2012) Inhibition of MMP-9 by a selective gelatinase inhibitor protects neurovasculature from embolic focal cerebral ischemia. Mol Neurodegener 7:21

33. Shen S, Yu S, Binek J, Chalimoniuk M, Zhang X, Lo SC, Hannink M, Wu J, Fritsche K, Donato R, Sun GY (2005) Distinct signaling pathways for induction of type II NOS by IFNgamma and LPS in BV-2 microglial cells. Neurochem Int 47(4):298–307

Chapter 14

Two-Dimensional Zymography of Proteases from Steatotic Duck Liver

Jeff Wilkesman, María Fernanda Padrón, Liliana Kurz, and Hervé Rémignon

Abstract

Protease activity present in liver cells with steatosis can be electrophoretically characterized. Zymographic techniques allow semi-quantitative results, successfully detecting cathepsin and metalloprotease activity using polyacrylamide gels copolymerized with gelatin and quantified by densitometry. By using specific inhibitors, the identity of the proteases can be confirmed. 2D zymography allows the determination of both *Mr.* and p*I* of the metalloprotease and cathepsin activity present in the homogenates. The analysis of liver proteases activities in force fed ducks may elucidate the mechanisms behind steatosis development.

Key words Electrophoresis, MMP2, Cathepsins, Steatosis, Zymography

Abbreviations

2DE	Two dimensional electrophoresis
2DZ	Two dimensional zymography
DTT	Dithiothreitol
EDTA	Ethylenediaminetetraacetic acid
IEF	Isoelectric focusing
MMP	Matrix metalloproteases
PAG	Polyacrylamide gel
PAGE	Polyacrylamide gel electrophoresis
p*I*	Isoelectric point
PMSF	Phenylmethyl sulfonyl fluoride
SDS	Sodium dodecyl sulphate
TCA	Trichloroacetic acid

Jeff Wilkesman and Liliana Kurz (eds.), *Zymography: Methods and Protocols*, Methods in Molecular Biology, vol. 1626,
DOI 10.1007/978-1-4939-7111-4_14,

1 Introduction

The accumulation of triglycerides in the hepatic cells is known as fatty liver or steatosis, usually as the result of a rich carbohydrates diet force feeding [1]. In palmipeds, the liver is the place where most of the fat is synthesized and can be stored in case of excessive production. In force fed birds, due to the large amount of carbohydrates brought by the daily diet, a huge quantity of fat is produced and stored, leading to the development of a nutritional steatosis of the liver [2]. In any case, a steatotic liver in ducks is not a pathological organ, but one that presents an enhanced metabolic activity [1].

In the regulation of a wide range of biological processes are present many proteolytic activities, as many possess high specificity associated with the hydrolysis of the peptide bond [3, 4]. Numerous techniques are used to determine the presence of proteases in biological samples. The application of electrophoresis represents a simple alternative in the study of many types of proteases and its inhibitors [5, 6]. Zymography is an important tool for the detection of proteolytic activity on an electrophoretic gel matrix, employing a protein substrate [7].

In this chapter, we describe the detection of proteolytic profiles present in hepatic cells under steatosis, employing 1D and 2D zymography. To accomplish this, an appropriate extraction procedure was performed. For quantitative purposes, it is advised to measure in parallel the total proteolytic activity by some other method, e.g. photometry.

2 Materials

All chemicals used were of analytical grade, and deionized water was used. As main equipment an electrophoretic chamber is needed (many commercial providers offer different types and models) along with a power supply.

2.1 Extraction Buffers

1. *Extraction buffer without inhibitors*: 50 mM Tris–HCl pH 7.5, 150 mM NaCl, 1 mM $CaCl_2$.
2. *Extraction buffer selective for metalloproteases*: 50 mM Tris–HCl pH 7.5, 150 mM NaCl, 1 mM $CaCl_2$, 10 μg/mL leupeptin, 1 mM PMSF.
3. *Extraction buffer selective for cathepsins*: 50 mM Tris-HCl pH 7.5, 150 mM NaCl, 1 mM $CaCl_2$, 20 mM EDTA.

2.2 Electrophoresis Stock Solutions

1. 1.5 M Tris pH 8.8.
2. 0.5 M Tris pH 6.8.

3. 40% Acrylamide.
4. 2% Bis-acrylamide.
5. *5% gelatin stock solution*: weigh 0.5 g and dissolve in hot water until 10 mL. Aliquot 1 mL fractions and keep refrigerated.
6. *10% APS*: 0.0520 g in 520 μL. Prepare fresh.
7. *TEMED*.
8. *4× sample buffer*: 200 mM Tris–HCl pH 6.8, 4% (w/v) SDS, 40% (v/v) glycerol, 0.02% (w/v) bromophenol blue.
9. *Running buffer*: 25 mM Tris–HCl, 192 mM glycine, 0.1% (w/v) SDS.

2.3 Staining Solutions

1. *Coomassie staining solution*: 0.125% (w/v) Coomassie Brilliant Blue R-250, 40% (v/v) 2-propanol, 10% (v/v) acetic acid.
2. *Coomassie staining solution for zymograms*: 0.5% (w/v) Coomassie Brilliant Blue R-250, 40% (v/v) 2-propanol, 10% (v/v) acetic acid.
3. *Destaining solution*: 40% (v/v) 2-propanol, 10% (v/v) acetic acid.

2.4 2D Electrophoresis

1. *Rehydration buffer*: 8 M urea, 1% (w/v) CHAPS, 10 mM DTT, 0.25% (v/v) Biolytes 3-10.
2. *Equilibration buffer I*: 8 M urea, 0.375 M Tris–HCl pH 8.8, 2% (w/v) SDS, 20% (v/v) glycerol, 130 mM DTT.
3. *Equilibration buffer II*: 8 M urea, 0.375 M Tris–HCl pH 8.8, 2% (w/v) SDS, 20% (v/v) glycerol, 270 mM IAA.
4. *Agarose solution*: 1% agarose, 0.025% bromophenol blue.

2.5 Zymography Solutions

1. *Metalloprotease renaturing buffer*: 100 mM glycine pH 8.0, 2.5% (v/v) Triton X-100.
2. *Cathepsin renaturing buffer*: Tris–HCl 50 mM pH 7.4, 20% (v/v) glycerol.
3. *Metalloprotease activation buffer*: 100 mM glycine pH 8.0.
4. *Cathepsin activation buffer*: 0.1 M sodium acetate pH 5.5, 1 mM EDTA, 2 mM DDT.

2.6 Other Standards

1. *Bovine serum albumin solution*: BSA (1 g/L), weigh 1 mg protein and dissolve in 1 mL extraction buffer without inhibitor.
2. *Molecular weight standards* (commercially available, for SDS-PAGE and for zymograms).
3. *Bradford protein determination kit*.

3 Methods

3.1 Biological Samples

1. Collect the liver samples from ducks or other source [we studied Muscovy ducks (*Cairina moschata*)].
2. Treat the sample according to the protocol reported by Awde [8].
3. Briefly, livers are removed from carcasses, weighted, and samples are harvested, immediately frozen in liquid nitrogen and stored at −80 °C until analysis (*see* **Note 1**).

3.2 Protein Extraction

1. Weigh ~400 mg tissue and homogenize at 4 °C in a potter with a specific buffer.
2. Choose a specific extraction buffer, either: (a) without inhibitors; (b) selective for metalloproteases, or (c) selective for cathepsin.
3. Centrifuge suspension at 10,000 × *g* for 20 min at 4 °C. If necessary remove the fat cake due to lipids precipitation and redo the centrifugation.
4. Resulting supernatant is collected and either used immediately or alternately stored at −20 °C until assayed.

3.3 Protein Quantification

1. Determine the total amount of protein according to the Bradford method [9] (*see* **Note 2**).
2. Perform a calibration curve by triplicate employing bovine serum albumin as standard protein.
3. Assay samples by duplicate, measuring absorbance at λ595 nm.

3.4 SDS–PAGE

1. This electrophoresis is a modified procedure described by Laemmli [10]. Mind that non-reducing conditions are present, i.e., absence of 2-mercaptoethanol and samples are not heated.
2. Prepare solutions in order to polymerize 10% resolving gels and 5% stacking gel (Table 1).
3. Polymerize gels.
4. Dissolve ~30 μg protein in 4× sample buffer and apply samples into the well.
5. Apply the appropriate amount of molecular weight standards (consult the info-sheet of your product).
6. Add running buffer to the chamber.
7. Begin electrophoresis by setting on the power supply at 90 V for the first 15 min, and then increase to 120 V. Perform the run at 4 °C (*see* **Note 3**).
8. Run until the blue front reaches the bottom of the gel.
9. Remove gels from the glasses, and place them in container with Coomassie staining solution.

Table 1
Resolving and stacking gel preparation

	10% SDS-PAGE resolving gel	10% Zymography resolving gel	5% Stacking gel
Stock component	**Volume (mL)**		
H_2O	3.35	3.15	1.2
1.5 M Tris pH 8.8	2.5	2.5	–
0.5 M Tris pH 6.8	–	–	0.5
40% Acrylamide	2.5	2.5	0.2
2% Bis-Acrylamide	1.6	1.6	0.1
5% Gelatin	–	0.2	–
10% APS	0.05	0.05	0.01
TEMED	0.005	0.005	0.002
Final volume (for two mini-gels)	*10*	*10*	2

10. Remove staining solution and place the gel in destaining solution until patterns are observed.
11. Digitalize the gels with a transilluminator and save images as jpeg files.
12. Determine MW by using a calibration curve with the standard proteins, and relate the log MW vs. relative mobility.

3.5 2DE

1. Add rehydration buffer to an aliquot of liver extract containing between 100 and 400 μg of total protein such that the final volume is under 150 μL (*see* **Note 4**). Centrifuge briefly.
2. Place the IPG over the sample and leave 30 min for absorption.
3. Add 500 μL mineral oil.
4. Program the focusing run according to Table 2.
5. Submit strips to rehydration under active conditions: 50 V, 20 °C, 15 h, without pause after rehydration, and with 50 μA/strip.
6. Take notice that the final V × h displayed on the screen should be >10,000 V·h.
7. During the last focusing step, prepare the equilibration buffer I and II according to Table 3 (*see* **Note 5**).
8. Also during this time, polymerize a 10% resolving gel (Table 1) and a 10% zymogram *without* stacking gel.

Table 2
Program conditions for focusing

Step	Total time (h)	Voltage (V)
Rehydration	15	50
Focusing (rapid ramp)	3.5	50 → 4000
Slope down (linear ramp)	5	4000 → 500

Table 3
Composition of equilibration buffers

	Equilibration buffer I	Equilibration buffer II	
Final concentration of the component	**Amount**	**Amount**	**Stock**
6 M urea	1.8 g	1.8 g	60 g/mol
0.375 M Tris pH 8.8	1.25 mL	1.25 mL	1.5 M
2% SDS	1 mL	1 mL	10%
20% Glycerol	1 mL	1 mL	100%
2% DTT	0.1 g	–	
2.5% IAA	–	0.125 g	
BrPhBlue		25 μL	1%
Final volume	5 mL	5 mL	–

9. After focusing, equilibrate strips for 20 min in equilibration buffer I. Place strips in the special equilibration/rehydration tray and add 2.5 mL buffer per strip.
10. Now, equilibrate strips in equilibration buffer II.
11. Finally, wash strips briefly with water and then place one of them over the 10% polyacrylamide gels, and the other over the zymogram containing 0.1% gelatin substrate.
12. Seal the strip by overlaying it with warm agarose solution.
13. Run gels as described under Subheading 3.4, **step 6**.
14. When run is finished, stain the gel without substrate with Coomassie staining solution. Depending on the type of protease to be detected, proceed as stated under Subheading 3.8 for the zymogram.

3.6 1DZ for Metalloproteases

1. Prepare a 10% SDS-PAGE (Table 1) copolymerized with gelatin (*see* **Note 6**).
2. Pre-run the gel in running buffer at 4 °C for 20 min at 125 V.
3. Apply ~50 μg total protein per lane.
4. Run the system at 4 °C for 120 min at 125 V.
5. After run, incubate gels twice in renaturing buffer for 30 min each.
6. Discard solution and place gel in activating buffer at 37 °C for 18 h.
7. Finally, stain gels with Coomassie staining solution for zymograms.
8. Add destaining solution until pale bands under a deep blue background appear.

3.7 1DZ for Cathepsins

1. According to Platt et al. [11] for the case of cathepsins, polymerize 10% PAG (Table 1) containing 0.1% gelatin (*see* **Note 7**).
2. Prepare the samples in sample buffer. Set up the chamber and add running buffer.
3. Run at 100 V, 4 °C, until the blue front reaches the bottom of the gel (~2 h).
4. Once the run is completed, wash gel twice for 15 min with 0.1 M sodium acetate.
5. Place gel in cathepsin renaturing buffer under mild agitation overnight.
6. Now incubate gels with cathepsin activation buffer at 37 °C for 18 h.
7. Afterward, soak gels briefly in water (dd) and stain as described in Subheading 3.6, **step 7**. Figure 1 shows the protein profile of a duck liver sample in a 10% SDS-PAGE and a 10% zymogram.

3.8 2DZ

1. After the 2DE run, place gels in 1% Triton-X 100 solution, for 1 h.
2. Incubate in a suitable activation buffer, depending on the protease activity to be identified.
3. Wash and incubate gels with the same solutions used for the 1DZ. An example of a 2DE and 2DZ is given in Fig. 2, for the cathepsin analysis.
4. Perform densitometric analysis for p/ and MW determination. Spot intensities can be correlated with proteolytic activities (*see* **Note 8**) [12–15].

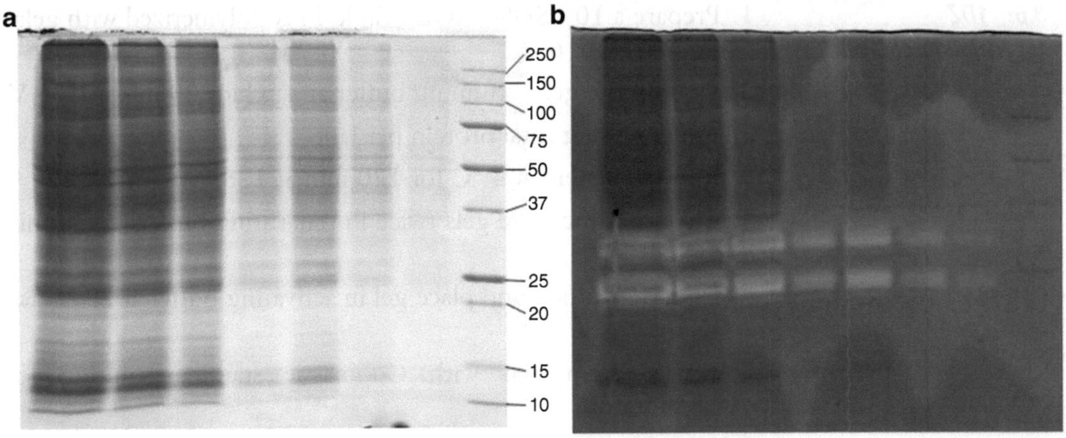

Fig. 1 10% SDS PAGE (**a**) and 10% zymogram (**b**) of duck liver homogenate extracted for cathepsin detection

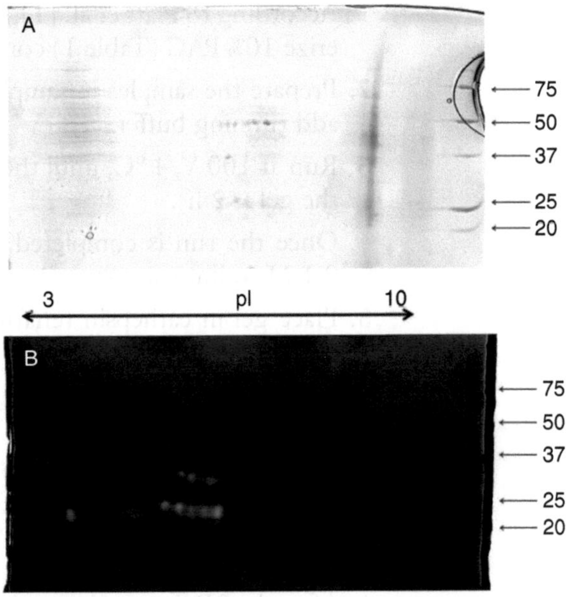

Fig. 2 2DE and 2DZ examples. First dimension run with IEF strips from pH 3–10. Second dimension run in SDS-PAGE 10%. (**a**) 2DE of the protein extract from duck liver. (**b**) 2DZ of the cathepsin activity

4 Notes

1. Before performing any experiments, consult the bioethical regulations of your laboratory, institution or country. The experiment described here fully complies with legislation on research involving animal subjects according to the European Communities Council Directive of November, 24 1986 (86/609/EC), the Venezuelan Law of Science & Technology (2010) and the Venezuelan Bioethic Code (2010). In the case of France, French Investigators must be certificated by the

French governmental authority for carrying out these experiments [16–18].

2. Total protein content (μg/μL) in the sample needs to be determined before submitting samples to electrophoresis, in order to know the amount of sample (in our case ~30 μg) needed for correct visualization of the activity on the zymograms.
3. The chamber may be placed directly inside a refrigerator, or a cold room.
4. For two strips, for example, we prepare a final volume of 250 μL. Our total protein content in the extract was 7.6 μg/μL. We took a 20 μL aliquot and combined it with the rehydration buffer. Each strip was covered with 125 μL solution.
5. Equilibration buffer I and II are freshly prepared.
6. We have also used 7.5% SDS-PAGE to obtain a better band resolution. The adequate % will depend on the MW of the protease and its polypeptide profile.
7. We have also used 12% SDS-PAGE to obtain a better band resolution. The adequate % will depend on the MW of the protease and its polypeptide profile.
8. Proteolytic activity can be measured by the Kunitz method [19].

Acknowledgments

We thank INP-ENSAT for supporting the stay of Dr. Wilkesman at GenPhySE, Université de Toulouse, as well as the *Consejo de Desarrollo Científico y Humanísitico de la Universidad de Carabobo*, Venezuela, for partial funding of this research (CDCH-AM-030-11, CDCH-187-2015). The authors thank the staff of GenPhySE lab for technical assistance and Dr. Molette for her useful comments.

References

1. Zhu LH, Meng H, Duan XJ, Xu GQ, Zhang J, Gong DQ (2011) Gene expression profile in the liver tissue of geese after overfeeding. Poult Sci 90:107–117
2. Hermier D, Salichon MR, Guy G, Peresson R (1999) Differential channelling of liver lipids in relation to susceptibility to hepatic steatosis in the goose. Poult Sci 78:1398–1406
3. Bugg T (2012) Introduction to enzyme and coenzyme chemistry, 3rd edn. John Wiley & Sons, UK, pp 77–102
4. Rawlings N, Barrett A (1993) Evolutionary families of peptidases. Biochem J 290:205–218
5. Wilkesman J, Kurz L (2009) Protease analysis by zymography: a review on techniques and patents. Recent Pat Biotechnol 3:175–184
6. Shiju J, Sudhakaran P (2003) Changes in the activity of matrix metalloproteinases in regenerating rat liver after CCl_4-induced injury. Indian J Biochem Biophys 40:324–329
7. Snoek P, Von den Hoff J (2005) Zymographic techniques for the analysis of matrix metalloproteinases and their inhibitors. Biotechniques 38:73–83
8. Awde S, Marty-Gasset N, Wilkesman J, Rémignon H (2013) Proteolytic activity alterations resulting

from force-feeding in Muscovy and Pekin. Poult Sci 92:2997–3002
9. Bradford MB (1976) A rapid and sensitive method for the quantitation of micrograms quantities of protein utilizing the principle of protein-dye binding. Anal Biochem 72:248–254
10. Laemmli UK (1970) Cleavage of structural proteins during the assembly of the head of bacteriophage T4. Nature 227:680–685
11. Platt M, Randall A, Hanjoong J (2006) Laminar shear stress inhibits cathepsin L activity in endothelial cells. Arterioscler Thromb Vasc Biol 26:1784–1790
12. Brauner JM, Groemer TW, Stroebel A, Grosse-Holz S, Oberstein T, Wiltfang J, Kornhuber J, Maler JM (2014) Spot quantification in two dimensional gel electrophoresis image analysis: comparison of different approaches and presentation of a novel compound fitting algorithm. BMC Bioinformatics 15:181
13. Morris JS, Gutstein HB (2016) Detection and quantification of protein spots by pinnacle. Methods Mol Biol 1384:185–201
14. Natale M, Caiazzo A, Ficarra E (2016) A novel Gaussian extrapolation approach for 2-D gel electrophoresis saturated protein spots. Methods Mol Biol 1384:203–211
15. Chen S, Meng F, Chen Z, Tomlinson BN, Wesley JM, Sun GY et al (2015) Two-dimensional zymography differentiates gelatinase isoforms in stimulated microglial cells and in brain tissues of acute brain injuries. PLoS One 10(4):e0123852. doi:10.1371/journal.pone.0123852
16. Código de Bioética y Bioseguridad (2008) 3era Edic, pp 1–63
17. European Communities Council Directive (1986) (86/609/EC). On line: http://ec.europa.eu/food/fs/aw/aw_legislation/scientific/86-609-eec_en.pdf. 13 July 2016
18. Ley Orgánica de Ciencia, Tecnología e Innovación. (2010) Gaceta Oficial N° 39.575. 16 Dec 2010
19. Kunitz M (1947) Crystalline soybean trypsin inhibitor II. General properties. J Gen Physiol 30:291–310

Part V

Special Zymography Cases

Chapter 15

Simultaneous Detection of Activity and Relative Molecular Mass of Adenylate Kinases After SDS-PAGE and Blotting

Silvia Ravera and Isabella Panfoli

Abstract

Adenylate kinases (AKs) are ubiquitous monomeric phosphotransferases, which play a pivotal role in the energetic metabolism. At the present, nine isoforms are known. AKs catalyze the following reversible reaction: ATP + AMP ↔ 2 ADP, even though isoform 3 uses GTP instead ATP. For many years, the activity of AKs has been detected only after native polyacrylamide gel separations, i.e. in the absence of sodium dodecyl sulfate or methanol. In this work, we report the possibility to detect the activity of the isoforms able to use ATP as substrate, directly onto gel or nitrocellulose sheets, after denaturing SDS-PAGE and electroblotting. This method is innovative because it allows to determine simultaneously the activity and the molecular weight of AKs, especially onto nitrocellulose where bands are sharper, thanks to absence of protein diffusion.

Key words Adenylate kinase, SDS-PAGE, Electroblotting

1 Introduction

Adenylate kinase (AK; EC 2.7.4.3) isoenzymes are involved in energy metabolism and homeostasis of cellular adenine nucleotide ratios in different intracellular compartments [1–3]. AKs belong to the cellular nucleotide synthetic machinery, catalyzing the reversible reaction AMP + MgATP ↔ ADP + MgADP [2]. Nucleotides make up the structure of nucleic acids and are also important in cell metabolism either as a source of chemical energy or as activated intermediates in many biosynthetic pathways, in cell signaling and as components of coenzymes [2]. The preferred substrate of all AKs is AMP and their main phosphate donor is ATP, although some can phosphorylate other substrates and use other phosphate donors, such as GTP [2].

At the present, in vertebrates, nine AK isoforms are known: ubiquitous and cytoplasmic AK1, AK7 and AK8 (all around 21 kDa) [2, 4]; AK2, located in the outer compartment of mitochondria (26 kDa) [5]; AK3, mitochondrial matrix GTP:AMP

Jeff Wilkesman and Liliana Kurz (eds.), *Zymography: Methods and Protocols*, Methods in Molecular Biology, vol. 1626,
DOI 10.1007/978-1-4939-7111-4_15,

phosphotransferase (26 kDa) [6]; AK4, expressed in the pyramidal hippocampus cells (21 kDa) [7]; brain-specific AK5, expressed during differentiation (21 kDa) [8] and AK6 (21 kDa) expressed at nuclear level [9]. AK has also been identified in bacteria, and is called AK9 (22 kDa) [10]. Until now, the AKs have been categorized into two major subgroups: the short-type AK1 group and the long-type AK2–AK3 group, based on their *Mr.* [11]. Metabolically, the several isoforms can be sorted into two groups: those that utilize ATP (AK1, 2, 4, 5, 6, 7, 8, and 9), and those that use GTP as a phosphate group donor (AK3) [2].

AKs activity can be assayed using several techniques. Principally, it is monitored spectrophotometrically at 340 nm, on the basis of the enzyme coupling method, which follows either the NADP reduction or the NADH oxidation [12]. However, this method does not allow distinguishing the several isoforms, because most of them catalyze the same reaction, using ATP.

Another way to assay the AKs activity is represented by zymography, performed without denaturing agents [13]. However, also in this case is not possible to discriminate the different isoforms. More recently, isoelectric focusing has been proposed to determine the intracellular distribution of the AK isoforms in mammalian tissues, discriminating only two AK isoforms, AK1 and AK2, on the basis of the different isoelectric points [14].

Here, we report a method to detect AKs activity after denaturing SDS-PAGE and/or blotting on nitrocellulose (NC), allowing to detect concurrently both the activity and the *Mr.* It is important to note that, even though, in this method, the sample is denatured by heat (95–100 °C for 5 min) and β-mercaptoethanol is present, the AKs activity can be restored, because the enzyme is stable to heat [15] and does not contain Cys residues in the active site (as reported on Uniprot Data Bank: http://www.uniprot.org/uniprot/P00568). However, to detect AKs activity after SDS-PAGE is fundamental to perform the run at very low amperage (20 mA) and to remove the Sodium dodecyl sulfate (SDS) from the gel, with casein buffer for *on gel* assay [1]. However, it is not applicable to observe the activity of the isoform 3, the GTP:AMP phosphotransferase. Moreover, the different isoforms can be identified either by using a specific antibody after the activity detection on NC, or by performing the assay on a specific cellular fraction, which contains only one of the cited isoforms.

2 Materials

Prepare all the solutions using ultrapure water (prepared by purifying deionized water to attain a sensitivity of 18 MΩ cm at 25 °C) and analytical grade reagents. Take the appropriate safety precautions for chemical hazards in carrying out the experiments. Diligently follow all waste disposal regulations when disposing waste materials.

2.1 Sample Preparation

1. In this work, we used liver and muscle, withdrawn from male, 10 weeks old ICRCD1 mice anaesthetized with ether and decapitated (*see* **Note 1**). However, this method is apt for several types of samples, from bacterial cultures to mammalian tissues. Homogenize the samples in 0.25 M sucrose, 5 mM HEPES buffer, 1 mM EDTA, pH 7.2.
2. Prepare 50 mL homogenization buffer, by mixing 4.27 g sucrose, 0.06 g HEPES and 0.03 g EDTA with 40 mL ultrapure water. When the salts are completely dissolved, add 5 M HCl until pH 7.2 is reached (*see* **Note 2**), and make up the solution to 50 mL, with ultrapure water. The solution can be conserved at −20 °C for several months.

2.2 Electrophoresis

For this analysis, a polyacrylamide gel at 14% was employed. Polymerize gel using Mini PROTEAN® Tetra Cell Casting Module and Mini PROTEAN® 3 System glass plates. Perform the electrophoresis using Mini-PROTEAN® tetra cell system and a universal power supply. For gel polymerization, prepare the following solutions:

1. Resolving gel buffer composed by 1.5 M Tris–HCl pH 8.8, 10 mM EDTA, 14 mM SDS. For 0.5 L buffer, dissolve 91.2 g Tris and 1.5 g EDTA in 400 mL ultrapure water. When the salts are dissolved, adjust the pH to 8.8 with 5 M HCl (*see* **Note 2**). Make up the solution to 0.5 L and add 1.5 g SDS (*see* **Note 3**). Store an aliquot at 4 °C, and stock the remainder at −20 °C.
2. Stacking gel buffer, composed by 0.5 M Tris–HCl, pH 6.8 (corresponding to 30.4 g for 0.5 L ultrapure water), 10 mM EDTA and 14 mM SDS. Prepare the buffer as described for the resolving gel buffer. Store an aliquot at 4 °C, and stock the remaining buffer at −20 °C.
3. 30% Acrylamide/Bis-acrylamide solution (*see* **Note 4**). The ratio among acrylamide and bis-acrylamide is 37.5:1 (2.7% cross-linker) (*see* **Note 5**).
4. Ammonium persulfate: 10% solution in water. Store the solution at −20 °C in the dark.
5. *N*, *N*, *N*′, *N*′-tetramethyl-ethylenediamine (TEMED) (*see* **Note 6**).
6. SDS-PAGE running buffer 5× (stock solution): 0.25 M Tris–HCl, pH 8.3, 1.5 M glycine, 11 mM EDTA and 0.1% SDS. Prepare the solution by dissolving 30 g Tris, 144 g glycine and 3.38 g EDTA in 1 L ultrapure water. Once salts are dissolved, add 5 g of SDS (*see* **Note 3**). The solution can be stored at 4 °C for several months.
7. Sample buffer 4× (stock solution): 25% stacking gel buffer (pH 6.8), 10% SDS, 25% 2-mercaptoethanol, 0.1% bromophenol

blue, 0.012 M EDTA and 1.5 M sucrose. For 20 mL, mix 5 mL stacking gel buffer with 8 g sucrose, 0.25 mL 2-mercaptoethanol, 0.02 g bromophenol blue, and 0.07 g EDTA in a 50 mL conical flask. Made up to 20 mL with ultrapure water. Add 1.6 g SDS and mix gently, to avoid bubble formation. 0.5 mL aliquots were prepared and stored at −20 °C (*see* **Note 7**).

2.3 Blotting

1. Nitrocellulose membranes with 0.45 μm pore size.
2. Wypall*X-60 reinforced paper.
3. Blotting buffer: 0.025 M Tris, 0.192 M glycine, and 20% methanol. Prepare 1 L of solution, adding 14.4 g glycine to 200 mL methanol plus 600 mL ultrapure water, in a 1 L beaker. Adjust the pH to 8.3 with Tris (*see* **Note 8**) and make up to 1 L with ultrapure water. Store the solution at 4 °C (*see* **Note 9**).
4. Perform the blotting using Mini Trans-Blot Module (*see* **Note 10**) with a magnetic stirrer and a universal power supply.

2.4 In Gel Detection of AKs Activity

1. 1 M Tris–HCl pH 8.6: dissolve 12.1 g Tris in 80 mL ultrapure water; when the salts are completely dissolved add 5 M HCl, until pH 8.6 is reached (*see* **Note 2**) and make up the volume to 100 mL with ultrapure water. The solution can be stored at 4 °C for several months.
2. 1 M KCl: dissolve 7.45 g KCl in 100 mL ultrapure water. The solution can be stored at 4 °C for several months.
3. 1 M $MgCl_2$: dissolve 9.52 g $MgCl_2$ in 100 mL ultrapure water.
4. 2% agarose: dissolve 1 g agarose in 50 mL ultrapure water (*see* **Note 11**) and heat it in a microwave oven, until complete melting.
5. SDS removal buffer, containing 2% casein (w/v), 0.04 M Tris–HCl pH 8.6 and 2 mM EDTA, pH 8.0. For each gel, prepare 100 mL of solution, dissolving 2 g casein and 0.058 g EDTA in 80 mL ultrapure water, add 4 mL Tris–HCl pH 8.6 and make up to 100 mL with water (*see* **Note 12**).
6. 20% Ethanol solution: dissolve 20 mL ethanol in 80 mL ultrapure water.

2.4.1 Reaction Mixture to Assay AKs Activity, Following ADP Formation

The reaction mixture contains 0.5 M Tris–HCl pH 8.6, 4 mM AMP, 4 mM ATP, 10 mM phosphoenolpyruvate (PEP), 1.2 mM NADH, 0.167 M KCl, 0.2 M $MgCl_2$.

For one gel, 15 mL solution is needed. Dissolve 0.02 g AMP, 0.033 g ATP, 0.025 g PEP and 0.0119 g NADH in 7.5 mL 1 M Tris–HCl pH 6.8 stock solution, 0.25 mL 1 M KCl stock solution, 3 mL $MgCl_2$ stock solution and 4.25 mL ultrapure water (*see* **Note 13**).

For this assay, a solution of purified pyruvate kinase (PK) + lactate dehydrogenase (LDH) is needed.

2.4.2 Reaction Mixture to Assay AKs Activity, Following ATP Formation

The reaction mixture contains 0.2 M Tris–HCl pH 8.0, 0.2 M $MgCl_2$, 5 mM ADP, 100 mM glucose, 1.6 mM NADP, 0.2 mM phenazine methosulfate (PMS), 0.5 mM nitro blue tetrazolium (NBT). For one gel, 15 mL of solution are needed. Dissolve 0.032 g ADP, 0.27 g glucose, 0.018 g NADP, 9 mg PMS and 6 mg NBT in 3 mL 1 M Tris–HCl pH 6.8 stock solution, 3 mL $MgCl_2$ stock solution and 11 mL ultrapure water (*see* **Note 14**).

For this assay, a solution of purified hexokinase (HK) + glucose 6 phosphate dehydrogenase (G6PD) is needed.

2.5 Detection of AKs Activity on Nitrocellulose

1. 1 M Tris–HCl pH 7.4 stock solution: dissolve 12.1 g Tris in 80 mL ultrapure water; when the salt is completely dissolved, add 5 M HCl until pH 7.4 is reached (*see* **Note 2**) and make up the volume to 100 mL with ultrapure water. This solution can be stored at 4 °C for several months.
2. Tris buffered saline (TBS):1.5 M NaCl, 0.1 M Tris–HCl pH 7.4. For 0.5 L, dissolve 87.66 g NaCl in 100 mL Tris–HCl pH 7.4 and make up to 0.5 L with ultrapure water.
3. Blocking solution: 5% milk in TBS (*see* **Note 15**).
4. The reaction mixtures are the same described for the in-gel detection (*see* Subheading 2.4).

3 Methods

3.1 Sample Preparations

All operations were carried out at 4 °C.

1. Skeletal muscle S9 fraction: homogenate 3 g of mouse muscle by Potter-Evehjem system in 10 mL homogenization buffer. Centrifuge the sample for 10 min at 500 × *g*, discard the pellet and centrifuge the supernatant for 20 min at 20,000 × *g*. Retain the resulting supernatant and store it at −80 °C.
2. Liver nuclei enriched fractions: homogenate 5 g of mice liver by Potter-Evehjem system in 27 mL of homogenization buffer (*see* **Note 16**). Centrifuge the sample at 800 × *g* for 10 min. Resuspend the resulting pellet in 10 mL homogenization buffer and repeat the centrifugation. Resuspend the pellet in ultrapure water. Aliquot sample and store it at −80 °C.
3. Liver mitochondria enriched fractions: centrifuge the supernatant resulting by the first centrifugation of nuclei isolation at 15,000 × *g* for 20 min. Dissolve the pellet in 5 mL homogenization buffer and repeat the centrifugation. Resuspend the pellet in ultrapure water. Aliquot sample and store it at −80 °C.
4. Brain homogenates: Homogenate 5 g of mice brain by Potter-Evehjem system in 27 mL homogenization buffer (*see* **Note 16**). Aliquot sample and store it at −80 °C.

5. For each sample, measure the protein concentration. For example, using the Bradford method [16], the protein concentration should be assessed around 20–30 mg/mL. It is important to load the same amount of total protein in gel.

3.2 SDS-PAGE Electrophoresis

Considering the low *Mr.* of AKs, we have used a 14% acrylamide gel. For 1 gel it is necessary to prepare 10 mL resolving gel and 2.5 mL stacking gel.

1. Prepare the resolving gel mixing: 4.67 mL 30% acrylamide, 2.5 mL resolving buffer, 50 μL APS, 6 μL TEMED and 2.77 mL ultrapure water in a 15 mL conical flask (*see* **Note 17**). Cast gel within a 7 × 8.3 × 0.1 cm3 gel cassette. Allow space for stacking the gel and gently overlay with isobutanol (*see* **Notes 18** and **19**).
2. Wait 20–30 min to allow the polymerization of resolving gel (*see* **Note 20**).
3. Discard isobutanol.
4. Prepare the stacking gel by mixing 0.256 mL 30% acrylamide, 0.5 mL stacking buffer, 40 μL APS and 2 μL TEMED and 1.7 mL ultrapure water in a 10 mL conical flask (*see* **Note 16**). Insert a 10-well gel comb immediately without introducing air bubbles.
5. Wait 10–20 min to allow the polymerization of stacking gel (*see* **Note 19**).
6. For each sample, use 40 μg of total proteins. Warm the sample at 95–100 °C for 5 min with 1/4 of total volume of Sample Buffer 4× solution.
7. Perform gel run at 20 mA/gel for 120–150 min with Running Buffer at 4 °C or over ice (*see* **Note 21**).

Alternatively, a precasted gel commercially available can be used, provided it has the same percentage of acrylamide that we recommend.

3.3 Detection of Adenylate Kinase Activity on Gel

Before proceeding with the AKs activity assay, SDS must be removed from the gel.

1. Prepare the SDS removal buffer and warm it at 37 °C.
2. Following electrophoresis, separate the glass plates with a spatula. Gel remains on one of the glass plates. Rinse gel with water and transfer it carefully to a plastic container bearing the same dimensions as the gel.
3. Wash the gel with the pre-heated SDS removal buffer for 1 h at 37 °C, with gentle agitation on a shaker rotating at 50–60 rpm, changing the buffer each 15 min (*see* **Note 22**).

4. Prepare one of the two assay solutions, put it on the gel, add 15 mL of the 2% agarose solution and incubate at 37 °C in the dark.
5. Check the activity band formation: (a) If the ATP formation is followed, a violet band will appear on a light yellow background (*see* Fig. 1, Panel B). (b) If the ADP formation is followed, the gel must be observed using a UV lamp. The AK activity appears as a dark band on a fluorescent background (*see* Fig. 1, Panel A) (*see* **Note 23**).
6. When the bands are visible, reaction can be stopped by adding 25 mL 20% ethanol solution.

3.4 Detection of Adenylate Kinase Activity After SDS-PAGE and Nitrocellulose-Blotting

3.4.1 Blotting

1. After the electrophoresis run, separate the glass plates, using a spatula. Gel remains on one of the glass plates. Rinse gel with water and transfer it carefully to a container filled with blotting buffer.
2. Cut a nitrocellulose membrane of the gel size of and rinse it twice in blotting buffer.
3. Transfer the proteins on nitrocellulose membrane by the classic electroblotting technique: 400 mA, for 1 h, at 4 °C, in blotting buffer (*see* **Note 24**).

3.4.2 AKs Activity Detection

After transfer, wash twice the nitrocellulose membrane with 10 mL TBS.

1. Incubate the nitrocellulose with the blocking solution for 1 h, at room temperature with a gentle agitation on a shaker rotating at 100–150 rpm.
2. Wash twice the nitrocellulose with 10 mL TBS.
3. Incubate the nitrocellulose with one of the two assay solutions, in the dark at 37 °C.
4. Check the activity band formation. (a) If the ATP formation is followed, a violet band will appear on a light yellow background

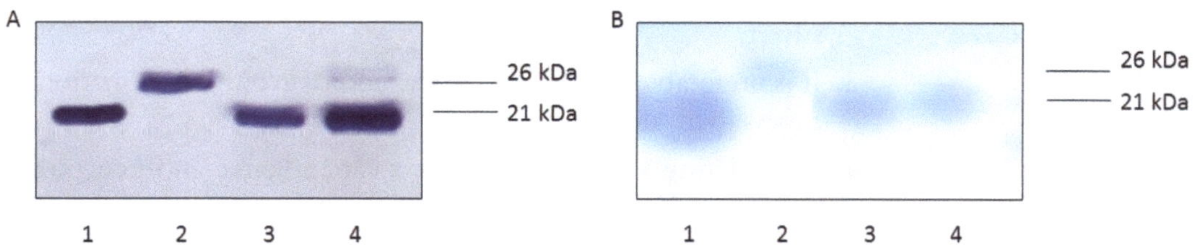

Fig. 1 On-gel assay of adenylate kinase activity, following ATP (*Panel A*) or ADP (*Panel B*) formation. The loaded samples are: *Lane 1*, skeletal muscle cytosolic fraction, containing the isoform 1 (21 kDa); *Lane 2*, liver mitochondria-enriched fraction, containing the isoform 2 (26 kDa); *Lane 3*, liver nuclei-enriched fraction, containing isoform 6 (21 kDa); *Lane 4*, brain homogenate, containing isoform 5 (21 kDa). The different isoforms have been identified considering the subcellular fraction employed. The figure is representative of at least ten experiments

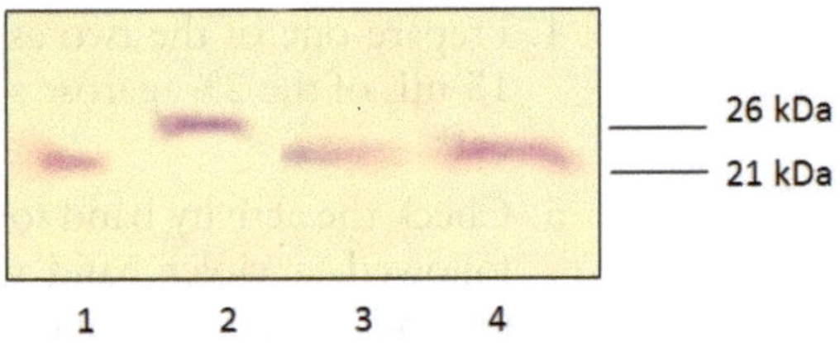

Fig. 2 Assay of AKs for ATP formation onto NC. The loaded samples are: *Lane 1*, skeletal muscle cytosolic fraction, containing the isoform 1 (21 kDa); *Lane 2*, liver mitochondria-enriched fraction, containing the isoform 2 (26 kDa); *Lane 3*, liver nuclei-enriched fraction, containing isoform 6 (21 kDa); *Lane 4*, brain homogenate, containing isoform 5 (21 kDa). The different isoforms have been identified considering the subcellular fraction employed. The figure is representative of at least ten experiments

(*see* Fig. 2). (b) If the ADP formation is followed, the nitrocellulose membrane must be observed in a close cabinet, equipped with an UV lamp. The AK activity appears as a dark band on a fluorescent background.

5. In both cases, stop the reaction, discarding the assay solutions and rinse the membrane with TBS.

4 Notes

1. Approval to conduct the experiments was obtained from the Italian ministry of Health in compliance with animal care requirements that are requested by Italian law (law D.L. 27.1.1992 n. 116, in agreement with the Council Directive 2010/63EU of the European Parliament and the Council of 22 September 2010 on the protection of animals used for scientific purposes). All efforts were made to minimize the number of animals used and their suffering.
2. It is important that 5 M HCl is added drop by drop, to avoid a sudden change in pH below the required one. Moreover, it is recommended that the solution is prepared with a magnetic stirrer.
3. It is preferable to add SDS at the end to avoid the formation of excess foam during mixing.
4. Considering that acrylamide is a carcinogenic and neurotoxic molecule, the use of a ready-made solution minimizes the danger of its use.
5. This type of solution was chosen because it is ideal for high molecular weight protein electrophoresis.
6. Considering the pungent smell, the solution must be used under fume hood.

7. SDS precipitates at 4 °C. Therefore, the lysis buffer needs to be warmed prior to use.
8. Normally, 4–5 g Tris are necessary to reach a pH 8.3. The Tris quantity varies on the basis of quality and purity of glycine and on the starting pH of ultrapure water.
9. The solution can be reused several times, as long as it remains transparent. When the solution color varies to yellow, it must be discarded. It is important that the solution remains at 4 °C until its use, to help to maintain a low temperature during the blot.
10. This apparatus contains also a plastic tank to be filled with ice to keep the temperature low during the blot. For convenience, we recommend to fill the tank with water and freeze it. This will render temperature control more effective.
11. This stock solution can be preserved at 4 °C for 1 week. However, in these conditions agarose becomes solid. Therefore, it is necessary to warm-up the solution in a microwave oven, until complete melting of agarose, letting it cool up to 45 °C just prior use.
12. The solution must be freshly prepared each time and heated at 37 °C to dissolve completely the casein.
13. The solution must be prepared freshly each time, just prior to the assay.
14. The solution must be prepared freshly each time, just prior to the assay, and maintained in the dark.
15. The solution must be prepared freshly each time.
16. Sample is considered well homogenized when the tissue debris are no longer visible. To improve the quality of the homogenate, it is recommended to filter sample through gauze.
17. TEMED and APS must be added last, just before transferring the solution in the gel cast system, to avoid that the solution polymerizes in the conical flask.
18. This overlay prevents contact with atmospheric oxygen (which inhibits acrylamide polymerization), moreover it helps to level the resolving gel solution.
19. During the resolving gel polymerization, it is necessary to avoid the formation of air bubbles.
20. To accelerate the gel polymerization, the system can be put in an incubator at 37 °C.
21. To maintain AKs activity during SDS page, it is fundamental to perform the run with less than 20 mA. Moreover, is also important to keep the system at low temperature, around 4 °C.
22. This step is fundamental to assay the AKs activity after SDS-PAGE. AKs activity could not be recovered on-gel when

detection was performed without previous casein washing. The casein buffer removes the SDS, improving the extent of enzyme renaturation.

23. In this case, the band visualization is more difficult with respect to that following ATP formation. For this reason, the signal may appear most blurred.
24. Blotting must be performed in cold room (at 4 °C) or in an ice-bath, to maintain temperature low.

References

1. Ravera S, Calzia D, Panfoli I, Pepe IM, Morelli A (2007) Simultaneous detection of molecular weight and activity of adenylate kinases after electrophoretic separation. Electrophoresis 28:291–300
2. Panayiotou C, Solaroli N, Karlsson A (2014) The many isoforms of human adenylate kinases. Int J Biochem Cell Biol 49:75–83
3. Notari L, Morelli A, Pepe IM (2003) Studies on adenylate kinase isoform bound to disk membranes of the rod outer segment of bovine retina. Photochem Photobiol Sci 2:1299–1302
4. Kubo S, Noda LH (1974) Adenylate kinase of porcine heart. Eur J Biochem 48:325–331
5. Tomasselli AG, Noda LH (1980) Mitochondrial ATP:AMP phosphotransferase from beef heart: purification and properties. Eur J Biochem 103:481–491
6. Tomasselli AG, Schirmer RH, Noda LH (1979) Mitochondrial GTP-AMP phosphotransferase. 1. Purification and properties. Eur J Biochem 93:257–262
7. Yoneda T, Sato M, Maeda M, Takagi H (1998) Identification of a novel adenylate kinase system in the brain: cloning of the fourth adenylate kinase. Brain Res Mol Brain Res 62:187–195
8. Van Rompay AR, Johansson M, Karlsson A (1999) Identification of a novel human adenylate kinase. cDNA cloning, expression analysis, chromosome localization and characterization of the recombinant protein. Eur J Biochem 261:509–517
9. Ren H, Wang L, Bennett M, Liang Y, Zheng X, Lu F, Li L, Nan J, Luo M, Eriksson S, Zhang C, Su XD (2005) The crystal structure of human adenylate kinase 6: an adenylate kinase localized to the cell nucleus. Proc Natl Acad Sci U S A 102:303–308
10. Amiri M, Conserva F, Panayiotou C, Karlsson A, Solaroli N (2013) The human adenylate kinase 9 is a nucleoside mono- and diphosphate kinase. Int J Biochem Cell Biol 45:925–931
11. Fukami-Kobayashi K, Nosaka M, Nakazawa A, Go M (1996) Ancient divergence of long and short isoforms of adenylate kinase: molecular evolution of the nucleoside monophosphate kinase family. FEBS Lett 385: 214–220
12. Bergmeyer HU (1983) Method of enzymatic analysis. Verlag Chemie, Germany
13. Bockelmann W, Ritter H (1968) Tissue variability of the phosphotransferases adenylate kinase (EC: 2.7.4.3.) and pyruvate kinase (EC: 2.7.1.40.) Hum Genet 6:373–376
14. Kurokawa Y, Takenaka H, Sumida M, Oka K, Hamada M, Kuby SA (1990) Multiforms of mammalian adenylate kinase and its monoclonal antibody against AK1. Enzyme 43:57–71
15. Calzia D, Panfoli I, Ravera S, Dazzi E, Gandolfo S, Pepe IM, Vergani L, Morelli A (2009) Structural modification of proteins by direct electric current from low voltage. J Biochem Mol Toxicol 23:309–317
16. Bradford MM (1976) A rapid and sensitive method for the quantitation of microgram quantities of protein utilizing the principle of protein-dye binding. Anal Biochem 72:248–254

Chapter 16

Silver-Stained Fibrin Zymography: Separation of Proteases and Activity Detection Using a Single Substrate-Containing Gel

Chang-Su Park, Dae-Ook Kang, and Nack-Shick Choi

Abstract

Silver-stained fibrin zymography for separation of protease bands and activity detection using a single substrate gel was designed. The method takes advantage of the nano-scale sensitivity of both zymography and silver staining. After sodium dodecyl sulfate (SDS) polyacrylamide gel electrophoresis (PAGE) in a gel containing fibrin (protease substrate), the gel was incubated in enzyme reaction buffer and the zymogram gel was silver-stained. Bands with protease activity were stained with silver in clear areas where the protein substrate had been degraded. The molecular sizes of proteases were accurately determined.

Key words Protease, Electrophoresis, Silver staining, Nano-scale detection, Coomassie Brilliant Blue, Zymography, Binding mode

1 Introduction

Proteases (EC 3.4._._) are essential constituents of all living cells, including prokaryotes, fungi, plants, and animals, and play a critical role in cell growth and differentiation [1]. Proteases have many commercial applications, particularly in the food, leather, detergent, pharmaceutical, diagnostics, waste management, and silver-recovery industries [2].

Substrate-containing polyacrylamide gel electrophoresis (PAGE) (zymography) is valuable for the analysis of proteolytic systems. Zymography is a versatile two-stage technique involving protein separation by sodium dodecyl sulfate (SDS) PAGE, followed by detection of enzyme activity in polyacrylamide gels under non-reducing (no treatment of reducing agent DTT or β-mercaptoethanol) and non-denaturing (no heating) conditions. Protease activity in zymograms is visualized as a clear band of proteolysis that appears against the dark-blue background [after Coomassie Brilliant Blue (CBB) staining] of undigested

Jeff Wilkesman and Liliana Kurz (eds.), *Zymography: Methods and Protocols*, Methods in Molecular Biology, vol. 1626,
DOI 10.1007/978-1-4939-7111-4_16, © Springer Science+Business Media LLC 2017

protein substrate [3–5]. The technique is simple and sensitive, and protease activities may be quantified; the approach is thus widely used to study enzyme function in complex mixtures and to search for specific enzyme types. In the study, we compare silver staining of zymogram gels with protease detection using CBB [6]. Silver staining of polyacrylamide gels allows most proteins to be detected in the nanogram range and the technique is thus 100-fold more sensitive than CBB staining. Using this method, we succeed in overcoming some limitations of zymography. These are, first, differences in protease migration patterns when various enzymes are examined; second, poor resolution of high molecular weight proteins; and third, problems with detection of proteases that remain near the origins of substrate-containing gels (binding mode) [7, 8].

2 Materials

Prepare all solutions using distilled water and analytical grade reagents. We used the Bio-Rad Mini-Protean III apparatus mini-gel system (Bio-Rad, Hercules, CA, USA) due to the savings in materials and time. SDS-PAGE was carried out based on the Laemmli system [9].

2.1 Sample Preparation

1. Two crude culture supernatants from *Bacillus* sp. strains were used (*see* **Note 1**). If necessary the samples were concentrated using organic solvent [10, 11] (*see* **Note 2**).

2.2 SDS Fibrin Polyacrylamide Gel Components

1. Resolving gel buffer: 1.5 M Tris–HCl, pH 8.8. Weigh 181.7 g Tris and add to 800 mL distilled water. Mix and adjust pH with concentrated HCl (*see* **Note 3**). Add distilled water to a total volume of 1 L. Store at 4 °C.
2. Stacking gel buffer: 0.5 M Tris–HCl, pH 6.8. Weigh 30.3 g Tris and add to 400 mL distilled water. Mix and adjust pH with concentrated HCl. Add distilled water to a total volume of 500 mL. Store at 4 °C.
3. Bovine fibrinogen and bovine thrombin: 1.2% fibrinogen solution and 10 NIH units thrombin in each 1 mL distilled water (*see* **Note 4**).
4. 30% acrylamide (acrylamide:bis-acrylamide 29.2:0.8): Weigh 29.2 g of acrylamide and 0.8 g of bis-acrylamide and add distilled water about 80 mL. Stir until completely dissolved and make up to 100 mL with distilled water and filter through a 0.45 μm filter (*see* **Note 5**). Store in a dark bottle at 4 °C.
5. 10% SDS: Weigh 10 g and add distilled water to make 100 mL. Store at room temperature.

6. 10% ammonium persulfate (APS): Dissolve 0.1 g APS in 1 mL distilled water (*see* **Note 6**).
7. Electrophoresis buffer (Tris-glycine buffer): 25 mM Tris base, 192 mM glycine, 0.1% SDS, pH 8.3 (*see* **Note 7**). Store at room temperature.
8. Zymogram SDS sample buffer (5×): 0.5 M Tris–HCl, 10% SDS, 20% glycerol, and 0.1% bromophenol blue (BPB), pH 6.8. Store at −20 °C (*see* **Note 8**).
9. 1% BPB: Dissolve 0.1 g BPB in 10 mL distilled water. Filter to remove impurity. Store at room temperature.

2.3 Zymogram Gel Activation

1. Renaturation buffer: 50 mM Tris–HCl (pH 7.4) with 2.5% (v/v) Triton X-100 (*see* **Note 9**). Store at room temperature.
2. Zymogram reaction buffer: 30 mM Tris–HCl, 200 mM NaCl, 0.02% NaN_3, pH 7.4 (*see* **Note 10**). Store at room temperature.

2.4 Coomassie Brilliant Blue Staining Components

1. Coomassie staining solution: 0.1% (w/v) Coomassie Blue R-250, 40% methanol, 10% acetic acid, 50% distilled water. Store at room temperature and can be stored for several weeks.
2. Coomassie destaining solution: 40% methanol, 10% acetic acid, 50% distilled water. Store at room temperature and can be stored for several weeks.

2.5 Silver Staining Components

1. Fixing solution: 40% ethanol, 10% acetic acid, 50% distilled water. Store at room temperature and can be stored for several weeks.
2. Sensitizing solution (0.02% sodium thiosulfate): Dissolve 0.02 g sodium thiosulfate in 100 mL distilled water.
3. Staining solution (0.2% silver nitrate): Dissolve 0.2 g silver nitrate (Bio-Rad) in 100 mL distilled water and add 74 μL 35% formaldehyde just before use (*see* **Note 11**).
4. Developing solution (6% sodium carbonate): Dissolve 6 g sodium carbonate in 100 mL distilled water and add both 50 μL 35% formaldehyde and 4 mL sensitizing solution (**item 2**) just before use (*see* **Note 11**).
5. Stop and storage solution: The same as fixing solution (**item 1**). Store at room temperature for several weeks.

3 Methods

Carry out sample preparation and electrophoresis at 4 °C (cold room) and silver staining and other procedures at room temperature.

3.1 Sodium Dodecyl Sulfate Fibrin Zymogram Gel Electrophoresis (12%)

1. Prepare an Erlenmeyer flask and mix the fibrin containing separation gel stated in Table 1 (*see* **Note 5**). Degas it for 10 min to get rid of dissolved oxygen. Add SDS, APS, and TEMED (Table 1), and cast gel within a mini-gel cassette (7.25 × 10 × 0.10 cm3). Allow space for stacking gel and overlay the cast gel with isobutanol or distilled water (*see* **Note 12**).
2. After polymerization of the separating gel pour off the isobutanol or distilled water and wash the space with distilled water (*see* **Note 13**).
3. Prepare an Erlenmeyer flask and mix the stacking gel stated in Table 1 (*see* **Note 5**). Degas it for 10 min to get rid of dissolved oxygen. Add SDS, APS, and TEMED (Table 1). Pipet stacking gel solution onto separating gel and insert comb immediately without trapping air bubbles.

Table 1
Preparation of substrate-containing SDS zymogram gel

Component	Volume (mL)
Separating gel (12%)	
Acrylamide-bisacrylamide mixture	4.0
1.5 M Tris–HCl	2.5
Bovine fibrinogen (0.012 g/mL)	1.0
Bovine thrombin (1 NIH unit/mL)	0.1
Distilled water	2.2
Degas	
10% SDS	0.1
10% APS	0.1
TEMED	0.04
Total volume	10.0
Stacking gel (5%)	
Acrylamide-bisacrylamide mixture	0.33
0.5 M Tris–HCl	0.25
Distilled water	1.38
Degas	
10% SDS	0.02
10% APS	0.02
TEMED	0.002
Total volume	2.0

4. After polymerization of the stacking gel remove the comb carefully and rinse the wells with electrophoresis buffer (Tris-glycine buffer) using a microsyringe to remove unpolymerized acrylamide.
5. Two crude culture supernatants from soil microorganisms (*see* **Note 1**) were diluted fivefold with zymogram sample buffer. Samples were centrifuged and loaded to the two types of gel (SDS gel and zymogram gel) under non-reducing (no treatment of reducing agent DTT or β-mercaptoethanol) and non-denaturing (no heating) conditions (*see* **Note 14**).
6. The gel was electrophoresed at 12 mA constantly in the cold room or in ice till the dye front has reached the bottom of the separating gel (*see* **Note 15**).

3.2 Renaturation and Enzyme Activation of Fibrin Zymogram Gel

1. After electrophoresis, disassemble the glass plate sandwich and remove the stacking. Rinse the gel with distilled water and transfer carefully to a container.
2. Soak the gel with renaturation buffer for 30 min with gentle agitation (*see* **Note 9**).
3. Rinse 10 min with distilled water three times to remove Triton X-100.
4. Incubate in zymogram reaction buffer at 37 °C for specified time periods (*see* **Note 10**).

3.3 Coomassie Brilliant Blue (CBB) Staining

1. After incubation, zymogram gel was stained with CBB staining solution (*see* **Note 16**). Agitate 15–20 min on slow rotary shaker.
2. Pour out staining solution and add destaining solution (Fig. 1b) (*see* **Note 17**).

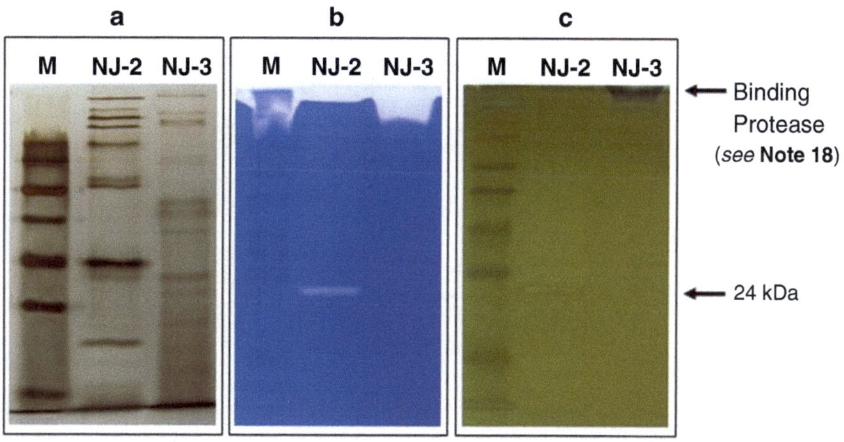

Fig. 1 SDS-PAGE and fibrin zymography of culture supernatants from two *Bacillus* sp. strains NJ-2 and NJ-3. SDS-PAGE was performed and fibrin (0.12%, w/v) zymograms (12% polyacrylamide) were run. After electrophoresis, the SDS-PAGE gel was silver-stained (**a**). Zymograms (**b**, **c**) were incubated in enzyme reaction buffer for 16 h at 37 °C. One gel was stained with CBB (**b**) and the other gel was silver-stained (**c**) (reproduced from [6] with permission from Springer)

3.4 Silver Staining

1. Transfer SDS gel (Fig. 1a) and zymogram gel (Fig. 1c) to a clean small container (*see* **Note 19**).
2. Soak the gel in fixing solution for at least 20 min with gentle agitation (*see* **Note 20**).
3. Wash the gel in 50% methanol for at least 10 with 2–3 changes (*see* **Note 21**).
4. Sensitize the gel in sensitizing solution for only 1 min (*see* **Note 22**).
5. Wash the gel in distilled water for 20 s with three changes (*see* **Note 23**).
6. Incubate the gel in silver staining solution for 20 min (*see* **Note 11**).
7. Wash the gel in distilled water for 20 s with two changes (*see* **Note 24**).
8. Develop the gel in developing solution (*see* **Note 25**).
9. Stop development SDS gel (Fig. 1a) and zymogram gel (Fig. 1c) by rinsing in stop and storage solution.
10. Repeat this procedure with pre-stained protein size markers (a PageRuler prestained protein ladder) to compare silver-stained SDS gel (Fig. 2a), CBB-stained common zymogram gel (Fig. 2b), and silver-stained zymogram gel (Fig. 2c) (*see* **Note 26**).

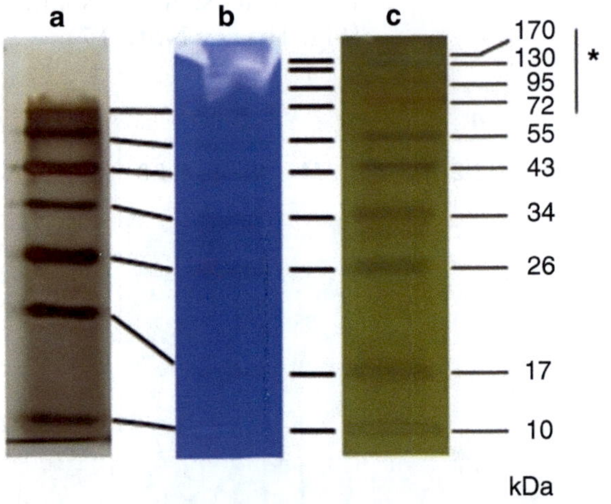

Fig. 2 Pre-stained protein size markers (a PageRuler pre-stained protein ladder) on the SDS-PAGE gel (**a**) and zymograms (**b**, **c**) were stained with silver (**a**, **c**) or with CBB (**b**) (reproduced from [6] with permission from Springer)

4 Notes

1. Two *Bacillus* sp. strains were grown at 37 °C in a tryptic soy broth (TSB, Difco, Detroit, USA) for 2 days. The cells were precipitated by centrifugation at 10,000 × *g* for 10 min. The supernatants were used as protease samples.
2. The supernatants (300 μL amounts) were concentrated by adding 3 volumes of cold acetone (900 μL amounts) for 3 h at −20 °C. The mixture was centrifuged for 5 min in an Eppendorf centrifuge. Remove supernatant and air dry pellet. Resuspend pellet in zymogram sample buffer for zymography.
3. Concentrated HCl (6 M) can be used at first and then it would be better to use a diluted HCl (1 or 2 M) to avoid a sudden drop in pH below the required pH.
4. Separating gel solution containing 0.12% (w/v) fibrinogen and thrombin was centrifuged to remove insoluble impurities which may be induced when SDS stock solution was mixed.
5. Unpolymerized acrylamide is a neurotoxin and a skin irritant and care should be exercised to avoid skin contact. Always handle with gloves and mask. Transfer the weighed acrylamide to the beaker inside the fume hood.
6. We recommend the solution is best to prepare freshly each time.
7. It is convenient to make a 10× stock solution. Prepare 10× native buffer (0.25 M Tris, 1.92 M glycine) and 10% SDS. Weigh 30.3 g Tris and 144 g glycine and add distilled water to 1 L. To make 1× electrophoresis buffer dilute 100 mL of 10× native buffer to 990 mL with distilled water and add 10 mL of 10% SDS.
8. Non-reducing sample buffer, do not add reducing agents (DTT or β-mercaptoethanol). Store for weeks at 4 °C and for months at −20 °C. Zymogram sample buffer needs to be prewarmed prior to use (SDS precipitates at 4 °C).
9. After electrophoresis, detergent SDS is removed from the zymogram gel by washing in 2.5% Triton X-100. This allows the protease to renature and the substrate is degraded during incubation in the zymogram reaction buffer.
10. In order to detect proteolytic activity, it is necessary to work the protease in a pH of reaction buffer. Therefore, it is best to check the optimal pH for activity of your protease prior to applying zymography and then the buffer needs to be changed if necessary (optimal pH-conditioned buffer).
11. We recommend the solution is best to prepare freshly each time (add formaldehyde just before use). Vapors of formaldehyde are

very irritating. Thus, store well-closed at room temperature (according to the manufacturer's datasheet).

12. The separating gel will fill approximately 80–85% of volume between the glass plates. Isobutanol or distilled water keeps the separating gel surface flat. Gels can be stored for up to a week at 4 °C wrapped in wet plastic wrap.
13. Polymerization of the separating gel is completed when a distinct layer is formed between the upper edge of the separating gel and isobutanol or distilled water.
14. Centrifuging the samples prior to preparation for loading helps remove insoluble material which may interfere with electrophoresis [12]. Filling the unused wells with 1× sample buffer is good for avoiding edge effects. When using unknown samples to the first attempt to zymography we recommend non-denaturing systems. Some proteases are inactive under reducing condition and many proteases denatured by heat treatment.
15. It is advisable to cool the gel during electrophoresis to keep the inactive state of proteases in sample and maintain substrate in the gel not to be degraded during the electrophoresis by the proteases.
16. Use gloves to prevent staining hands. Cover container with lid during staining and destaining.
17. Activity in zymograms is visualized as clear bands or binding mode (*see* **Note 18**) against blue background, where active proteases have degraded the fibrin substrate.
18. We found two types of binding modes. One was formed because of binding between the enzyme and substrate [7, 8] and the other was due to a high p*I* value (8.8) of the enzyme [13]. These two types were identified using reverse zymography and diagonal electrophoretic zymography.
19. Wear gloves at all times when handling the gel and use a clean (acid-washed) staining tray with lid to avoid contamination of the gel. Agitate the gel and carry out in a fume hood. An alternative silver staining methods or commercial silver staining kits [ProteoSilver Plus (Sigma) or Dodeca Silver Stain (Bio-Rad)] are available.
20. Gel may be stored overnight at this step.
21. Washing with 2–3 changes will remove acetic acid in fixing solution, reduce background staining and increase sensitivity.
22. Longer time will decrease small protein (peptide) recovery from the gel.
23. Make sure gel is completely immersed. Sensitizer should be completely removed. Residual sensitizer increase dark uniform background.

24. Transfer the gel in a new container. Residual silver nitrate on the gel and staining container will increase background staining.
25. Change developing solution immediately when it turns yellow. Stop the step when the staining is sufficient.
26. We recommend Pre-stained size marker in zymogram gel, which can be seen before and after staining. Plus, it is possible to determine the band against the dark background after silver staining.

Acknowledgements

This research was supported by the Basic Science Research Program through the National Research Foundation of Korea (NRF) funded by the Ministry of Education (grant number 2013R1A1A2062475).

References

1. Gupta R, Beg QK, Lorenz P (2002) Bacterial alkaline proteases: molecular approaches and industrial applications. Appl Microbiol Biotechnol 59:15–32
2. Godfrey T, West S (1996) Introduction to industrial enzymology. In: Godfrey T, West S (eds) Industrial enzymology, 2nd edn. Macmillan, London, pp 1–8
3. Kim SH, Choi NS, Lee WY (1998) Fibrin zymography: a direct analysis of fibrinolytic enzymes on gels. Anal Biochem 263:115–116
4. Kim SH, Choi NS (1999) Electrophoretic analysis of protease inhibitors in fibrin zymography. Anal Biochem 270:179–181
5. Choi NS, Kim SH (2000) Two fibrin zymography methods for analysis of plasminogen activators on gels. Anal Biochem 281:236–238
6. Chung DM, Kim KE, Ahn KH et al (2011) Silver-stained fibrin zymography: separation of proteases and activity detection using a single substrate-containing gel. Biotechnol Lett 33:1663–1666
7. Brown TL, Yet MG, Wold F (1982) Substrate containing gel electrophoresis: sensitive detection of amylolytic, nucleolytic, and proteolytic enzymes. Anal Biochem 122:164–172
8. Lantz MS, Ciborowski P (1994) Zymographic techniques for detection and characterization of microbial proteases. Methods Enzymol 235:563–594
9. Laemmli UK (1970) Cleavage of structural protein during the assembly of the head of bacteriophage T4. Nature 227:680–685
10. Choi NS, Kim BH, Park CS et al (2009) Multiple-layer substrate zymography for detection of several enzymes in a single sodium dodecyl sulfate gel. Anal Biochem 386:121–122
11. Choi NS, Choi JH, Han YJ et al (2009) Mixed-substrate (glycerol tributyrate and fibrin) zymography for simultaneous detection of lipolytic and proteolytic enzymes on a single gel. Electrophoresis 30:2234–2237
12. Hames BD (1981) Gel electrophoresis of proteins. In: Hames BD, Rickwood D (eds) A practical approach. IRL Press, London, p 290
13. Park CS, Kang DO, Lee WY et al (2015) Identification of two types binding modes using reverse or diagonal electrophoretic zymography. Acad J Biotechnol 3:52–55

Chapter 17

Zymography Methods to Simultaneously Analyze Superoxide Dismutase and Catalase Activities: Novel Application for Yeast Species Identification

Esther Gamero-Sandemetrio, Rocío Gómez-Pastor, and Emilia Matallana

Abstract

We provide an optimized protocol for a double staining technique to analyze superoxide dismutase enzymatic isoforms Cu-Zn SOD (Sod1) and Mn-SOD (Sod2) and catalase in the same polyacrylamide gel. The use of NaCN, which specifically inhibits yeast Sod1 isoform, allows the analysis of Sod2 isoform while the use of H_2O_2 allows the analysis of catalase. The identification of a different zymography profiling of SOD and catalase isoforms in different yeast species allowed us to propose this technique as a novel yeast identification and classification strategy.

Key words SOD, Catalase, Zymogram, Electrophoretic isoforms, Yeast

1 Introduction

Separation of enzymes by polyacrylamide gel electrophoresis (PAGE) followed by zymography is a widely used technique for identifying enzymatic activities [1]. The term *zymogram* refers to the analysis of an enzymatic activity [2] and the term *zymography* (introduced in 1962, by Gross and Lapière [3]) involves the electrophoretic technique for the detection of a particular enzyme, based on the substrate repertoire for such enzyme. With zymography, the enzyme substrate is converted into a product that can literally be visualized. The biochemical reaction is measured by detection methods that analyze either appearance of the reaction product or disappearance of the substrate [4]. This technique provides reliable identification of enzymes based on the electrophoretic mobility of their active forms after gel electrophoresis.

The conventional zymography technique was created for the identification of enzymes with protease activity and it was based on a sodium dodecyl sulfate gel impregnated with a protein substrate,

Jeff Wilkesman and Liliana Kurz (eds.), *Zymography: Methods and Protocols*, Methods in Molecular Biology, vol. 1626,
DOI 10.1007/978-1-4939-7111-4_17, © Springer Science+Business Media LLC 2017

which is degraded by the proteases resolved during the incubation period. Coomassie blue staining of the gel reveals sites of proteolysis as white bands on a dark blue background [1]. Over the last 4 decades, zymography has been transformed into different forms, such as chromogenic substrate autography (substrate indicator gel), reverse zymography, 2D-zymography, in situ zymography, in vivo zymography [5] and native PAGE-coupled zymography [6] allowing to identify not only hundreds of proteases, but also other enzymes such as lipases, esterases, catalase, and superoxide dismutase (SOD). Moreover, the advantage of this technique is that any kind of biological sample can be analyzed, including cell and tissue extracts, whole blood, plasma, and other complex body or lavage fluids.

The analysis of antioxidant enzymes like SOD and catalase has important implications in cosmetic, chemical, food, and pharmaceutical industries because of their capability to catalyze the destruction of reactive oxygen species and prevent oxidation. Currently, the most popular method to detect SOD and catalase activity is by native PAGE-coupled zymography [7, 8]. This method has the advantage of detecting only active enzymes across the spectrum of different types of SOD and catalase and discriminating between different enzyme isoforms.

SODs are a class of highly conserved enzymes that catalyze the dismutation of superoxide in oxygen and hydrogen peroxide [9]. Different SOD isoenzymes are currently classified according to the metal cofactor in their active site. Therefore, we can classify SODs into four main types: CuZn-SOD, Mn-SOD, Fe-SOD, and Ni-SOD. In the yeast *Saccharomyces cerevisiae*, there are two main distinct superoxide dismutases: CuZn-SOD (Sod1) and Mn-SOD (Sod2). Sod1 is mainly located in the cytosol although a small percentage can be found in the intermembrane space of the mitochondria [10], while Sod2 is located in the mitochondria. In *Candida albicans*, an irregular cytosolic Mn-SOD (Sod3) has been reported [11]. In addition, this yeast has other three additional genes for expressing Sod1-related GPI-anchored extracellular proteins Sod4, Sod5, and Sod6. Interestingly, it has been recently showed that Sod5 represents a unique class of Cu-only SOD, required for pathogen defense [12]. On the other hand, there are two described catalases, Cta1, which localizes in the peroxisome, and Ctt1, which is cytosolic. They reduce hydrogen peroxide using the redox properties of a heme group complexed to the polypeptide [13].

The two different SOD isoforms (Sod1 and Sod2) of *S. cerevisiae* can be separated and visualized in PAGE zymograms by using nitro blue tetrazolium (NBT) staining [7, 14], where SOD bands are visualized as clear bands on a blue background. Inhibition of Sod1 can be achieved by using NaCN, which allows the specific detection of Sod2 isoforms [15]. In the case of catalase isoforms, they can be separated on starch gel zymograms although their similar molecular weights in yeast make it difficult to separate them on normal PAGE zymograms [16]. However, the activity of Ctt1, which is vital in protecting yeast against exogenous H_2O_2, is much higher than the activity of Cta1 under physiological conditions [17]. Therefore, the catalase activity observed in a PAGE zymogram can be attributed mainly to Ctt1, without the need to separate both isoforms. When a 3,3′-diaminobenzidine (DAB) peroxidase stain is applied to a polyacrylamide gel, followed by a ferricyanide negative stain, catalase appears as clear bands on a green background [18].

Despite the high sequence conservation of SOD and catalase across the evolution, polymorphic forms of these enzymes have been reported under different conditions. Interestingly, those polymorphisms alter the total charge of the protein, affecting their electrophoretic mobility and creating a unique electrophoretic profile in different organisms. In human Sod1 mutant alleles carrying amino acidic changes E100G and E100K, the lack of one negative charge or the increase in total positive charge respectively, resulted in a diminished Sod1 zymogram mobility compared to non-mutated SOD [19]. Unique SOD and catalase profiles are also exhibited by different *Deinococcus* and *Naegleria* species and the zymogram patterns of theses enzymes have been used to differentiate these bacteria and ameba [7, 20]. This differentiation was possible by a double-staining zymogram technique to visualize SOD and catalase bands in the same gel [7]. However, the polymorphic characterization of these enzymes in different yeast species has never been explored.

In this chapter, we describe an optimized double staining protocol for Sod1, Sod2, and catalase zymogram detection to identify new enzyme polymorphic forms in different yeasts. This strategy provides a unique SOD-catalase zymogram profiling, which can be used as a novel approach to discriminate and classify different *Saccharomyces* and non-*Saccharomyces* yeast species.

2 Materials

Prepare all solutions using ultrapure water (obtained by purifying deionized water to attain a sensitivity of 18 MΩ cm at 25 °C) and analytical grade reagents.

2.1 Buffers, Solutions, and Reagents

1. Extraction buffer: 0.05 M Tris–HCl, pH 8.0. Weight 6.06 g Tris and dissolve in about 900 mL of water. Mix and adjust pH with HCl. Make up to 1 L with water. Store at 4 °C.

2. Resolving gel buffer: 1.5 M Tris–HCl, pH 8.8. Weigh 181.7 g Tris and prepare 1 L solution. Store at 4 °C.
3. Stacking gel buffer: 0.5 M Tris–HCl, pH 6.8. Weigh 60.6 g Tris and prepare 1 L solution. Store at 4 °C.
4. SS4X load buffer: sucrose 20% (m/v), 0.25% bromophenol blue (m/v).
5. Running buffer: Weight 3.03 g Tris, 14.41 g glycine. Add water to a volume of 500 mL. Mix and adjust pH to 8.0 with HCl. Make up to 1 L with water. Store at 4 °C.
6. NaCN Inhibitor solution: 1 mM NaCN. Weigh 25 mg NaCN and mix in 500 mL water. Store at 4 °C.
7. SOD staining solution: 80 mL of 0.05 M Tris–HCl pH 8.0 containing 10 mg MTT [3-(4,5(dimethylthiazol-2-yl)-2,5-diphenyltetrazolium bromide], 6 mg PMS (phenazine methosulfate) and 15 mg $MgCl_2$ (*see* **Note 1**).
8. H_2O_2 solution: 100 mL of 50 mM phosphate buffer (pH 7.8) containing 0.04 M H_2O_2.
9. Catalase staining solution: 20 mL 2% (m/v) ferric chloride and 20 mL 2% (m/v) potassium ferric cyanide in water (*see* **Note 2**).
10. Acrylamide/Bis-acrylamide (29:1), 30% solution.
11. Bradford Reagent.
12. Bovine serum albumin (BSA) stock 1 mg/mL.
13. Ammonium persulfate: 10% solution in water.
14. *N,N,N′,N′*-Tetramethylethylenediamine (TEMED).

2.2 Special Equipment

1. FastPrep-24 5G cell homogenizer.
2. Protein electrophoresis equipment: XCell SureLock™ Mini-Cell Electrophoresis System.
3. Image analyzer: Luminescent Image Analyzer LAS-1000plus and Image Gauge software.

3 Methods

The method described in this section allows the identification of different polymorphic isoforms for Sod1, Sod2 and catalase in different yeast species:

1. Collect 50 mg of yeast cell samples for protein extraction by centrifuging at 2200 × *g* for 3 min at 4 °C.
2. Wash cells in 50 mL water and centrifuge at 2200 × *g* for 3 min at 4 °C.

3. Resuspend cells in 500 μL extraction buffer containing the mixture of protease inhibitors 1× (Roche, Switzerland) and 0.4 g of glass beads (*see* **Note 3**).
4. Cells are lysed in a FastPrep homogenizer by two 30 s breaking cycles at 5.0 m/s and then centrifuged at 1620 × *g* for 3 min at 4 °C.
5. Determine protein concentration in the extracts by using the Bradford assay and a BSA calibration curve (10–70 μg/mL).
6. All samples are diluted to 20 μg/μL in 0.05 M Tris–HCl pH 8.0, and resuspended in SS4X load buffer to a final concentration of 1×.
7. 80 μg of total protein of each sample were separated in 10% native-PAGE (Table 1) developed in running buffer for 1 h at 15 mA and 4 °C. Two gels were performed in parallel, one was used to analyze Sod1 and Sod2 isoforms (Gel A) and the other was used to identify Sod2 isoform (Gel B).
8. **Gel A** was immersed in distillated water and **Gel B** was immersed in 1 mM NaCN, which specifically inhibits Sod1p isoform (*see* **Note 4**). Both gels were incubated in their corresponding solutions for 4 min at approximately 20 °C.
9. Immediately, both gels were immersed in 80 mL of SOD staining solution.
10. The gels were exposed to sunlight for 5–10 min to visualize the SOD bands as white bands on a blue-black background (*see* **Note 5**).

Table 1
Composition of 10% native acrylamide gel

Stacking gel	
Tris–HCl (pH 6.8)	125 mM
Acrylamide/Bis-acrylamide (29:1) (m/m)	5% (v/v)
Ammonium persulfate	0.05% (m/v)
TEMED	0.025% (v/v)
Resolving gel	
Tris–HCl (pH 8.8)	375 mM
Acrylamide/Bis-acrylamide (30:0.8) (m/m)	10% (v/v)
Ammonium persulfate	0.05% (m/v)
TEMED	0.025% (v/v)

11. **Gel A** (corresponding to Sod1 and Sod2 isoforms) and **Gel B** (corresponding to Sod2 isoforms) images are captured with Luminescent Image Analyzer LAS-1000plus Camera for further analysis (*see* **Note 6**).
12. Afterward, **Gel A** is washed with distilled water for 5 min and immersed in 100 mL of 50 mM phosphate buffer (pH 7.8) containing 0.04 M H_2O_2 for 20 min at room temperature in darkness and shaking (**Gel A + H2O2**).
13. Then, the gel is washed with distilled water for 5 min and immersed in catalase staining solution, shaking it under sunlight for 5 min to visualize the yellow bands of catalase.
14. Stop reaction by transferring the gel into distilled water.
15. Gel image acquisition was performed in a Luminescent Image Analyzer LAS-1000plus Camera.
16. Finally, the combined analysis of the three different captured images **Gel A** (Sod1, Sod2), **Gel B** (Sod2), and **Gel A + H2O2** (Catalase) shows a specific zymographic profiling for every yeast species (*see* Figs. 1, 2, and Table 2). *See* also Ref. 16 for published results.

Figure 1 shows a schematic description of the SOD-catalase double staining technique for the identification of specific yeast polymorphic isoforms and an example of previously published results is shown in Fig. 2 and Table 1 [19].

4 Notes

1. Weigh each reagent in separate tubes and dissolve them in 1 mL 0.05 M Tris–HCl, pH 8.0. MTT resuspension is difficult so it is recommended to cut the extreme of the tip. Keep PMS and MTT in the darkness prior to prepare the final solution. Add the reagents to the final volume solution in the following order: $MgCl_2$, PMS, and MTT.
2. Prepare 2% (m/v) ferric chloride and 2% (m/v) potassium ferric cyanide solutions separately and mix them into the final catalase solution right before use.
3. Add 1 tablet of protease inhibitor to 50 mL 0.05 M Tris–HCl, pH 8.0, at the time of use.
4. NaCN inhibits specifically Sod1 allowing the identification of bands belonging to Sod2. The combined analysis of treated and untreated gels allows the identifications of the different Sod bands in each yeast species. We have also used another specific inhibitor, diethyldiothiocarbamate (DDC) but it does not efficiently inhibit Sod1 in zymogram analysis.

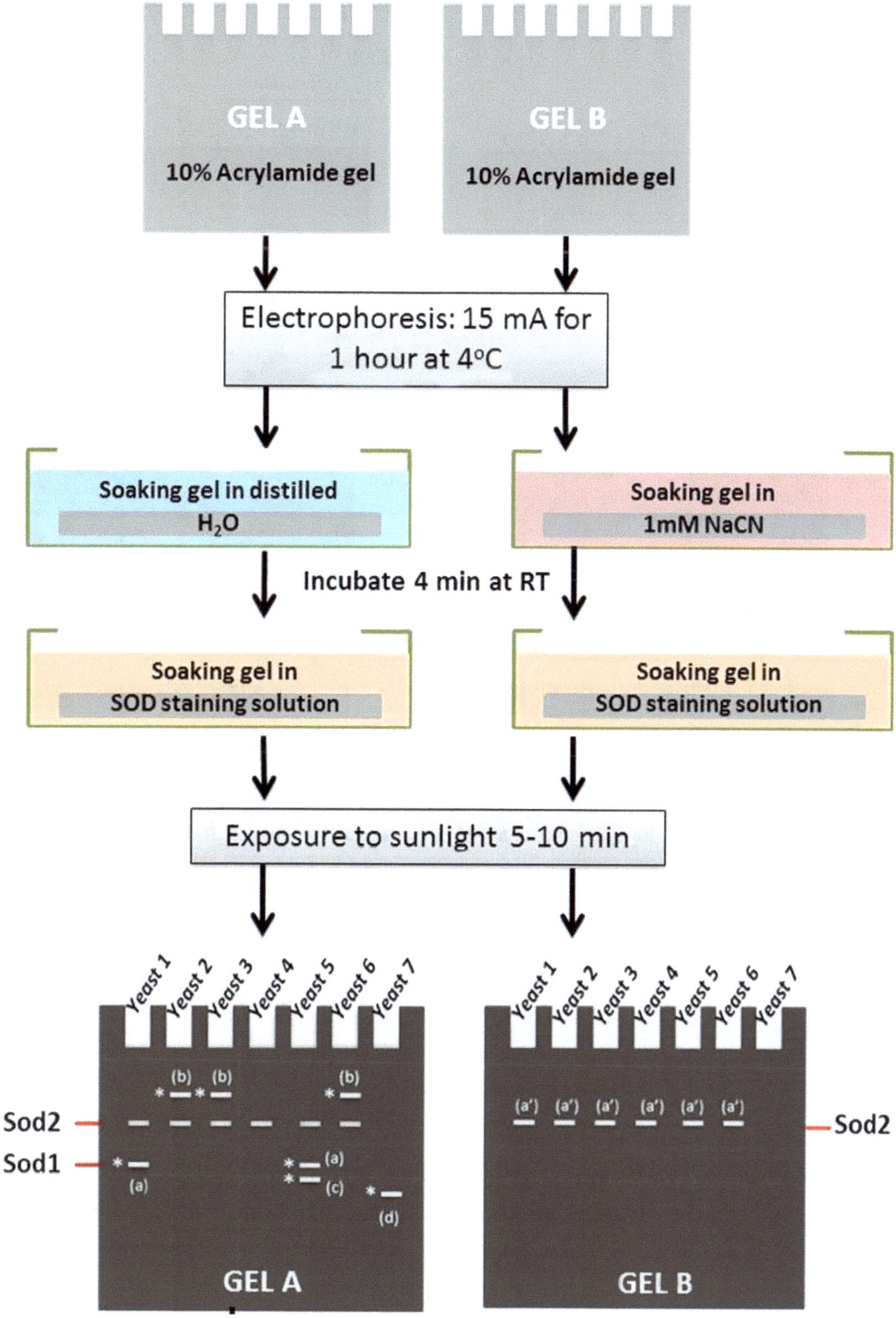

Fig. 1 Diagram of the catalase and SOD double staining zymogram technique used for the resolution of different electrophoretic isoforms in several yeast species. In the zymogram profile of superoxide dismutases Sod1 and Sod2, the different electrophoretic Sod1 isoforms (*a* and *b*) are indicated with *white asterisks* and the unique Sod2 isoforms is indicated as (*a'*). In the zymogram profile of catalase the different isoforms are labeled as *a*, *b*, and *c*. The inserted table shows the zymogram profile of SOD and catalase for the hypothetical yeast strains

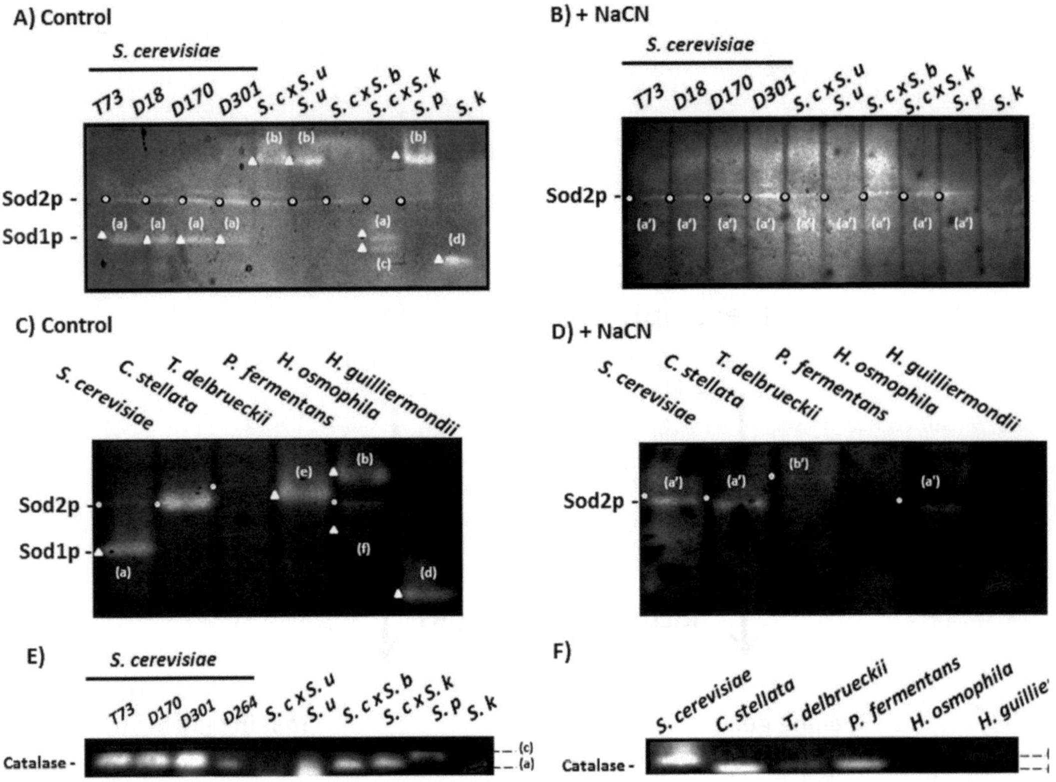

Fig. 2 Identification of SOD and catalase isoforms from *Saccharomyces* and non-*Saccharomyces* species. (**A**) Zymogram profile of superoxide dismutase Sod1 and Sod2 for different *Saccharomyces* yeast strains after 24 h of growth in molasses. (**B**) Zymogram profile after soaking an identical acrylamide gel in NaCN to specifically inhibit the Sod1 isoform. (**C**) Zymogram profile of Sod1 and Sod2 for different non-*Saccharomyces* wine yeast strains cultured in molasses medium. (**D**) Zymogram profile after soaking an identical acrylamide gel in NaCN. In all the panels, Sod1 isoforms (*a–f*) are indicated with *white triangles* and the different electrophoretic isoforms of Sod2 (*a′–b′*) are indicated with *white spots*. (**E**) Catalase zymogram profile for yeast belonging to the genus *Saccharomyces*. (**F**) Catalase zymogram profile for other oenological non-*Saccharomyces* yeast species. Catalase (*a*), (*b*), and (*c*) correspond to different electrophoretic isoforms

5. Although the Sod zymogram can be developed with a white light lamp, the use of sunlight provides faster developing and better bands resolution.
6. These gels can also be used for quantification of the SOD activity. For this, the gel has to be stained with Coomassie blue to calculate the total loaded protein (Pi) and this value is used to normalize the signal data (Ci/Pi). In parallel, Sod1p and Sod2p western blots allow to calculate the specific activity [(Ci/Pi)/Sodp].

Table 2
SOD and catalase zymographic isoforms

Strain	Sod1p isoform	Sod2p isoform	Catalase isoform
S. cerevisiae (*S.c*)	a	a	a
S. bayanus (var. uvarum) (*S.u*)	b	a	b
S.c x *S.u*	b	a	b
S.c x *S.b* (var. bayanus)	–	a	a
S. paradoxus (*S.p*)	b	a	c
S. kudriavzevii (*S.k*)	d	–	–
S.c x *S.k*	a,c	a	a
Candida stellata	–	a	b
Candida albicans	c	b	b
Torulaspora delbruekii	–	b	b
Pichia fermentans	e	–	b
Pichia pastoris	e	–	a
Hanseniospora osmophila	b,f	–	–
Hanseniospora guilliermondii	d	–	–

Acknowledgements

This work has been supported by grants AGL2011-24353 and AGL2014-52985-R from the Spanish Ministry of Economy and Competitiveness (MINECO) to E.M., and it has been performed within the Program VLC/Campus, Microcluster IViSoCa (Innovation for a Sustainable Viticulture and Quality), and Microcluster BBLM (Model Yeasts in Biomedicine & Biotechnology). E.G.-S. was a predoctoral fellow of the JAE program from the CSIC (Spanish National Research Council). R.G.-P. was a postdoctoral researcher at Universitat de València.

References

1. Budic M, Kidric M, Meglic V, Cigic B (2009) A quantitative technique for determining proteases and their substrate specificities and pH optima in crude enzyme extracts. Anal Biochem 388:56–62
2. Wilkesman J, Kurz L (2009) Protease analysis by zymography: a review on techniques and patents. Recent Pat Biotechnol 3:175–184
3. Gross J, Lapière CM (1962) Collagenolytic activity in amphibian tissues: a tissue culture assay. Proc Natl Acad Sci U S A 48:1014–1022
4. Vandooren J, Geurts N, Martens E, Van den Steen PE, Opdenakker G (2013) Zymography methods for visualizing hydrolytic enzymes. Nat Methods 10:175–184

5. Michaud D (1998) Gel electrophoresis of proteolytic enzymes. Anal Chim Acta 372:173–185
6. Marokházi J, Lengyel K, Pekár S, Felföldi G, Patthy A, Gráf L, Fodor A, Venekei I (2004) Comparison of proteolytic activities produced by entopathogenic *Photorhabdus* bacteria: strain- and phase-dependent heterogeneity in composition and activity of four enzymes. Appl Environ Microbiol 70:7311–7320
7. Shukla MR, Yadav R, Desai A (2009) Catalase and superoxide dismutase double staining zymogram technique for *Deinococcus* and *Kocuria* species exposed to multiple stresses. J Basic Microbiol 49:593–597
8. Niyomploy P, Srisomsap C, Chokchaichamnankit D, Vinayavekhin N, Karnchanatat A, Sangvanich P (2014) Superoxide dismutase isozyme detection using two-dimensional gel electrophoresis zymograms. J Pharm Biomed Anal 90:72–77
9. Fridovich I (1978) Superoxide dismutase: defence against endogenous superoxide radical. Ciba Found Symp 65:77–93
10. Sturtz LA, Diekert K, Jensen LT, Lill R, Culotta VC (2001) A fraction of yeast Cu,Zn-superoxide dismutase and its metallochaperone, CCS, localize to the intermembrane space of mitochondria: a physiological role for Sod1 in guarding against mitochondrial oxidative damage. J Biol Chem 276:38084–38089
11. Tsang CK, Liu Y, Thomas J, Zhang Y, Zheng FZ (2014) Superoxide dismutase 1 acts as a nuclear transcription factor to regulate oxidative stress resistance. Nat Commun 5:3446
12. Gleason JE, Galaleldeen A, Peterson RL, Taylor AB, Holloway SP, Waninger-Saroni J, Cormarck BP, Cabelli DE, Hart PJ, Culotta VC (2014) *Candida albicans* SOD5 represents the prototype of an unprecedented class of Cu-only superoxide dismutase required for pathogen defense. Proc Natl Acad Sci U S A 111:5866–5871
13. Herrero E, Ros J, Bellí G, Cabiscol E (2007) Redox control and oxidative stress in yeast cells. Biochim Biophys Acta 1780:1217–1235
14. Luk EE, Culotta VC (2001) Manganese superoxide dismutase in *Saccharomyces cerevisiae* acquires its metal co-factor through a pathway involving the Nramp metal transporter, Smf2p. J Biol Chem 276:47556–47562
15. Gregory EM, Goscin SA, Fridovich I (1974) Superoxide dismutase and oxygen toxicity in a eukaryote. J Bacteriol 2:456–460
16. Wayne LG, Diaz GA (1986) A double staining method for differentiating between two classes of mycobacterial catalase in polyacrylamide electrophoresis gels. Anal Biochem 157:89–92
17. Martins D, English AM (2014) Catalase activity is stimulated by H_2O_2 in rich culture medium and is required for H_2O_2 resistance and adaptation in yeast. Redox Biol 2:308–313
18. Gamero-Sandemetrio E, Gómez-Pastor R, Matallana E (2013) Zymogram profiling of superoxide dismutase and catalase activities allows *Saccharomyces* and non-*Saccharomyces* species differentiation and correlates to their fermentation performance. Appl Microbiol Biotechnol 97:4563–4576
19. Carroll MC, Girourard JB, Ulloa JL, Subramaniam JR, Wong PC, Valentine JS, Culotta VC (2004) Mechanisms for activating Cu- and Zn-containing superoxide dismutase in the absence of the CCS Cu chaperone. Proc Natl Acad Sci U S A 101:5964–5969
20. Yun EJ, Lee YN (2000) Production of two different catalase-peroxidases by *Deinococcus radiophilus*. FEMS Microbiol Lett 184:155–159

Chapter 18

Detection of Guaiacol Peroxidase on Electrophoretic Gels

Diana Castro, Lellys M. Contreras, Liliana Kurz, and Jeff Wilkesman

Abstract

It is possible to analyze peroxidase (POD) from different vegetable sources by electrophoresis. Zymography, i.e., a SDS-PAGE method to detect enzyme activity, is used to specifically detect POD activity and to visualize the total protein profile. For this purpose, we describe how a radish homogenate is prepared and submitted first to electrophoresis, and then, the POD activity present in the gel is reactivated and selectively stained using guaiacol as substrate. After scanning the gel, the same gel is further stained with Coomassie blue to determine the whole protein profile of the sample.

Key words Guaiacol peroxidase, Electrophoresis, Zymography

1 Introduction

The large family of enzymes named peroxidases (EC 1.11.1.x), typically catalyze the reaction type ROH+HOR′:

$$\mathrm{ROOR'} + 2e^- + 2H^+ \xrightarrow[\text{Peroxidase}]{} \mathrm{ROH} + \mathrm{HOR'}$$

Many of these enzymes use hydrogen peroxide as substrate, yet others are specific towards organic hydroperoxides such as lipid peroxides. Peroxidases may contain a heme cofactor in their active sites, or alternately, redox-active cysteine or selenocysteine residues.

POD from radish (*Raphanus sativus*) belongs to the ferroprotoporphyrin group of peroxidases. It has been described as a glycoprotein conformed by two polypeptide chains, having a *Mr* of 31,965.5 each, containing four disulfide bridges with an average *Mr* of 66,000 Da. POD catalyzes H_2O_2 decomposition, yet the resultant [HRP–H_2O_2] intermediate complex can oxidize a variety of chromogenic hydrogen donors. The use of chemiluminescent substrates such as luminol or isoluminol, and fluorogenic substrates such as 4-hydroxyphenyl acetic acid, homovanillic acid, and tyramine, is also possible. Though many different kinds of chromogenic substrates for peroxidases are reported (*o*-dianisidine, benzidine,

Jeff Wilkesman and Liliana Kurz (eds.), *Zymography: Methods and Protocols*, Methods in Molecular Biology, vol. 1626, DOI 10.1007/978-1-4939-7111-4_18,

diaminobenzidine, etc.), the procedure proposed employs guaiacol, a non-carcinogenic substrate [1]. Visualization of the enzyme activity is performed after a reddish product is produced in minutes, and the molecular mass, subunits conformation, enzymatic activity, and densitometry can be conveniently further analyzed.

The detection of enzyme activity on electrophoretic gels, or zymography, has become an important tool to evaluate and analyze many enzymes [1–3]. Herein, we describe the analysis of radish peroxidase (POD) using as substrate guaiacol, employing a safe, economical, and effective zymography experiment. With this technique, enzymatic activity is visualized as reddish bands on a translucent gel. Afterwards, the total protein content may be visualized by staining with Coomassie blue.

2 Materials

2.1 Solutions for Gel Preparation

1. *Acrylamide: bis-acrylamide 30:0.8*: Stock solution is prepared by weighing 30 g acrylamide and 0.8 g *bis*-acrylamide. Dissolve in 100 mL water and filter. Solution is stable at 4 °C for months (*see* **Note 1**).
2. *10% (w/v) APS*: weigh 0.1 g APS and dissolve in 1 mL (*see* **Note 2**).
3. *1.5 M Tris–HCl buffer pH 8.8*: Prepare 100 mL of solution (*see* **Note 3**).
4. *1 M Tris buffer pH 6.8*: Prepare 100 mL of solution (*see* **Note 3**).

2.2 Buffers and Solutions

1. *Sample buffer 4×*: 200 mM Tris pH 6.8, 4% (w/v) SDS, 40% (v/v) glycerol, 0.02% (w/v) bromophenol blue.
2. *Running buffer 10×*: 250 mM Tris–HCl pH 8, 0.1% (w/v) SDS, 192 mM glycine.
3. *Reactivation solution*: 1% (v/v) Triton X-100 (*see* **Note 4**).
4. *Solution A*: pipette 10 μL 25% (v/v) guaiacol to 1.5 mL ethanol, and complete final volume to 10 mL with 50 mM phosphate buffer pH 7.
5. *POD staining solution*: Mix 10 mL of solution A with 9 mL 50 mM phosphate buffer pH 7 and add 1 mL 0.3% (v/v) hydrogen peroxide (*see* **Note 5**).
6. *Coomassie staining solution*: 0.018 mg/mL Coomassie blue R-250, 1% (v/v) acetic acid, 24% (v/v) methanol (*see* **Note 6**).
7. *Destaining solution*: 1% (v/v) acetic acid, 24% (v/v) methanol.

3 Methods

3.1 POD Extraction

1. Rinse and cut one or two pieces of fresh radish (or any other vegetable source to be tested) into small pieces.
2. Weigh ~25 g of the pieces and place them in a blender with a minimal amount (~20 mL) of cold 50 mM phosphate buffer pH 7.
3. Filter the resulting mixture through cheesecloth, so that the big particles are retained and only the liquid extract (homogenate) is obtained.
4. Spin briefly. Aliquot 1 mL of the homogenate and keep on ice until electrophoresis (*see* **Note** 7).
5. If the effect of concentration is to be analyzed, aliquot different quantities of homogenate (i.e., 50, 25, 10, 5, 1 μL) into sample buffer 4× (*see* **Note 8**).

3.2 Electrophoresis

1. Choose small gels (8 × 10 cm^2) as they run faster and clean glasses thoroughly (*see* **Note 9**).
2. Prepare the resolving and stacking gels according to Table 1 (*see* **Notes 10** and **11**).
3. Run gels at 200 V (~ 40 mA for two mini-gels) at 4 °C. Run until the blue front reaches the bottom of the gel (*see* **Notes 12** and **13**).

Table 1
Preparation of SDS-PAGE mini-gels (8 × 6 cm^2)

		Volume[a]/(mL)	
Stock solution component	**Stock concentration**	**Resolution gel (10%)**	**Stacking gel (4.5%)**
H_2O (d)	–	3.28	2.2
Acrylamide mix	30%	2.67	0.6
Tris-HCl pH 8.8	1.5 M	2.00	–
Tris-HCl pH 6.8	1 M	–	1.0
APS[b]	10%	0.08	0.03
TEMED[c]	100%	0.008	0.001
	Final volume:	~8	~4

[a]Volume given for two mini-gels, 4 mL each
[b]Ammonium persulfate
[c]*N*,*N*,*N*′,*N*′ Tetramethylethylenediamine

3.3 Gel Staining

1. Immediately after the run, place the gel in a Petri dish, previously filled with reactivation solution.
2. Incubate twice with gentle shaking for 15 min.
3. Discard reactivation solution and add POD staining solution. Shake gently until reddish bands in a clear background appear (*see* **Note 14**).
4. After staining, gel is scanned or photographed, since color fades with time, and also because after the next staining step, the specificity for peroxidase-stained bands will be lost (*see* **Notes 15** and **16**).
5. Discard the solution in the Petri dish. Add Coomassie solution to the gel. And leave with gentle shaking, typically overnight (*see* **Note 17**).
6. Discard the staining solution (*see* **Note 18**), and remove the excess of dye from the gel with destaining solution (*see* **Note 19**). Leave shaking gently.
7. Stop destaining once blue bands with a clear or bluish background are obtained.
8. Gels can be now scanned, photographed, or dried (*see* **Note 20**). Figure 1 presents a typical result. According to Wilkesman et al. [4], the sensitivity limit of the assay is around 4 μg of total

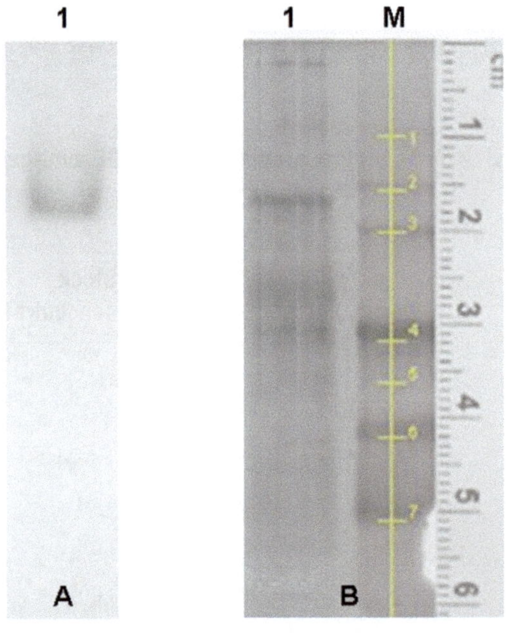

Fig. 1 Analysis of radish POD. (**a**) 10% Zymogram. (**b**) 10% SDS-PAGE stained with Coomassie blue. Lanes: (*1*) 4.4 μg total protein; (M) Molecular weight marker (in kDa): (*1*) Myosin (229), (*2*) phosphorylase (116), (*3*) BSA (82.2), (*4*) ovalbumine (49.1), (*5*) Carbonic anhydrase (32), (*6*) Trypsin inhibitor (25.7), and (*7*) Lysozyme (17.7)

protein per lane, in order to visualize efficiently POD activity. Figure 1a shows the zymogram with POD activity and Fig. 1b shows the protein profile stained with Coomassie blue.

4 Notes

1. Acrylamide is a proven neurotoxin, with carcinogenic and mutagenic effects; extreme care must be taken. Wear gloves when handling Acrylamide powder and solutions. Wear a safety mask when weighing the acrylamide solid. Another option is to buy the already made solution. In this case, it is just necessary to aliquot the liquid in small portions for better handling.
2. Prepare solution fresh. Normally 1 mL is prepared in a micro vial. Solution is stable for 2 weeks at 4 °C.
3. If stored at 4 °C, check pH before use.
4. Normally, 50 mL/gel is enough. As it will be used twice, preparation of 100 mL is enough for one gel.
5. Some guaiacol preparations are quite smelly. Maintain guaiacol solutions in the fume hood. Normally, 50 mL is enough to cover the gel placed in a Petri dish.
6. Methanol is a volatile, flammable, and toxic organic solvent.
7. For POD detection, radish homogenate must always be fresh. Refrigeration of the samples does not assure POD activity throughout time. Additionally, a parallel and complementary enzymatic activity test is suggested to be performed, e.g., photometry test at 470 nm [5–7].
8. It is strongly recommended to measure the total protein concentration of the homogenate by some of the traditional methods reported in the literature (Bradford, BCA, Lowry, UV, etc.) [8–11]. In this manner, it is possible to determine the amount (μg) of proteins introduced in each lane of the gel. Moreover, it is advised to reserve a lane for a commercial peroxidase to be used as a positive control.
9. Wear gloves when manipulating the glasses.
10. To optimize time, during gel polymerization, the POD extract can be prepared.
11. Commercial precasted gels may be used as well.
12. The amperage values used during electrophoresis are potentially lethal. During the run, the electrophoretic unit must not be touched.
13. Typical running times are between 35 and 50 min.
14. Bands should be visualized in the next 5 min. However, time for band apparition varies from 30 s until 10 min. Longer incubation times normally suggest enzyme inactivation.

15. Record (scan, photograph) the results immediately after band apparition as bands do fade away with time.
16. Densitometry analysis may be performed at any time later, as long as the file containing the gel image has been properly saved.
17. Staining times may vary from a couple of hours until overnight. Other Coomassie recipes may be used.
18. Coomassie staining solution may be reused several times. Have ready an empty glass bottle in order to recover the solution after staining the gels.
19. Destaining solution may be recovered too in an amber glass bottle filled with some active charcoal in the bottom. After 1 month, the solution is clear and may be reused.
20. If the polyacrylamide gel is to be discarded, do it appropriately.

Acknowledgments

We thank the *Consejo de Desarrollo Científico y Humanístico, Universidad de Carabobo* and *LOCTI* for partial funding of this work (CDCH-UC-1080-07).

References

1. Manchenko G (2003) Handbook of detection of enzymes on electrophoretic gels, 2nd edn. CRC Press LLC, USA, pp 160–162
2. Wilkesman J, Kurz L (2009) Protease analysis by zymography: a review on techniques and patents. Recent Pat Biotechnol 3:175–184
3. Wilkesman J, Kurz L (2012) Advances in zymography techniques and patents regarding protease analysis. Recent Pat Biotechnol 6:106–114
4. Wilkesman J, Castro D, Contreras LM, Kurz L (2014) Guaiacol peroxidase zymography for the undergraduate laboratory. Biochem Mol Biol Educ 42:420–426
5. Ghamsari L, Ezzatollah K, Shokoofeh G (2007) Kinetics properties of guaiacol peroxidase activity in *Crocus sativus* L. corm during rooting. Iran Biomed J 11:137–146
6. Mika A, Lüthje S (2003) Properties of guaiacol peroxidase activities isolated from corn root plasma membranes. Plant Physiol 132: 1489–1498
7. Doerge DR, Divi RL, Churchwell MI (1997) Identification of the colored guaiacol oxidation product produced by peroxidases. Anal Biochem 250:10–17
8. Westermeier R (1997) Electrophoresis in practice, 2nd edn. Wiley-VCH, Weinheim, pp 5–36. 165–186
9. Mikkelsen S, Cortón E (2004) Bioanalytical chemistry. John Wiley and Sons, USA, pp 167–184
10. Rosenberg IM (2005) Protein analysis and purification, 2nd edn. Birkhäuser, Boston, pp 63–79
11. Walker JM (2009) The protein protocols handbook, 3rd edn. Humana Press, New York, pp 177–186

Chapter 19

In Situ Demonstration and Characteristic Analysis of the Protease Using Substrate Immersing Zymography

HaiLun He, Hao Li, and Dan Liu

Abstract

Zymography, the detection of proteolytic activities on the basis of protein substrate degradation, has been a technique described in the literature for at least in the past 50 years. In this study, we used substrate immersing zymography to analyze proteolysis of proteases. Instead of being directly added into a sodium dodecyl sulfate-polyacrylamide gel electrophoresis (SDS–PAGE) gel, the substrates were added into the immersing solution after electrophoresis. With substrate immersing zymography, some characters of proteases, such as enzyme forms, potential proteolytic activity, molecular weights, presence of complexes, and potentially active enzyme fragments in complex biological samples, can be determined.

Key words Zymography, Substrate immersing zymography, Protease, SDS–PAGE

1 Introduction

Zymography is known as an electrophoretic technique for the study of proteolytic activities on the basis of substrate degradation [1]. Zymography was originally established for detecting collagen degradation and describing a matrix metalloproteinase (MMP) in 1962 by Gross and Lapière [2]. In the traditional zymography, proteins are separated by nonreducing sodium dodecyl sulfate-polyacrylamide gel electrophoresis (SDS-PAGE) in which the acrylamide is copolymerized with a protein substrate. After the electrophoretic run, the SDS is soaked out from the gel by incubation in a non-buffered Triton X-100, followed by incubation under conditions that are conducive for protease activity before staining. The zymogram is subsequently stained, and light-stained bands contrasted with the dark-stained background would indicate the protease activity for hydrolyzing the substrate [3].

In recent years, many new techniques have been developed based on zymography, such as two-dimensional (2-D) zymography, real-time zymography, multiple-layer substrate zymography, and in situ zymography [4–7]. Vandooren and his colleagues

Jeff Wilkesman and Liliana Kurz (eds.), *Zymography: Methods and Protocols*, Methods in Molecular Biology, vol. 1626, DOI 10.1007/978-1-4939-7111-4_19, © Springer Science+Business Media LLC 2017

compared different types of zymography experiments and listed the advantages, disadvantages, and limitations of different zymography methods [8]. With the zymography technique, numerous kinds of biological samples, including cells and tissue extracts, secretion of cells or tissues, whole blood, plasma, serum, and even whole organisms can be analyzed [9, 10]. Zymography strategy has been mainly used to detect potential proteolytic activity or to analyze the number of active components in a multi-enzyme system. Furthermore, zymography can be used to analyze the molecular weights of proteases. Unfortunately, the conventional zymography often determines molecular sizes of some proteases inaccurately, due to decreased mobility of proteins in an acrylamide gel copolymerized with protein substrate [11]. Moreover, the protein transfer of traditional zymography was of low sensitivity due to the loss of the protein during the transfer process [12].

In order to overcome these disadvantages of traditional zymography, a substrate immersing zymography strategy, which allows accurate detection of activity and molecular weight of the protease, was developed in our laboratory. Of course, in the substrate immersing zymography there is no need to transfer the enzyme to a substrate gel; therefore, the protease lost is avoided. In this study, a substrate immersing zymography strategy was performed to detect proteolytic activity. Moreover, the molecular weight of protease, the trend of protease production, and components of multi-protease system of marine bacteria were also analyzed by the modified zymography method. This technique will provide a useful and rapid method for detecting potential hydrolytic activities, enzymatic characteristics, and bacterial multi-protease system.

2 Materials

All solutions were prepared using ultrapure water (prepared by purifying deionized water to attain a sensitivity of 18.25 MΩ cm at 25 °C) and analytical grade reagents. All reagents were prepared and stored at room temperature (unless indicated otherwise). Waste disposal instructions were strictly performed while disposing waste materials.

2.1 SDS Polyacrylamide Gel

2.1.1 Stock Solutions

1. 2 M Tris–HCl, pH 8.8, 100 mL. Weigh 24.2 g Tris and transfer to the cylinder. Add water to a volume of 50 mL (*see* **Note 1**). Mix and adjust pH with HCl (*see* **Note 2**). Make up to 100 mL with water. Store at 4 °C.
2. 1 M Tris–HCl, pH 6.8, 100 mL. Weigh 12.1 g Tris and prepare a 100 mL solution as described in previous step. Store at 4 °C.
3. 10% (w/v) SDS, 100 mL. Weigh 10 g SDS and make up to 100 mL with water. Store at room temperature (~25 °C).

4. 50% (v/v) glycerol, 100 mL. Measure 50 mL 100% glycerol and make up to 100 mL with water. Store at room temperature (~25 °C).
5. 1% (w/v) bromophenol blue (BPB), 10 mL. Weigh 100 mg BPB and add water to 10 mL, stir until dissolved.

2.1.2 Working Solutions

1. Solution A: 30% acrylamide/Bis solution (29.2:0.8 acrylamide: Bis): Weigh 29.2 g of acrylamide monomer and 0.8 g Bis (cross-linker) and transfer to a 100 mL graduated cylinder containing about 50 mL of water, slowly stir until the acrylamide was completely dissolved. Make up to 100 mL with water and filter through filter paper (*see* **Note 3**). Store at 4 °C, in a bottle wrapped with aluminum foil (*see* **Note 4**).
2. Solution B: 4× Resolving gel buffer, 100 mL. 75 mL 2 M Tris–HCl (pH 8.8); 4 mL 10% SDS; Add about 21 mL water; store at 4 °C.
3. Solution C: 4× Stacking gel buffer, 100 mL. 50 mL 1 M Tris–HCl (pH 6.8); 4 mL 10% SDS; Add about 46 mL water; store at 4 °C.
4. Ammonium persulfate (APS): 10% (w/v) solution in water, Store at 4 °C (*see* **Note 5**).
5. *N*,*N*,*N*′,*N*′-Tetramethyl-ethylenediamine (TEMED). Store at 4 °C and wrapped with aluminum foil.
6. 10× SDS-PAGE running buffer, 1 L. Weigh 30 g Tris, 144 g glycine, 10 g SDS, add water to 1 L, pH 8.3. Store at room temperature (*see* **Note 6**).
7. 5× Loading buffer, 10 mL. 0.6 mL 1 M Tris–HCl (pH 6.8); 5 mL 50% glycerol; 2 mL 10% SDS; 1 mL 1% bromophenol blue (BPB). Add about 1.4 mL water. Leave one aliquot at 4 °C for current use and store remaining aliquots at −20 °C (*see* **Note 7**).

2.2 Substrate Immersing Zymography

1. Triton X-100: 2.5% (v/v) solution in 50 mM Tris–HCl buffer, pH 7.5. Store at 4 °C (*see* **Note 8**).
2. 0.2% Gelatin (w/v): dissolved in 50 mM Tris–HCl buffer, pH 7.5.
3. 0.2% Casein (w/v): dissolved in 50 mM Tris–HCl buffer, pH 7.5 (*see* **Note 9**).
4. Coomassie blue staining solution, 1 L. Mix 1.5 g of Coomassie brilliant blue R-250, 350 mL methanol, 100 mL acetic acid, and 550 mL water. Mix thoroughly and store solution at room temperature.
5. Coomassie blue destaining solution, 1 L. Mix 250 mL methanol, 100 mL acetic acid, and 650 mL water. Mix thoroughly and store solution at room temperature (*see* **Note 10**).

3 Methods

Carry out all procedures at room temperature unless otherwise specified.

3.1 Sodium Dodecyl Sulfate-Polyacrylamide Gel

1. Prepare the resolving monomer solution by combining all reagents as specified in Table 1 (*see* **Note 11**).
2. Pour the solution between the glass plates. Continue to pour until 2 cm below the top of the short plate is reached. Gently overlay the gel with isobutanol or water.
3. Vacuum the water out. Prepare the stacking monomer solution by combining all reagents except the APS and the TEMED (Table 2). Vortex for 1 min. Total volume 4 mL.
4. Pour the solution between the glass plates. Continue to pour until the top of the short plate is reached. Insert a 10-well gel comb immediately without introducing air bubbles.
5. Gently remove the comb and rinse the wells thoroughly with distilled water or running buffer.

Table 1
Amounts for the preparation of separating gel

Separating gel	20%	15%	12%	10%	7.5%
Solution A/mL	6.7	5	4	3.3	2.5
Solution B/mL	2.5				
ddH_2O/mL	0.8	2.5	3.5	4.2	5
10%AP/μL	50				
TEMED/μL	5	5	5	5	10
Total/mL	10				

Table 2
Amounts for the preparation of stacking gel

Stacking gel	5%
Solution A/mL	0.67
Solution C/mL	1.0
10% AP/μL	30
TEMED/μL	5
ddH_2O	2.3
Total	4 mL

3.2 Sample Preparing and Electrophoresis

1. Assemble the gel cassette. Fill the assembly with buffer to just under the edge of the outer gel plate.
2. Protein samples with proper dilution or concentration are mixed with the loading buffer at the ratio of 4:1 and placed directly in the wells (*see* **Note 12**).
3. Carry out electrophoresis at 4 °C with constant voltage (100 V) until the sample has entered the gel and then continue at 150 V until the bromophenol blue dye reached the bottom (*see* **Note 13**).

3.3 Substrate Immersing Zymography

1. Following electrophoresis, pry the gel plates open with the use of a spatula and then wash the gels three times with Triton X-100 (2.5%) for 15 min at room temperature with constant shaking to remove SDS.
2. Subsequently wash gel three times for 5 min with 50 mM Tris–HCl buffer.
3. The resolved gel is immersed in a pre-warmed substrate solution (0.1% casein or gelatin) and kept at 37 °C for 2 h.

3.4 Coomassie Stain

1. After washing, the gels are fixed in 50 mL Bio-Safe Coomassie stain (or enough to completely cover gel). Gently shake for 3 h.
2. Gels are destained with Coomassie blue destaining solution for at least 30 min until clear bands indicating proteolytic activity become visible. Steps are shown in Fig. 1 [13]. Stained gels can be stored in water (*see* **Note 14**).

4 Notes

1. Having water at the bottom of the beaker helps to dissolve Tris relatively easily. If using a glass beaker, Tris can be dissolved faster provided the water is warmed to about 37 °C. However, the downside is that care should be taken to bring the solution to room temperature before adjusting pH.
2. Concentrated HCl (approximately 4 mL) can be used at first to narrow the gap from the starting pH to the required pH. From then on it would be better to use dilute HCl (1 M) to avoid a sudden drop in pH below the required pH.
3. Unpolymerized acrylamide is a neurotoxin and care should be taken to avoid skin contact. Wear disposable gloves when handling solutions of acrylamide and a mask when weighing out powder. Polyacrylamide is considered to be nontoxic, but polyacrylamide gels should also be handled with gloves due to the possible presence of free acrylamide.

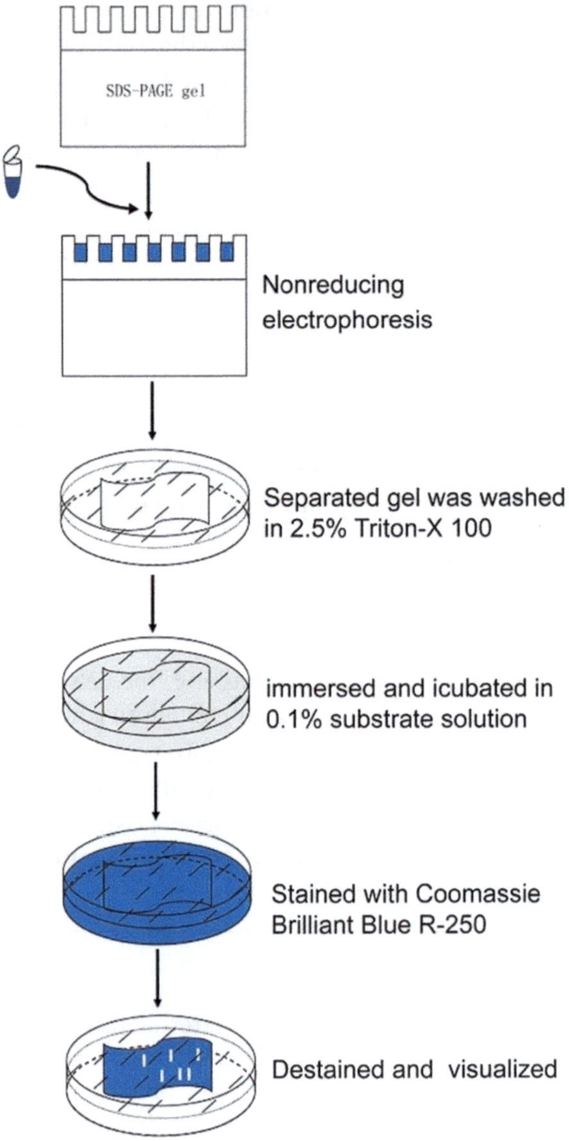

Fig. 1 The process of substrate immersing zymography. In immersing gel zymography, samples are separated on a conventional SDS-PAGE gel. Then, the separating gel is washed with Triton X-100, and then the gel is immersed in a substrate solution and incubated for the hydrolytic reaction about 1 h at a certain temperature. The gel is afterwards stained with Coomassie Brilliant Blue R-250, followed by destaining

4. A well-sealed acrylamide solution can be stored at 4 °C for up to 1 month.
5. We find that it is best to prepare this solution fresh each time.
6. Dilute 100 mL of 10× SDS buffer to 900 mL with water. Care should be taken to add water, since SDS makes bubbles.

7. SDS precipitates at 4 °C. Therefore, the lysis buffer needs to be warmed prior to use.
8. Triton X-100 would be dissolved faster provided the Tris–HCl buffer is warmed to about 37 °C. However, the solution should be taken to 4 °C before using. It is best to prepare this fresh each time.
9. The buffer temperature is gradually increased with gentle stirring to 100 °C for about 10 min until a homogenous dispersion is achieved. The pH is then adjusted if necessary with NaOH and HCl. Under acid condition, casein is insoluble.
10. The Coomassie Blue destaining solution could be recycled and reused.
11. Allow space for stacking the gel and gently overlay with isobutanol or water. Add water very, very slowly! Do not mix water with the resolving gel. This overlay prevents contact with atmospheric oxygen (which inhibits acrylamide polymerization) in addition to helping to level the resolving gel solution. The gel composition must be selected depending on the size of the protease of interest.
12. Centrifuging the samples prior to the run helps remove insoluble debris, which could produce streaks in the protein lanes (revealed when stained with Coomassie blue). Load samples slowly to allow them to settle evenly on the bottom of the well. Be careful not to puncture the bottom of the well with the syringe needle or pipette.
13. 13 If the current is too large, great heat will denature protein during electrophoresis, while if the current is too low, protein bands will disperse due to the extension of electrophoresis time.
14. Rinsing the gel extensively in water after staining will remove background and allow proper visualization of the bands.

Acknowledgments

The work was supported by National Natural Science Foundation of China (31070061, 31370104), Hunan Provincial Natural Science Foundation of China (13JJ9001), and National Sparking Plan Project (2013GA770009).

References

1. Lantz MS, Ciborowski P (1994) Zymographic techniques for detection and characterization of microbial proteases. Methods Enzymol 235:563–594
2. Gross J, Lapiere C (1962) Collagenolytic activity in amphibian tissues: a tissue culture assay. Proc Natl Acad Sci U S A 48: 1014–1022

3. Heussen C, Dowdle E (1980) Electrophoretic analysis of plasminogen activators in polyacrylamide gels containing sodium dodecyl sulfate and copolymerized substrates. Anal Biochem 102:196–202
4. Dodia MS, Rawal CM, Bhimani HG et al (2008) Purification and stability characteristics of an alkaline serine protease from a newly isolated Haloalkaliphilic bacterium sp. AH-6. J Ind Microbiol Biotechnol 35:121–131
5. Wilder CL, Park K, Keegan PM et al (2011) Manipulating substrate and pH in zymography protocols selectively distinguishes cathepsins K, L, S, and V activity in cells and tissues. Arch Biochem Biophys 516:52–57
6. Rossano R, Larocca M, Riccio P (2011) 2-D zymographic analysis of Broccoli (*Brassica oleracea* L. var. Italica) florets proteases: follow up of cysteine protease isotypes in the course of post-harvest senescence. J Plant Physiol 168:1517–1525
7. Hattori S, Fujisaki H, Kiriyama T, Yokoyama T, Irie S (2002) Real-time zymography and reverse zymography: a method for detecting activities of matrix metalloproteinases and their inhibitors using FITC-labeled collagen and casein as substrates. Anal Biochem 301:27–34
8. Choi N, Kim B, Park C et al (2009) Multiple-layer substrate zymography for detection of several enzymes in a single sodium dodecyl sulfate gel. Anal Biochem 386:121–122
9. Kaberdin RV, McDowall JK (2003) Expanding the use of zymography by the chemical linkage of small, defined substrates to the gel matrix. Genome Res 13:1961–1965
10. Vandooren J, Geurts N, Martens E et al (2013) Zymography methods for visualizing hydrolytic enzymes. Nat Methods 10:211–220
11. Hummel KM, Penheiter AR, Gathman AC, Lilly WW (1996) Anomalous estimation of protease molecular weights using gelatin-containing SDS–PAGE. Anal Biochem 233:140–142
12. Pan D, Hill AP, Kashou A et al (2011) Electrophoretic transfer protein zymography. Anal Biochem 411:277–283
13. Liu D, Yang XH, Huang JF, Wu RB, Wu CL, He HL, Li H (2015) In situ demonstration and characteristic analysis of the protease components from marine bacteria using substrate immersing zymography. Appl Biochem Biotechnol 175:489–501

Chapter 20

Use of Zymography in Trypanosomiasis Studies

Jéssyka Fernanda Santiago Monte, Cláudia Jassica Gonçalves Moreno, Joana Patrícia Molato Figueiredo Lopes Monteiro, Hugo Alexandre de Oliveira Rocha, Aline Rimoldi Ribeiro, and Marcelo Sousa Silva

Abstract

Zymography assay is a semiquantitative technique, very sensitive, and commonly used to determine metalloproteinase levels in different types of biological samples, including tissues, cells, and extracts of protein. Samples containing metalloproteinases are loaded onto a polyacrylamide gel containing sodium dodecyl sulphate (SDS) and a specific substrate (gelatin, casein, collagen, etc.). Then proteins are allowed to migrate under an electric current and the distance of migration is inversely correlated with the molecular weight. After migration, the gel is placed in a renaturing buffer to allow proteins to regain their tertiary structure, necessary for enzymatic activity (metalloproteinase activity). In the context of infections caused by trypanosomatids (*Leishmania* spp., *Trypanosoma cruzi*, and *Trypanosoma brucei*), the characterization of metalloproteinase by zymography can contribute to the comprehension of the pathogenesis mechanisms and host–parasite interaction.

Key words Zymography, Metalloproteinases, Trypanosomatids, *Trypanosoma cruzi*, *Trypanosoma brucei*, *Leishmania* spp.

1 Introduction

Zymography is a technique that has been widely used for identification and characterization of proteases, especially metalloproteinases. It consists of a proteolytical assay capable of detecting latent and active forms of enzymes in cells, tissues, biological fluids, and preparations (e.g., cell culture supernatants, crude protein extract, serum, etc.) [1]. Described since 1980 by Heussen and Dowdle [2], this technique provides a sensitive, quantifiable, and functional analysis of metalloproteinases, by gel electrophoresis under denaturing and nonreducing conditions, being possible to estimate the proteolytic activity of these proteases based on the degradation of a specific substrate, as well as their identification based on molecular weight [3].

Briefly, this technique consists of copolymerizing the substrate (gelatin, casein, collagen, etc.) on the same gel for use in

Jeff Wilkesman and Liliana Kurz (eds.), *Zymography: Methods and Protocols*, Methods in Molecular Biology, vol. 1626, DOI 10.1007/978-1-4939-7111-4_20,

electrophoresis. The sample buffer is designed to increase sample viscosity, provide a tracking dye, denaturing molecules, and control the pH of the sample. Thereafter, the sample is fractionated by electrophoresis in an electric field. Metalloproteinases present in the sample are denatured by SDS, but are not reduced. After the run, the proteases separated within the gel are renatured by repeated washing with a nonionic detergent such as Triton X-100, which replaces the SDS on the gel. The gel is then incubated in a suitable buffer, allowing the renatured protease perform digestion of the substrate in the surrounding region of its electrophoresis position. These areas are visualized by staining the gel with Coomassie blue, where the gel turns blue and digestion areas appear as clearer zones. Metalloproteinases are then identified by comparing the digested areas with molecular weight standards and/or by inhibition tests with specific inhibitors of certain classes of metalloproteinases [4].

Thus, zymography is by far the most common test used for the detection of metalloproteinases activity; besides being very effective, it has several advantages compared to other techniques of biological molecules, namely: (a) it does not require antibody, making it relatively inexpensive, (b) the protein separation by molecular mass and nonreducing electrophoretic migration confirms visually the identity of the enzyme, (c) densitometry may be used for quantitative analysis, (d) there is the ability to distinguish the presence of latent and active enzymes isoforms, (e) its relative ease of performance, (f) the development of different specific assay using different protein substrates for enzymatic cleavage, and (g) it provides a much more sensitive test than immunoblotting for quantifying these metalloproteinases [3, 5].

Using the zymography technique, it is possible to characterize metalloproteinases in a wide variety of cells, tissues, and parasites (Figs. 1 and 2). Given that a better understanding of the interactions established between parasites and their respective vertebrate host is needed, the biochemical characterization of metalloproteinases may be a way to achieve this. A set of enzymes is known to be common to Trypanosomatids, such as *Leishmania* spp. (etiological agent of Leishmaniasis in human and animal), *Trypanosoma cruzi* (etiological agent of Chagas diseases or American trypanosomiasis), and *Trypanosoma brucei* (responsible for the sleeping sickness or African trypanosomiasis). All these are involved in several aspects of pathogenesis of neglected tropical diseases.

In the context of infections caused by trypanosomatids, we can use the zymography assay for targeting the following issues: (a) What are the different zymography profiles of the different species and forms of trypanosomatids? (Fig. 1) (b) Which are the specific substrates of metalloproteinases present in different species and

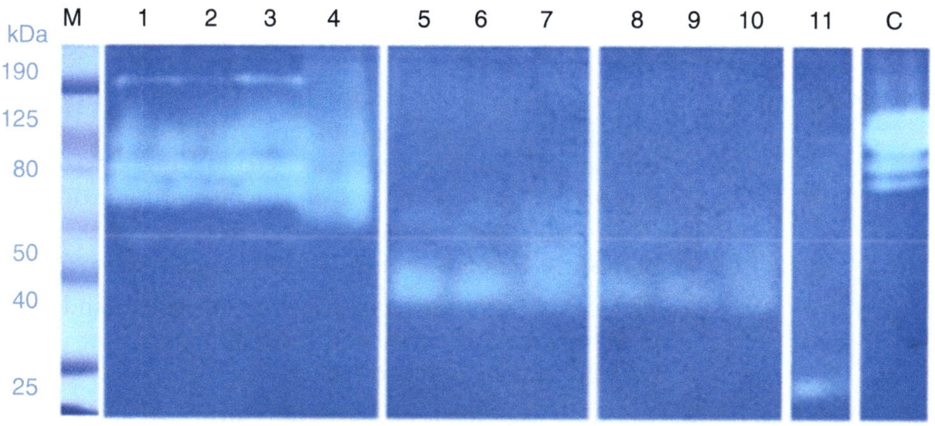

Fig. 1 Zimography profile of different species of Trypanosomatids (*Leishmania* spp., *Trypanosoma cruzi*, and *Trypanosoma brucei brucei*) in collagen substrate. M (marker), *1* (*Leishmania amazonensis*), *2* (*Leishmania guyanensis*), *3* (*Leishmania infantum*), *4* (*Leishmania shawi*), *5* (epimastigote *Trypanosoma cruzi* strain Y), *6* (epimastigote *Trypanosoma cruzi* strain Bol), *7* (epimastigote *Trypanosoma cruzi* strain QMM5), *8* (trypomastigote *Trypanosoma cruzi* strain Y), *9* (trypomastigote *Trypanososoma cruzi* Bol Strain), *10* (trypomastigote *Trypanosoma cruzi* strain QMM5), *11* (bloodstreams *Trypanosoma brucei brucei*), and positive control (purified collagenase). Image obtained by Monteiro, 2015 [6]

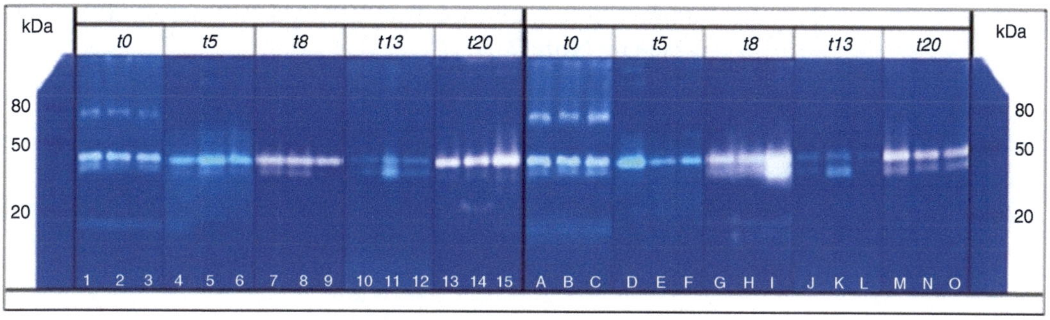

Fig. 2 Zimography profile (SDS-PAGE gelatin substrate) of spleens and livers samples obtained from infected mice by *Trypanosoma brucei brucei* parasites. Image obtained by Gonçalves, 2011 [7]. Days postinfection (from T0 to T20). Lanes from *1* to *15* (spleen samples) and from A to O (liver samples)

forms of trypanosomatids? (c) During the interaction of parasites with different cell types and tissues, which are the endogenous metalloproteinases produced as a consequence of host–parasite interaction? (Fig. 2) (d) Which types of molecules can be used to inhibit enzyme activity by zymography (i.e., screening for novel metalloproteinase inhibitors)?

Indeed, numerous other scientific questions can be answered by zymography regarding the activation process, regulation, and interaction of parasites metalloproteinase involved in the biology and pathogenesis of the parasite–host interaction (Fig. 2).

2 Materials

2.1 Zymography

1. Have prepared your electrophoresis system (casting mount, glass plates with 1.5 mm spacers, combs, power supply, and gel spatula); blotting paper; ice. Gel loading Pipette Tips and Pipettes; 15- and 50-mL polypropylene centrifuge tubes, sterile; Plastic gel wash containers. Agitator (lab shaker, rocking platform). Humidified incubator set to 37 °C.
2. Prepare all solutions using deionized water and analytical grade reagents. Prepare and store all reagents at room temperature (unless indicated otherwise).
3. Running Gel: Deionized water, substrate (gelatin, casein, collagen, etc.), 30% acrylamide-bis solution (30% acrylamide + 0.8% bis-acrylamide), 1.5 M Tris–HCl pH 8.8, 10% sodium dodecyl sulphate (SDS, w/v), 30% ammonium persulphate (APS) (w/v) and *N,N,N′,N′*-tetramethyl-ethylenediamine-TEMED (*see* **Note 1**).
4. Stacking Gel: Deionized water, bis-acrylamide (30% acrylamide + 0.8% Bis-acrylamide), 0.5 M Tris–HCl pH 6.8, 10% SDS (w/v), 30% APS (w/v), and TEMED.
5. Running buffer: 25 mM Tris, 192 mM glycine, and 0.1% SDS (w/v) (*see* **Note 2**).
6. Sample buffer: 62.5 mM Tris–HCl, pH 6.8; 5% glycerol (v/v); 4% (p/v) SDS; 0.01% bromophenol blue (w/v).
7. Renaturing solution: 2.5% Triton X-100 (v/v).
8. Development solution: 50 mM Tris–HCl pH 7.5, 200 mM NaCl, 5 mM $CaCl_2$, and 0.02% Brij-35 (v/v).
9. Staining solution: Coomassie Brilliant Blue (1 g of Coomassie Brilliant Blue; 40% ethanol and 10% acetic acid glacial per 1 L).
10. Destaining solution: 10% methanol and 5% acetic acid glacial.

3 Methods

3.1 Preparation of Gels

1. Each gel results from the juxtaposition of a stacking gel (without substrate) to running gel (with substrate). Thus, the preparation should be a 10% acrylamide gel with copolymerized substrate followed by a 4% acrylamide gel without substrate, stacking gel. Table 1 describes a resume of composition of each gel.

3.2 Ten Percent Acrylamide Gel with Gelatin Substrate at 4.1% (w/v)

1. Weigh 0.05 g of gelatin and dissolve completely in 5 mL of deionized water previously heated.
2. Add to a centrifuge tube, 4.1 mL of the solution prepared in previous point, 3.3 mL of bis-acrylamide, 2.5 mL of 1.5 M

Table 1
Amounts for the preparation of running and stacking gels

Running gel		Stacking gel	
10% acrylamide		**4% acrylamide**	
4.1 mL	Deionized water + substrate	Deionized water	2.5 mL
3.3 mL	30% acrylamide + 0.8% Bis-acrylamide		450 μL
2.5 mL	1.5 M Tris–HCl pH 8.8	0.5 M Tris–HCl pH 6.8	333 μL
100 μL	10% SDS		33 μL
50 μL	30% APS		16.6 μL
5 μL	TEMED		4.7 μL

Tris–HCl pH 8.8, 100 μL of 10% SDS, 50 μL of APS 30%, and 5 μL TEMED (*see* **Note 3**).

3. Transfer rapidly 8 mL of the above solution to the glass plates where it will be polymerize. Cover the surface of the gel with distilled water to eliminate air bubbles and isolate it from oxygen (in order to accelerate polymerization).
4. Remove the remaining water after polymerization and add the stacking gel.

3.3 Ten Percent Acrylamide Gel with Collagen Substrate at 10%

1. Using a 15 mL centrifuge tube, dilute 1 mL of collagen 0.04% (v/v) collagen (Collagen, type I solution from rat tail—commercially available) in 3.1 mL of deionized water.
2. Add, to the above solution, 3.3 mL of bis-acrylamide, 2.5 mL of 1.5 M Tris–HCl pH 8.8, 100 μL of 10% SDS, 50 μL of 30% of APS, and 5 μL TEMED (*see* **Note 3**).
3. Transfer rapidly 8 mL of the above solution to the glass plates, where it will be polymerized.
4. Cover the gel surface with deionized water to eliminate air bubbles.
5. Remove the remaining water and add the stacking gel.

3.4 Acrylamide 4% Without Substrate (Stacking Gel)

1. Mix 2.5 mL deionized water, 450 μL bis acrylamide, 333 μL of 0.5 M Tris–HCl pH 6.8, 33 μL of 10% SDS, 16.6 μL 30 μL 4.7% APS, and TEMED (*see* **Note 3**).
2. Add quickly this solution to the running gel already polymerized in glass plates.
3. Fit into the glass holder the plastic comb that will shape the wells (you may choose among single-well, 10-well, or 15-well comb) plunging into the stacking gel still in its liquid state.

3.5 Preparation of Samples

1. Samples must be thawed on ice. Prior to running the gel, adjust the concentration of each sample (1:1) in sample buffer for zymography.

3.6 Separation of Proteins by Electrophoresis

1. Proceed to fit the glass plates in the electrode support (*see* **Note 4**).
2. Complete assembly of the electrophoresis system.
3. Add running buffer at 4 °C, between the glasses. Fill the chamber, to the top surrounding the gel according to manufacturer specifications.
4. Remove carefully combs of the gels. Remove any air bubble present. Add samples into each well.
5. Complete the assembly of the electrophoresis system. Cover the tank with the lid. Connect to power source.
6. Run at 80 V and 4 °C (*see* **Note 5**).
7. After sample migration to the running gel, the voltage can be raised to 100 V (*see* **Note 6**).

3.7 Renaturing and Developing the Gel

1. After electrophoresis, gels are carefully removed from their glass plates using a gel knife (scalpel or a spatula) and place into a container with 50 mL renaturing solution. Cut a corner to mark the gel's direction and incubate during 1 h at 37 °C.
2. Remove the renaturing solution and add 50 mL development solution to the gel. Incubate overnight at 37 °C.
3. Remove the development solution and rinse gel briefly with deionized water. Proceed to staining with 50 mL Coomassie Brilliant Blue, for approximately 20 min under gentle agitation at room temperature.
4. Finally, the gels must be transferred into a destaining solution. At this stage, the colorant must remain on the entire surface of the gel except in areas where their enzymatic activity is capable of degrading the substrate incorporated into the gel. Thus, enzymes capable of decomposing the substrate should stand as white bands on a blue background. Figures 1 and 2 show the different zymogram profiles with collagen or with gelatine, using extract from different species of Trypanosomatids (*Leishmania* spp., *Trypanosoma cruzi*, and *Trypanosoma brucei brucei*).

All extracts were enzymatically active in colagen substrates and each parasite evidenced a specific zymographic profile. An enzyme with a molecular mass of 50–80 kDa, possibly the 63 kDa glycoprotein, was detected in Leishmania species. Bands of approximately 40 and 20 kDa with enzymatic activity were observed in extracts of

T. cruzi and *T. B. brucei*, respectively. Given the biological importance of metalloproteases, clarifying their role in the establishment of infection and in the survival of tripanosomatids may constitute an opportunity to identify new targets and, consequently, for the design and development of new therapeutic and prophylactic strategies.

4 Notes

1. Thirty percent acrylamide-bis solution is stored in a bottle wrapped with aluminum foil at 4 °C. TEMED and Tris–HCl pH 8.8 and pH 6.8 is stored at 4 °C. Ammonium persulfate, APS: 30% solution in water in large batches (10 mL) frozen in aliquots (for greater day-to-day reproducibility) and used indefinitely.
2. Simple method of preparing running buffer: Prepare 10× native buffer (0.25 M Tris, 1.92 M glycine, and 0.1% SDS) and make it to 1 L with water. Dilute 100 mL of 10× native buffer to 900 mL with water. Based on the manufacturer's recommendations, the pH of the running buffer must not be adjusted. Its correct preparation is, per se, enough to achieve the desired pH.
3. The higher the percentage of PSA and TEMED, the higher will be the speed of polymerization. Their concentrations should not change too much, so as not to alter the concentrations of the remaining components of the gel.
4. The separation of proteins in gels containing different substrates should be held in different electrophoresis runs.
5. If the dye front is yellow rather than blue, proceed to check the pH of all solutions, especially the running buffer solution.
6. Monitor the progress of the migration by using the bromophenol blue included in the loading buffer as an indicator. Let the gel run until the indicator dye reaches the bottom of the gels.

Acknowledgments

This work was supported by Capes—Brazil (*Bolsa Jovem Talento*—Grant 019/2013) and by Global Health and Tropical Medicine (GHTM-UID/multi/04413/2013). We wish to thank Dr. João Aristeu da Rosa from Faculdade de Ciências Farmacêuticas de Araraquara—Brazil, who isolated the QMM5 strain of *Trypanosoma cruzi* parasite.

References

1. Kleiner DE, Stetler-Stevenson WG (1994) Quantitative zymography: detection of picogram quantities of gelatinases. Anal Biochem 218:325–329
2. Heussen C, Dowdle EB (1980) Electrophoretic analysis of plasminogen activators in polyacrylamide gels containing sodium dodecyl sulfate and copolymerized substrates. Anal Biochem 102:196
3. Frankowski H, YH G, Heo JH, Milner R, del Zoppo G (2012) Use of gel zymography to examine matrix metalloproteinase (gelatinase) expression in brain tissue or in primary glial cultures. Methods Mol Biol 814:221–233
4. De Sousa KP, Atouguia J, Silva MS (2010) Partial biochemical characterization of a metalloproteinase from the bloodstream forms of *Trypanosoma brucei brucei* parasites. Protein J 29:283–289
5. Wilder CL, Park KY, Keegan PM, Platt MO (2011) Manipulating substrate and pH in zymography protocols selectively distinguishes cathepsins K, L, S, and V activity in cells and tissues. Arch Biochem Biophys 516:52–57
6. Monteiro JPMFL (2015) Caracterização bioquímica, molecular e propriedades biológicas de diferentes metaloproteinases de Tripansomatídeos (*Trypanosoma brucei* e *Leishmania* spp.). Dissertation, New University of Lisbon-IHMT
7. Gonçalves D (2011) Efeito da minociclina em *Mus musculus* infectados com *Trypanosoma brucei brucei*. Dissertation, New University of Lisbon-IHMT

Chapter 21

Zymography in Multiwells for Quality Assessment of Proteinases

Ambili Mechoor and Madathiparambil G. Madanan

Abstract

Zymography is a well-standardized protocol for the qualitative assessment and analysis of proteinases under specified conditions. However, analysis of a large number of samples simultaneously becomes a challenge when the zymography is carried out by the usual protocol of electrophoresis. This can be overcome by assaying the matrix-degrading proteinases in substrate-impregnated gels in multiwells. Enzymes are copolymerized with 300 mL of 10% acrylamide impregnated with gelatin substrate and incubated for 16 h. The gels are then stained with Coomassie blue, destained with water, and visualized with the naked eye. The intensity; if needed can be measured with a densitometer or gel documentation system. This method has been tested for bacterial collagenases as well as some matrix-degrading metalloproteinases that were purified from rat mammary gland. It can also be used to characterize the enzymes with respect to the type and concentration of the cations required for activity and the role of other regulatory molecules that may affect the enzyme activity. The added advantage of this method is that the electrophoresis set up and electricity is not needed for the procedure.

Key words Extracellular matrix, Proteinases, Collagenase, Multiwells, Zymography

1 Introduction

The extracellular matrix (ECM) plays an important role in regulating tissue-specific function apart from providing structural support and tensile strength to the tissue. This is brought about by providing substrates and pathways for cell adhesion, cell migration, and by regulating differentiated cellular function and metabolic activity directly or indirectly [1, 2]. Remodeling of the ECM occurs in adult tissues during mammary gland involution, hepatic regeneration, uterine involution, etc. The diverse forms of ECM include the tough elastic framework of tendons, cartilage and bone, the crystal like cornea of the eye, and supportive elements of epithelia as well. The various macromolecular components of the ECM include a wide variety of collagens, elastins, glycoproteins, proteoglycans, and complex carbohydrates. Different types of ECM are comprised of different sets of macromolecules [3–5]. ECM degradation and tissue

Jeff Wilkesman and Liliana Kurz (eds.), *Zymography: Methods and Protocols*, Methods in Molecular Biology, vol. 1626, DOI 10.1007/978-1-4939-7111-4_21,

remodelling take place both in normal physiological conditions and in pathological states. One of the major players in degradation and remodelling of the ECM are a group of zinc-dependent proteinases which are generally known as Matrix metalloproteinases (MMPs) [6].

The interaction of cells with their ECM is crucial in maintaining a differentiated phenotype. These interactions are altered during tissue regeneration, repair, and in pathological conditions. These interactions are governed by a group of enzymes called MMPs. They are zinc-containing endopeptidases that degrade extracellular matrix proteins during tissue morphogenesis and remodeling in mammary gland involution [7], liver regeneration, wound healing and are associated with tumor angiogenesis, invasion and metastasis, arthritis, and atherosclerosis [8–11]. MMP expression is regulated at the transcriptional level by growth factors, hormones [12], cytokines, and cell–cell matrix interactions [13]. Extracellular MMPs are activated by proteolytic cleavage of their NH_2 terminal domains and are inhibited by noncovalent 1:1 stochiometric interaction with tissue inhibitors of metalloproteinases (TIMPs) [14]. Many of these MMPs are not expressed constitutively in vivo, but rather are induced in response to cytokines, growth factors, hormones, oncogenes, etc. Apart from these, they are regulated at the transcriptional level as well as depending on the type of various tissues involved. Most of these MMPs are produced as proenzyme which needs activation to act on the specific substrate. Besides, TIMPs also play a key role in MMP regulation. Classification of MMPs is based on their specificity of substrate and is classified into collagenases, gelatinases, stromelysins, and matrilysins and membrane-type MMPs [15]. They show affinity to different metal ions based on which their activity can be inhibited.

Many pathogenic organisms synthesize and secrete a wide array of proteinases, of which the most common are metalloproteinases, particularly membrane-bound forms, which possess additional functional domains compared with secreted forms [16]. In pathogens, also metalloproteinases play key roles in degradation of ECM components during their invasion and pathogenesis. Some pathogens activate host proteinase to degrade the host ECM at their convenience [17, 18]. In such cases, these enzymes work against host proteinase cascades, cytokine networks, extracellular matrix components, and host enzyme inhibitor interactions [19]. But many pathogens produce their own proteinases [20]. Bacterial proteases may be secretory forms or membrane-bound forms [21]. These proteases are used for the degradation of ECM in order to facilitate their smooth invasion and makes pathogenesis [22]. They are considered to be having role in invasion of the pathogen and pathogenesis.

The conventional zymography method involves electrophoretic separation of proteins in a substrate-impregnated acrylamide

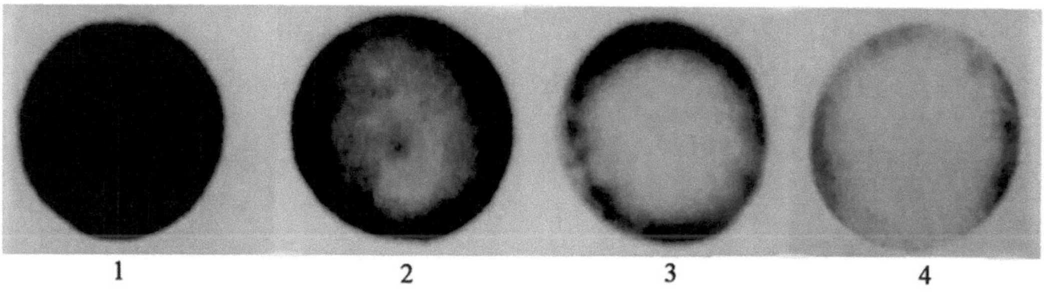

Fig. 1 Increasing concentration of bacterial collagenase in a total volume of 100 μL (200 mM phosphate buffer pH 7.5) was copolymerized with 300 μL of acrylamide in the wells of a 16 mm multiwell plate. The reaction was allowed to take place for about 1 h. The gels were released and treated with substrate buffer and analyzed as described in the procedure. Control without enzyme (*1*), 0.02 units (*2*), 0.2 unit (*3*), and 20 units (*4*) of collagenase was used

gel and the incubation in buffers which is necessary for the activity of these matrix-degrading enzymes. As many of these proteinase activities are specific to their ECM substrates and activators, requirement of metal ions for activity and pH conditions plays a significant role. Their characterization is also based on scaling their activity under such conditions. The analysis of a large number of samples simultaneously under different conditions becomes a challenge when the zymography is carried out by the usual protocol of electrophoresis. This is mainly due to the cumbersome procedure and limited number of samples which can be loaded on the gel. This can be overcome by assaying the matrix-degrading proteinases in substrate-impregnated gels in multiwells (Fig. 1). The effect of different ions and its concentration dependence can also be studied using this method by keeping the enzyme and substrate concentration at the optimal levels. This method can be further extended to convenient study of the effect of various regulatory molecules that affect the enzyme activity and their concentration dependence on the enzyme activity (Fig. 2) [23].

2 Materials

All solutions should be prepared using deionized water at 25 °C and analytical grade reagents. The following materials are needed: Gelatin from bovine skin, Triton X-100, Chondroitin sulphate A, Chondroitin sulphate C, Heparin, Hyaluronic acid, Sodium azide. The matrix-degrading enzymes, viz., Gelatinases, were purified from involuting mouse mammary gland in the laboratory. Besides, 16 mm multiwell plates, incubator (for incubating the gel trays for 18 h at 37 °C), and a shaker (rocking or rotary shaker with low speed for staining the gel).

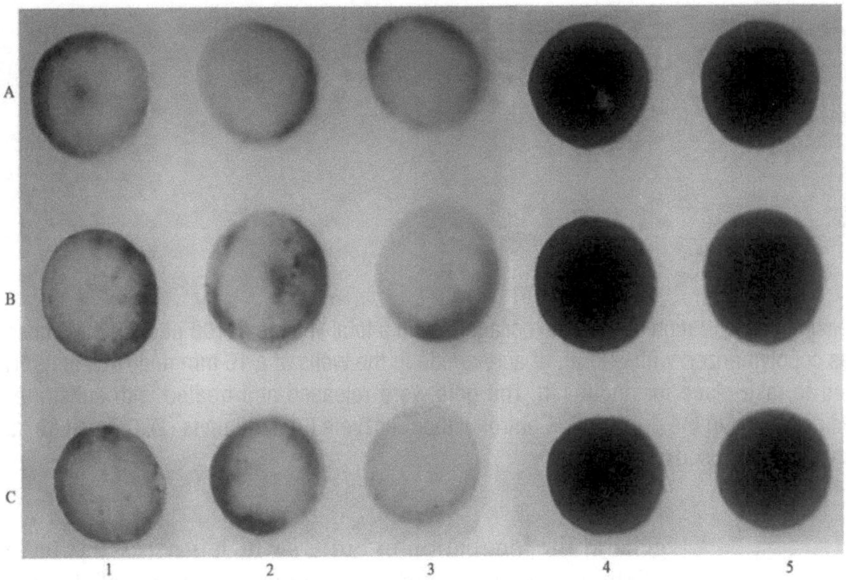

Fig. 2 Purified gelatinases *A*, *B*, and *C* were pretreated with 6 μg each of gylcosaminoglycans for 30 min at room temperature. The treated enzymes were then copolymerized with gelatine (2 mg/mL)-impregnated polyacrylamide in multiwell plates. After polymerization, the gels were incubated in substrate buffer, stained with Coomassie blue, and destained with distilled water. Untreated controls (enzyme only) (*1*), pretreated with heparin (*2*), hyaluronic acid (*3*), chondrotin sulphate A (*4*), and chondroitin sulphate C (*5*). Enzyme activity inhibition is noted in the case of samples treated with condroitin sulphate A and C

2.1 Buffers and Stock Solutions

1. Acrylamide-Bis Acrylamide Solution (30% T, 3% C): Dissolve acrylamide 29.1 g and bis-acrylamide 0.9 g in 40 mL water. Make up to 100 mL with water and filter through a Whatman® Grade 1 filter paper. Store the stock solution at 4 °C in an amber-colored bottle.
2. Sodium dodecyl sulfate 10%: Dissolve 10 g SDS in 80 mL of water and make up the volume to 100 mL and store at room temperature.
3. Ammonium persulfate (APS) 10%: Dissolve 100 mg ammonium persulfate in 1 mL of water and store at 4 °C. This stock is stable up to 2 weeks.
4. Substrate-impregnated gels: Acrylamide solution containing the substrate is prepared by mixing 200 mg gelatin in buffer/water, 30 mL of 30% acrylamide solution, 100 μL of TEMED, and 1 mL of APS for 100 mL.
5. Renaturing Buffer (2.5% Triton X-100 solution): Dissolve 2.5 mL of Triton X-100 in 97.5 mL of water.
6. Stock 500 mM Tris–HCl solution: 500 mM Tris–HCl pH 7.5 is prepared by dissolving 6.06 g Tris base in 80 mL of water and pH made up to 7.5 with HCl. Volume made up to 100 mL with water and store at 4 °C.

7. Stock 500 mM $CaCl_2$: Dissolve 5.54 g $CaCl_2$ in 100 mL water and store at room temperature.
8. Substrate Buffer (50 mM Tris–HCl pH 7.5, 1% Triton X-100, and 5 mM $CaCl_2$). Dissolve 1 mL of Triton X-100 in 79 mL of water. Add 10 mL of 500 mM Tris–HCl pH 7.5 and 1 mL of 500 mM $CaCl_2$. Make it up to 100 mL. The buffer was added with 0.02% NaN_3 in order to avoid microbial contamination.

2.2 Staining

1. Coomassie Blue Stain: To make 100 mL Coomassie blue stain mix (40 mL methanol, 10 mL acetic acid, 50 mL of water) dissolve 10 mg Coomassie brilliant blue R-250.
2. Destaining Solution: Pure double-distilled water.

3 Methods

3.1 Sample Preparation of Proteinases and Collagenases

1. Matrix-degrading proteinases are isolated from various stages of mammary gland involution by standard protocols of protein extraction [7].
2. The enzymes from the mammary epithelia cells in culture are isolated by substrate affinity using gelatin agarose beads [24]. The enzymes thus purified are used for all the experimental protocols.

3.2 Preparation of Substrate-Impregnated Gels in Multiwells

1. The required amount of enzyme (approximately 0.02–200 units) (MMPs/Collagenases/Gelatinases) in a maximum volume of 100 μL is pipetted out into each well of a 16 mm multiwell plate (*see* **Note 1**).
2. Gelatin or Collagen IV (2 mg/mL) is dissolved (by heating at 70 °C) in 50 mM Tris–Cl pH 7.5. Acrylamide 10% was made up to 40 mL with Tris–HCl buffer and mixed well. 300 μL of this mixture is pipetted out quickly into each well as a layer over the evenly spread enzyme (*see* **Note 2**). This leads to a final acrylamide concentration of 7.5%. The gels are allowed to polymerize for about 1 h. The gels are then carefully detached from the plastic and incubated with 1 mL of Tris–HCl buffer [50 mM Tris–HCl buffer pH 7.5, 5 mM $CaCl_2$, and 0.02% NaN_3] for 16 h at 37 °C.
3. The buffer is then carefully removed and the gels are stained with Coomassie brilliant blue for 15 min and destained with water for 2 h. The substrate degradation can be visualized with the naked eye and, if needed, can be quantitated using a densitometer.

3.3 Preparation of Multiwell Gel

1. Add different amounts (approximately 0.02–200 units) (concentration dependence as in legend from Fig. 1) of enzyme (MMPs/Collagenases/Gelatinases) dissolved in a final volume

of 100 μL of 50 mM Tris–HCl pH 7.5 per well. We have used different amounts of collagenase dissolved in the same buffer.

2. Add 300 μL substrate-containing acrylamide solution carefully over the evenly distributed enzyme in the well. This leads to a final acrylamide concentration of 7.5%.
3. The gels are then allowed to polymerize for about 1 h.
4. After polymerization, the gels are carefully detached from the plastic and incubated with 1 mL of substrate buffer for 16 h at 37 °C (*see* **Notes 1–3**).

3.4 Pretreatment of Enzymes and Quality Assessment

1. In order to study the effect of various parameters, viz., ions, other ECM components, activators, inhibitors, etc., the enzyme solution is incubated with the component for about 30 min at room temperature (25 °C) before subjecting it to polymerization with substrate-impregnated acrylamide solution (*see* **Note 4**).
2. The incubation buffer can be modified by adding different activators and inhibitors of choice to study their effect.
3. As shown in Fig. 2, we have assessed an untreated sample as negative control (enzyme only) (1), enzyme pretreated with heparin (2), hyaluronic acid (3), chondrotin sulphate A (4), and chondroitin sulphate C (5). Enzyme activity inhibition is noted in the case of samples treated with chondroitin sulphate A and C.

3.5 Quantification Using Densitometry or Gel Documentation Scanning

1. Densitometry or gel documentation scanning can be used to scale the activity between different samples.
2. The gels were kept in the scanner and the peak density across the diameter of round gels will be taken as measure of activity (for more accuracy, an average of diameters at multiple axis may be used) (*see* **Note 5**).
3. An interpolation against known peak density of standard samples with known activity or concentration may be used to measure the unknown sample (*see* **Note 6**).

4 Notes

1. The enzyme assay has to be done by other standard protocols, the optimal concentrations calculated, and has to be brought to the necessary concentration according to the experiment. Care has to be taken that the required concentration does not exceed a volume of 100 μL. One enzyme unit is defined as the amount that catalyzes the conversion of 1 μmol of substrate (gelatin) per min at 25 °C.
2. The multiwell plate has to be placed on an even surface and utmost care has to be taken for the even uniform spreading of the enzyme solution in each of the wells. Appropriate controls

are taken in the same volume. For instance, a standard proteinase whose activity has been established can act as a positive control. In our case, the standard taken was bacterial gelatinase.

3. Sodium azide in low concentrations ~0.01–0.02% is used as a preservative to prevent fungal and bacterial growth in buffers. Excess concentration can be toxic to the system and can possibly inactivate the enzyme.
4. The cations used in the study were 2 mM each of Ca^{2+}, Zn^{2+}, Mg^{2+}, and EDTA. The inhibitors/modulators used in this study were 1 mM each of FDNB (1-fluoro-2,4-dinitrobenzene) and iodoacetate. These molecules are known to inhibit MMP activity. In case of other ECM components, we had used heparin, hyaluronic acid, chondroitin sulfate A, and chondroitin sulfate C (1–6 μg each).
5. The gels are removed from the multiwell plate and scanned individually. A slight pressure with the end of the spatula releases the gels from the plate.
6. Among other applications and further advantages of multiwell zymography: similarly other purified proteinases from eukaryotes and bacteria may be used [25].

Acknowledgments

The author A.M. is grateful to Indian Council of Medical Research for receiving ICMR-JRF during the conduction of the study.

References

1. Adams JC, Watt FM (1989) Fibronection inhibits the terminal differentiation of human keratinocytes. Nature 34:307–309
2. Hay ED (1993) Extracellular matrix alters epithelial differentiation. Curr Opin Cell Biol 5:1029–1035
3. Yamada KM (1983) Cell surface interactions with extracellular materials. Annu Rev Biochem 52:761–799
4. Sudhakaran PR, Anilkumar N, Santhosh A (1997) Extracellular matrix in liver. In: Rana SVS, Taketa K (ed) Liver xenobiotics, pp. 1–18. Narosa Publishing House, New Delhi, India
5. Yurchenco PD, Schittney JC (1990) Heparin modulation of laminin by polymerisation. J Biol Chem 265:3981–3991
6. Matrisian LM (1992) The matrix-degrading metalloproteinases. Bioessays 14:455–463
7. Ambili M, Pillai MR, Sudhakaran PR (1997) Characteristics of a 60K gelatinase involved in rat mammary gland involution. Indian J Biochem Biophys 34:347–353
8. Nagase H (1996) Matrix metalloproteinases. In: Hooper NM (ed) Zinc metalloproteases in health and disease. Taylor and Francis, London, pp 153–204
9. Sternlicht MD, Lochter A, Sympson CJ, Huey B, Rougler JP, Gray JW, Pinkel D, Bissel MJ, Werb Z (1999) The stromal proteinase MMP3/stromelysin-1 promotes mammary carcinogenesis. Cell 98:137–146
10. Woessner JF (1998) The matrix metalloproteinase family. In: Perks WC, Mecham RP (eds) Matrix metalloproteinases. Academic Press, San Diego, pp 1–14
11. Mark DS, Werb Z (2001) How matrix metalloproteinases regulate cell behavior. Annu Rev Cell Dev Biol 17:463–516
12. Ambili M, Jayasree K, Sudhakaran PR (1998) 60K gelatinase involved in mammary gland

involution is regulated by beta-oestradiol. Biochim Biophys Acta 1403:219–231

13. Edwards DR, Waterhouse P, Holman ML, Denhardt DT (1986) A growth responsive gene (16 C8) in normal mouse fibroblast homologous to a human collagenase inhibitor with erythroid potentiating activity: evidence for inducible and sonstiturive transcripts. Nucleic Acids Res 14: 8863–8877
14. Matrisian LM (1990) Metalloproteinases and their inhibitors in matrix remodeling. Trends Genet 6:121–125
15. Verma RP, Hansch C (2007) Matrix metalloproteinases (MMPs): chemical–biological functions and (Q)SARs. Bioorg Med Chem 15:2223–2268
16. Travis J, Potempa J (2000) Bacterial proteinases as targets for the development of second-generation antibiotics. Biochem Biophys Acta 1477:35–50
17. Elkington PTG, O'Kane CM, Friedland JS (2005) The paradox of matrix metalloproteinases in infectious disease. Clin Exp Immunol 142:12–20
18. Behera AK, Hildebrand E, Scagliotti J, Steere AC, Hu LT (2005) Induction of host matrix metalloproteinases by *Borrelia burgdorferi* differs in human and murine lyme arthritis. Infect Immun 73:126–134
19. Murray GI, Duncan ME, Arbuckle E, Melwin WT, Fothergill JE (1998) Matrix metalloproteinases and their inhibitors in gastric cancer. Gut 43:791–797
20. Madathiparambil GM, Sandhana KC, Gayathri DM, Singh K, Sudhakaran PR, Subhash CS (2011) A zymography analysis of proteinase activity present in Leptospira. Curr Microbiol 62:917–922
21. Madathiparambil GM, Cattavarayane S, Perumana SR, Manickam GD, Sehgal SC (2011) Presence of 46 kDa gelatinase on the outer membrane of Leptospira. Curr Microbiol 62:1478–1482
22. Tsaplina O, Bozhokina E, Mardanova A, Khaitlina S (2015) Virulence factors contributing to invasive activities of Serratiagrimesii and Serratiaproteamaculans. Arch Microbiol 197:481–488
23. Ambili M, Sudhakaran PR (1998) Assay of matrix metalloproteinases in substrate impregnated gels in multiwells. Indian J Biochem Biophys 35:317–320
24. Ambili M, Sudhakaran PR (1999) Modulation of neutral matrix metalloproteinases of involuting rat mammary gland by different cations and glycosaminoglycans. J Cell Biochem 732:218–226
25. Lim DV, Jackson RJ, Pull-VonGruenigen CM (1993) Purification and assay of bacterial collagenases. J Microbiol Methods 18:241–253

Chapter 22

Visualization of Enzyme Activities in Earthworm Biopores by In Situ Soil Zymography

Bahar S. Razavi, Duyen Hoang, and Yakov Kuzyakov

Abstract

Earthworms produce biopores with strongly increased microbial and enzyme activities and consequently they form microbial hotspots in soil. In extremely dynamic microhabitats and hotspots such as earthworm biopores, the in situ enzyme activities are a footprint of process rates and complex biotic interactions. The effect of earthworms on enzyme activities inside biopores, relative to earthworm-free soil, can be visualized by in situ soil zymography. Here, we describe the details of the approach and discuss its advantages and limitations. Direct zymography provides high spatial resolution for quantitative images of enzyme activities in biopores.

Key words Earthworm, Drilosphere, Spatial distribution, Enzymes activity, Microbial hotspots, Zymography

1 Introduction

Extracellular enzymes are involved in innumerable biogeochemical processes and are central for the processing, stabilization, and destabilization of soil organic matter and nutrient cycling in terrestrial ecosystems [1, 2]. Extracellular enzyme activities are most strongly controlled by the concentration of enzymes, their confirmation, and corresponding substrates [3]. The abundance of C-, N-, and P-degrading enzymes in soils is controlled by numerous factors including microbial biomass, community composition, substrate availability, microclimate, and stoichiometric demands [4]. Most enzymes are assumed to originate from microorganisms; however, plant roots and soil animals can contribute strongly to enzyme abundance either directly, by enzyme production, or indirectly, by releasing organic substrates that stimulate microorganisms [5].

Earthworms, which are the most important soil dwelling animals, play an important role by mixing soil materials, aggregating soil particles, and digesting plant litter [6, 7]. The network of biopores formed by earthworms is termed drilosphere and is among

Jeff Wilkesman and Liliana Kurz (eds.), *Zymography: Methods and Protocols*, Methods in Molecular Biology, vol. 1626, DOI 10.1007/978-1-4939-7111-4_22, © Springer Science+Business Media LLC 2017

the most important microbial hotspots in soil [8]. The high microbial activity in the drilosphere is explained by the input of labile organic materials and their mechanical and microbial processing within earthworm gut in the well-aerated and stable biopore structure. High microbial activities, in turn, accelerate the transformation and mineralization of carbon (C) and nutrients, such as nitrogen (N) and phosphorus (P).

Determination of enzyme activities by fluorogenically labeled substrates is frequently applied in soil and rhizosphere studies. Visualization of enzyme activity inside burrow linings in undisturbed samples would strongly improve our understanding of earthworms' effects on enzyme activities [8, 9]. The zymography technique [10] has been applied in various scientific fields, such as medicine, physiology, and biochemistry. With zymography, the conversion of the substrate to a reaction product can be visualized nondestructively [11]. It yields spatially resolved quantitative and qualitative information about hydrolase activities on a surface [11]. Zymography has previously been adapted to visualize the spatial and temporal dynamics of enzyme activities in soil with living and dead roots [12–14]. However, visualization of enzyme activities inside earthworms' habitats was remained a challenge. Recently, this technique has been developed to be applied for quantitative imaging of enzyme activities in biopores [15].

The objectives of this chapter are to describe zymography application to visualize enzyme distribution inside earthworm biopores for nondestructive and quantitative assessment of enzyme activities both under laboratory and field conditions.

2 Materials

1. Dissolved each of substrates separately in universal buffer to a concentration of 12 mM. [This technique can be used for various hydrolytic enzymes (Table 1) (*see* **Notes 1–4**).]
2. Buffer preparation:
   (a) 4-Morpholineethane sulphonic acid (MES), (0.1 M), for 4-methylumbelliferone (MUF) substrate: Weigh 20.673 g MES (206.73 g/mol), put in a 1000 mL volumetric flask and add distilled water. Desired pH would be 6.5 [16].
   (b) TRIZMA (0.05 M) buffer for 4-methylcoumarin (AMC) substrate: Weigh 0.985 g Tris-Base (α-α-α-Tris-(hydroxymethyl)-methylamin) and 2.66 g Tris–HCl (Tris (hydroxymethyl) aminomethane-hydrochloride buffer, 0.036 M) and put them in one 500 mL volumetric flask and add distilled water, to reach 500 mL. Desired pH would be 7.2 [17] (*see* **Note 5**).

Table 1
Summary of selected enzymes, their main ecological functions, and substrate proxies

Enzyme	Synthetic substrate[a]	Enzyme function
β-Glucosidase	MUF-β-D-glucopyranoside	Releases glucose from cellulose
Cellobiohydrolase	MUF-β-D-cellobioside	Releases disaccharides from cellulose
Xylanase	MUF-β-D-xylopyranoside	Releases xylose from hemicellulose
Chitinase	MUF-*N*-acetyl-β-D-glucosaminide	Releases *N*-acetyl glucosamine from chitin
Phosphatase	MUF-phosphate	Releases attached phosphate groups
Leucine-aminopeptidase	L-Leucine-7-amido-AMC	Hydrolysis of the peptide bonds
Tyrosine-aminopeptidase	L-Tyrosine-7-amido-AMC	Hydrolysis of the peptide bonds

[a]For each synthetic substrate, the fluorescent dye 4-methylumbelliferone (MUF) and 4-methylcoumarin (AMC) are presented

3. Polyamide membrane (0.45 μm pore size) should be cut into pieces of biopore sizes and shapes (e.g., 5 × 10 mm) [15]. Thereafter, membrane should be soaked in prepared substrate solutions (*see* **Note 3**).

2.1 Experiment Under Lab Condition (Rhizobox Preparation)

1. Use a transparent (or opaque) plastic box (size variable depending on the objectives) for the experiment; with a removable front panel that could be easily opened without affecting the earthworm habitats (*see* **Notes 6** and 7).
2. Before filling the boxes with soil, lay a sand layer on the bottom of the pots for drainage, to prevent water saturation (*see* **Note 8**).
3. Regulate water content and keep it stable at 30% of soil dry weight during whole experiment.
4. Depending on the objectives, place number of mature earthworms (three or more), by different length (5–10 cm long) in each pot.
5. Prepare similar extra box without earthworm as a control box. Control box will comfort your further comparison between biopores and bulk soil.
6. If the effect of plant roots is also going to be considered in the study, plant seedling in the soil simultaneously with earthworms.
7. Keep the boxes under a stabled temperature of 18 ± 1 °C and a daily light cycle of 16 h, with light intensity of 300 μmol/m^2/s.
8. After 2 weeks that earthworms formed biopores, samples are ready for zymography (Fig. 1).

Fig. 1 An example of the rhizobox and biopores formed by earthworms during 2 weeks

2.2 Experiment Under Field Condition: Biopore-Window System (Rhizotrons Installation)

1. The biopore-window consists of a transparent acrylic sheet (3–5 mm thick).
2. The window can be 30 × 30 cm^2 or 20 × 20 cm^2.
3. It is recommended to install rhizotrons in field simultaneously with plant sowing (if effect of plant roots would be considered in the study).
4. Install rhizotrons at an angle of 90° to the soil surface, and fix it in place by two vertical steel rods and backfill it with soil to remove air gaps.
5. Place some earthworms (2, 3, to) behind the rhizotron to increase the number of burrows.
6. Backfill rhizotron with the soil (Fig. 2).
7. After 2–3 weeks, samples are ready for zymography (Fig. 3).

3 Methods

3.1 Experiment Under Lab Condition

1. Open rhizobox from one side.
2. The size of membranes should be adjusted to the rhizobox.
3. Additional membranes should be cut to pieces of the biopore size.

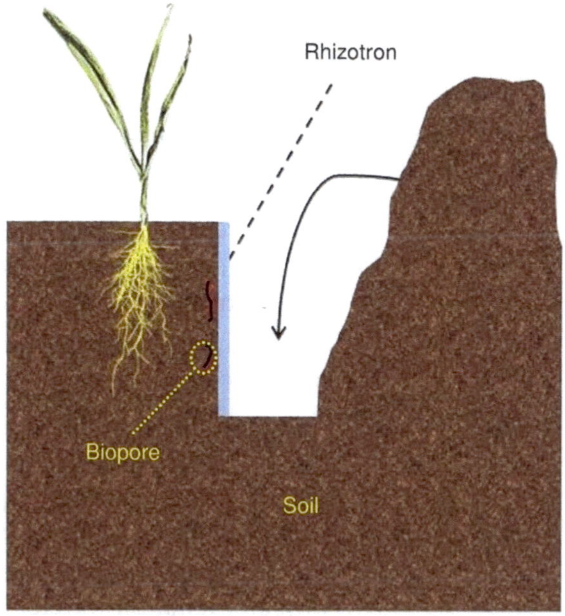

Fig. 2 Scheme of biopore-window under field conditions. Rhizotrone should be fixed, while the dug-out soil will be placed back

Fig. 3 An example of the biopores formed by earthworms under field condition within 4 weeks after biopore-window establishment

4. Substrate-soaked small pieces of membrane will be separately placed inside biopores.
5. Then they should be filled with flint glass beads (0.5 mm).
6. Soft plastic stuffing should be inserted between the glass beads and membrane to ensure the proper membrane attachment to the biopore wall.

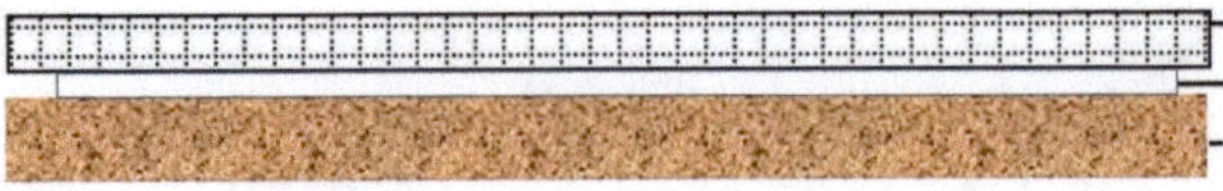

Fig. 4 The polyamide membrane is sandwiched between soil surface and glass plate (under lab condition) or rhizotrone (under field condition) and incubated for 1 h

7. At the same time, a large membrane (matching the box side) will be attached to the whole exposed soil surface.
8. The saturated membranes should be incubated at the soil surface for 1 h.
9. Cover membrane with a glass plate during incubation (to prevent evaporation) (Fig. 4).
10. After incubation, few soil particles attached to membranes should be gently removed using tweezers.
11. The membranes will be carefully placed under UV light with an excitation wavelength of 355 nm in a light-proofed (dark) room.
12. Thereafter, all membranes should be photographed under UV light.
13. If you will couple zymography with routine determination of enzyme activity, collect biopore samples carefully from inside earthworm burrows (*see* **Note 9**).

3.2 Experiment Under Field Condition

1. Dig out the soil that covered the rhizotron during biopore-window establishment.
2. Carefully, remove the rhizotron without any disturbance of biopores.
3. Cut the coated membranes into sizes adjusted for desired area.
4. Also, cut additional membranes into pieces associated with pore size.
5. Separately, place substrate-soaked small pieces of membrane inside biopores (Fig. 5).
6. Then fill them with flint glass beads (0.5 mm).
7. Insert soft plastic stuffing between the glass beads and membrane to ensure the proper membrane attachment to the biopore wall.
8. At the same time, attach a large membrane fitted for the larger area around the biopore to the soil surface (Fig. 5).
9. Incubate the saturated membranes to the soil surface for 1 h.
10. Cover membrane with a clean rhizotron (transparent acrylic sheet) during incubation (to prevent evaporation) (Fig. 4).

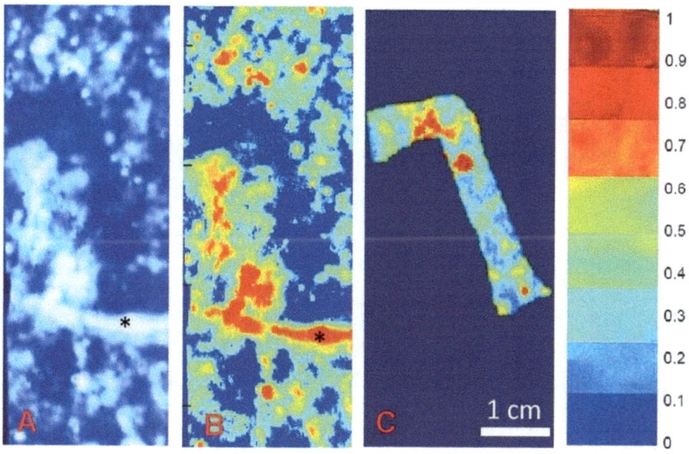

Fig. 5 An example of phosphatase activity, (**a**) initial zymogram of soil surface and biopore edges; (**b**) zymogram of soil surface and biopore edges after image processing; (**c**) zymogram of inside biopore after image processing. Side colormap is proportional to the calibration line of MUF substrate. *Red color* demonstrated high color intensity (high activity) and *blue* shows low color intensity (low activity). Asterisk (*) shows another source of hotspots, i.e., root

11. After incubation, gently remove few soil particles attached to membranes using tweezers.
12. Cover membranes from dust and to prevent evaporation and transport them to the laboratory.
13. Place the membranes under UV light with an excitation wavelength of 355 nm in a light-proofed (dark) room.
14. Thereafter all membranes should be photographed.
15. If you will couple zymography with routine determination of enzyme activity, collect biopore samples carefully from inside earthworm burrows (*see* **Note 9**).

3.3 Image Analysis

3.3.1 Calibration

For quantitative analysis of the zymogram images, calibration is required, which relates the enzyme activities to the gray-values projected onto the zymograms.

1. Eight to ten pieces of membranes should be cut to 3 or 4 cm^2 (Fig. 5).
2. Soak cut membranes in exact volume (e.g., 5 μL) of solution of MUF or AMC with the range of concentrations. For example: 0.01, 0.2, 0.5, 1, 2, 4, 6, 10 μM (This step is necessary to quantify your images.) (*see* **Note 10**).
3. Place the calibration membranes under UV light with an excitation wavelength of 355 nm in a light-proofed room.
4. All the membranes will be photographed.

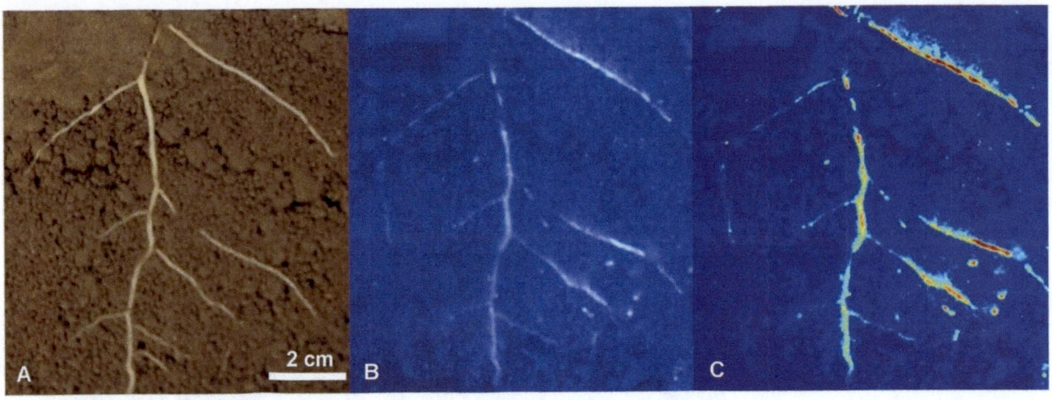

Fig. 6 (**a**) Example of roots grown in control box. (**b**) Initial zymogram of soil surface and roots. (**c**) Zymogram of soil surface and roots after image processing (AMC-leucine aminopeptidase). Hotspots in control soil, mainly, associated with roots

3.3.2 Image Processing

Image processing and analysis could be done by any image processor software. Here, image processing in Matlab environment would be explained briefly [14] (*see* **Notes 11** and **12**).

1. Transform the zymograms to 16 bit grayscale images as a matrix.
2. Make spatial referencing based on the gray-value received from a reference object, which is fixed in all the zymograms (*see* **Note 13**).
3. Calculate the average background gray-values through the calibration lines at zero concentration, and subtract this background value from all zymograms.
4. To illustrate the results, the values of the grayscale image could be depicted in color (Figs. 5 and 6) (*see* **Note 10**).
5. Convert the pixel-wise gray-values of the zymograms to enzyme activities using the enzyme-specific calibration function.

4 Notes

1. Sterile and autoclave all the plates, which will be used for preparation of buffer, substrate solution, and scaling function.
2. Wear gloves and use very precise scale to weight the all materials, especially substrates.
3. It is recommended to find the proper substrate amount in preliminary tests.
4. The substrate solution can be stored at 4 °C for 2 weeks. However, in our laboratory we make substrate solutions fresh. We find that it is better to prepare them fresh each time.

5. Adjust the pH, close to the soil natural condition. Above-maintained buffer is proposed for soil with pH: 7–7.5.
6. If the earthworms are collected manually from field, place them in a black pot with in situ soil at room temperature for 1 week, to adapt the earthworms to the new environment.
7. Cover the boxes with aluminum foil or black paper to protect them from the light and prevent algae growth.
8. Water content should be regulated and kept stable similar to the original habitat.
9. Have in mind that these burrows should not be used for zymography (because burrows cannot be considered undisturbed samples any more).
10. Soak cut membranes in different concentration of MUF solution to quantify the enzyme activities of substrates that are MUF-based, and soak cut membranes in different concentration of AMC solution for substrates that are AMC-linked (Table 1).
11. Interpretation of obtained images: Stronger color intensity (red color) compared to control soil illustrates the hotspots made by earthworms, which are concentrated mainly on the edges and wall of burrow linings and also inside burrow lines (Fig. 5). In comparison, hotspots (red color) in control soil are mainly associated with the root (rhizosphere) (Fig. 6).
12. Calculation of active sites (hotspots area): This step can be done after converting the pixel-wise gray-values of the zymograms to enzyme activities by using the enzyme-specific calibration function. Thereafter, segment the hotspot length and visible area based on different categories of activity (no activity, low, medium, high, and very high), [15]. To confirm the boundaries, you can apply one way analysis of variance (ANOVA) to assess the significant differences between independent variables (adjacent pixels). Calculate and compare these segments as the percentage of total hotspot area (percentage of hotspot area per cm^2) to the entire image (total surface of boxes, inside biopores, and control soil).
13. Blank sides of the sample could be considered as the referencing point (as background). Otherwise on each image, 2 cm^2 black color sheet could be placed, so that you will have same color scale for all zymograms.

Acknowledgments

We gratefully acknowledge the German Academic Exchange Service (DAAD) for supporting B.S.R. and Vietnamese government for supporting D.H. Authors thank Ali Feizi for linguistic help. The study was supported by the German Science Foundation (DFG) within project KU 1184/29-1.

References

1. Allison SD, Wallenstein MD, Bradford MA (2010) Soil-carbon response to warming dependent on microbial physiology. Nat Geosci 3:336–340
2. Chen R, Blagodatskaya E, Senbayram M et al (2012) Decomposition of biogas residues in soil and their effects on microbial growth kinetics and enzyme activities. Biomass Bioenergy 45:221–229
3. Sinsabaugh RL, Hill BH, Follstad SJJ (2009) Ecoenzymatic stoichiometry of microbial organic nutrient acquisition in soil and sediment. Nature 462:795–798
4. Sinsabaugh RL, Lauber CL, Weintraub MN et al (2008) Stoichiometry of soil enzyme activity at global scale. Ecol Lett 11:1252–1264
5. Burns RG (1982) Enzyme activity in soil: location and possible role in microbial ecology. Soil Biol Biochem 14:423–427
6. Jones CG, Lawton JH, Shachak M (1994) Organisms as ecosystem engineers. Oikos 69:373–386
7. Lavelle P, Bignell D, Lepage M et al (1997) Soil function in a changing world: the role of invertebrate ecosystem engineers. Eur J Soil Biol 33:159–193
8. Kuzyakov Y, Blagodatskaya E (2015) Microbial hotspots and hot moments in soil: concept & review. Soil Biol Biochem 83:184–199
9. Mackie KA, Schmidt HP, Müller T et al (2014) Cover crops influence soil microorganisms and phytoextraction of copper from a moderately contaminated vineyard. Sci Total Environ 500:34–43
10. Gross J, Lapiere CM (1962) Collagenolytic activity in amphibian tissues: a tissue culture assay. Proc Natl Acad Sci U S A 48:1014–1022
11. Vandooren J, Geurts N, Martens E, Van den Steen PE et al (2013) Zymography methods for visualizing hydrolytic enzymes. Nat Methods 10:211–220
12. Spohn M, Kuzyakov Y (2014) Spatial and temporal dynamics of hotspots of enzyme activity in soil as affected by living and dead roots – a soil zymography analysis. Plant Soil 379:67–77
13. Sanaullah M, Razavi BS, Blagodatskaya E et al (2016) Spatial distribution and catalytic mechanisms of β-glucosidase activity at the root-soil interface. Biol Fertil Soils 52:505–514
14. Razavi BS, Zarebanadkouki M, Blagodatskaya E et al (2016) Rhizosphere shape of lentil and maize: spatial distribution of enzyme activities. Soil Biol Biochem 96:229–237
15. Hoang TTD, Razavi BS, Kuzyakov Y et al (2016) Earthworm burrows: kinetics and spatial distribution of enzymes of C-, N-and P-cycles. Soil Biol Biochem 99:94–103
16. Razavi BS, Blagodatskaya E, Kuzyakov Y (2015) Nonlinear temperature sensitivity of enzyme kinetics explains canceling effect—a case study on loamy haplic Luvisol. Front Microbiol 6:1–13
17. Koch O, Tscherko D, Kandeler E (2007) Temperature sensitivity of microbial respiration, nitrogen mineralization, and potential soil enzyme activities in organic alpine soils: temperature sensitivity in alpine soils. Glob Biogeochem Cycles 21:1–11

Chapter 23

Multiplex Cathepsin Zymography to Detect Amounts of Active Cathepsins K, L, S, and V

Manu O. Platt

Abstract

Cysteine cathepsins are powerful proteases that can degrade other proteins, among which are the extracellular matrix proteins collagen and elastin. Multiplex cathepsin zymography is an assay that can quantify the amount of active cathepsins in a cell or tissue preparation. This method works for measuring the amounts of active cathepsins K, L, S, and V in a cell or tissue preparation without requiring the use of antibodies for specific identification which tremendously reduces cost. This chapter will demonstrate the utility and interpretation of this method with mammalian cells and tissue to quantify amounts of active cathepsins K, L, S, and V without complicating signals of the procathepsin. Multiplex cathepsin zymography has many advantages: (1) it separates cathepsins K, L, S, and V by electrophoretic migration distance, (2) allows visual confirmation of cathepsin identity, (3) does not detect procathepsins, and (4) can be quantified with densitometry.

Key words Cathepsins, Proteases, Cancer, Cardiovascular disease, Enzymology

1 Introduction

Cathepsins K, L, S, and V are members of the lysosomal cysteine cathepsin family which have been implicated in a number of pathological roles. Among the four, they share 60% sequence homology [1–3], but each has unique properties and different homeostatic functions. Cysteine cathepsins are a potent class of proteases that have been shown to be upregulated in a number of tissue-destructive diseases such as cancer, arthritis, osteoporosis, and atherosclerosis [1, 4–9].

Multiplex cathepsin zymography is based on gelatin zymography techniques that have been used to detect matrix metalloproteinases (MMPs) [10], but has been modified for specific members of the cysteine cathepsin family [11, 12]. It has also been referred to as substrate SDS-PAGE (sodium dodecyl sulfate polyacrylamide gel electrophoresis) due to polymerizing a gelatin substrate in the polyacrylamide gel meshwork. Proteins extracted from cells or

Jeff Wilkesman and Liliana Kurz (eds.), *Zymography: Methods and Protocols*, Methods in Molecular Biology, vol. 1626, DOI 10.1007/978-1-4939-7111-4_23, © Springer Science+Business Media LLC 2017

tissue can be resolved in a nonreducing electrophoresis with a standard Bio-Rad Mini-Protean chamber or equivalent apparatus to separate proteins by molecular size after being partially unfolded by SDS. Excluding beta-mercaptoethanol preserves disulfide bridges in the cathepsin native conformation [6, 13], preventing full linearization of the molecule, but enabling refolding after washing out the SDS in a renaturing buffer. The gel is then equilibrated to pH 6, which is optimal pH for their function, prior to overnight incubation in fresh assay buffer with a freshly added reducing agent, dithiothreitol (DTT), to provide an optimal reducing environment and pH for cathepsin activity. These refolded, active enzymes will then cleave the gelatin substrate in the polyacrylamide gel at the site to which they were separated by the electrophoresis such that after an overnight incubation period, the gel can be stained with Coomassie blue, which will stain the entire gel blue due to the cross-linked gelatin, except where the cathepsins have degraded the gelatin presenting as cleared white bands of proteolytic activity after destaining. Schematic of this entire protocol is presented in Fig. 1. Only refolded, active cathepsins will provide a signal that can be quantified by densitometry methods such as NIH ImageJ. This removes the confounding signals from proforms of cathepsins that would be detected by ELISA, Western blot, and other antibody-based methods.

Multiplex cathepsin zymography has been demonstrated with recombinant enzymes [14, 15], a number of different cell types including endothelial cells [8, 16–19], monocytes [17], macrophages [20, 21], breast cancer cells [20], prostate cancer cells [22], osteoclasts [21], and others, and also a number of homogenized tissues including tendons, breast, lung, and cervical tumors [23], endometriosis lesions [24], arteries [8, 18, 25–27], and heart valves [28, 29] among others.

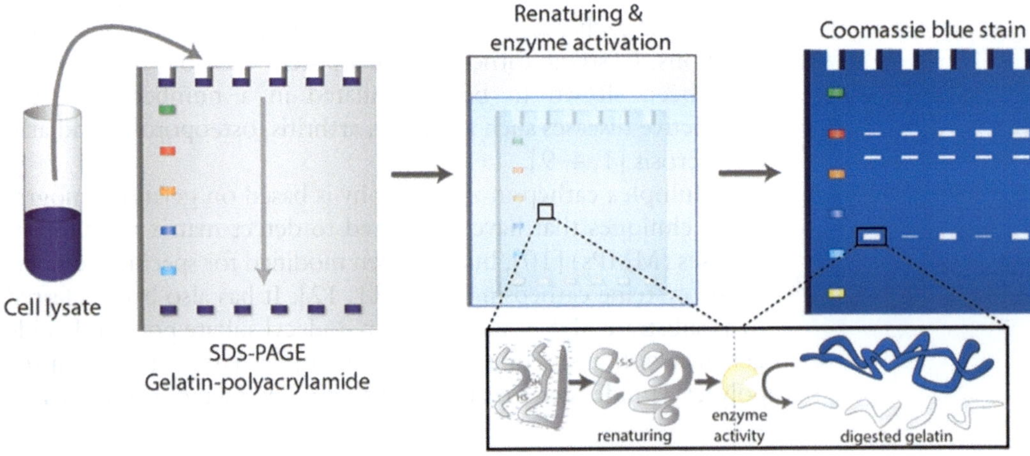

Fig. 1 Schematic representation of the cathepsin zymography protocol

As described, this protocol will require 2 h of attention and preparation prior to the overnight incubation step and usually requires 10–15 μg of protein material to obtain a signal. This technique has many benefits: (1) it does not require antibodies, (2) separation of proteins by molecular mass visually confirms cathepsin identity, (3) densitometry can be used for quantitative analysis, and (4) versatility of the assay enables inhibition with small molecules to corroborate the identity of the enzyme of interest.

2 Materials

2.1 SDS Polyacrylamide Gel Components

1. *Gelatin preparation*: Dissolve 0.150 g gelatin (from pig or bovine extract) in 30 mL of warmed, deionized water to obtain a final concentration of 5 mg/mL (*see* **Note 1**). Store this gelatin solution in a 50 mL conical tube labeled with the date and store at room temperature.
2. *Separating Gel Components*: Add about 200 mL of deionized water to a beaker with a magnetic stir bar and place this together on a stir plate. Weigh 113.3 g Tris base and transfer it to the beaker. Fill with water to 500 mL. Then adjust pH with HCl to 8.9. This solution can be stored at room temperature.
3. *Stacking Gel Components*: As above, prepare a beaker, stir bar, and stir plate with 200 mL of deionized water. Weigh 19 g Tris base and transfer it to the beaker. Fill with water to 500 mL. Then adjust pH to 6.7 with HCl. This solution can also be stored at room temperature.
4. *Acrylamide/bisacrylamide mixture at 37.5:1 ratio*: Commonly used version is ProtoGel available from National Diagnostics. To be stored in a dark jar protected from light.
5. *Sodium dodecyl sulfate (SDS)*: 10% solution in water (*see* **Note 2**).
6. *Ammonium persulfate*: 1.5% solution in water (*see* **Note 3**).
7. *TEMED N, N, N′, N′-tetramethyl-ethylenediamine (TEMED)*: store protected from light.
8. *SDS-PAGE running buffer (10× stock)*: In a large 5 L beaker, fill to about 2 L with deionized water, add a magnetic stir bar, and place on a stir plate. Measure out 120 g Tris base, 576 g glycine, and 40 g SDS and add to the stirring solution. Allow it to mix until all are completely dissolved. Fill with deionized water to 4 L. Then adjust pH to 8.3 with HCl or NaOH. This can be stored at room temperature. Dilute 100 mL of this 10× stock by adding 900 mL deionized water in a graduated cylinder to prepare a 1× working solution when ready to start electrophoresis.
9. *Prestained protein ladder*: Commercially available ladders with prestained bands from 10 kDa to at least 150 kDa.

2.2 Sample Preparation

1. *Zymogram lysis buffer*: To prepare 500 mL, dissolve the following components into deionized water. 20 mM Tris–HCl (1.21 g); 5 mM ethylene glycol tetra-acetic acid (EGTA, 951 mg); 150 mM sodium chloride (NaCl, 4.38 g); 20 mM beta-glycerol-phosphate (2.16 g); 10 mM sodium fluoride (NaF, 210 mg); and sodium orthovanadate (Na_3VO_4, 91.95 mg). Then add liquid detergents: 5 mL of Triton X-100 to get a final 1% v/v, and 500 μL of Tween-20 to a final 0.1% (v/v).
2. *Leupeptin*: To be freshly added to the zymo lysis buffer at a final concentration of 0.1 mM. Stock solutions of 20 mM can be aliquoted and stored at −20 °C. Leupeptin molecular weight is 475.59. Add 526 μL of deionized water to 5 mg of leupeptin to make the 20 mM solution. Then, aliquot into 20 μL aliquots that can be stored at −20 °C and thawed individually upon use. Then to prepare 1 mL of zymo lysis buffer, add 995 μL of zymo lysis buffer and 5 μL of freshly thawed 20 mM leupeptin to be diluted 1:200 to prepare a final concentration of 0.1 mM.
3. *5× Sample nonreducing loading buffer*: To prepare 100 mL, weigh out and measure 0.05 g Bromophenol Blue and 10 g SDS. Add this to 20.8 mL of 1.5 M Tris buffer, pH 6.8, and 25 mL deionized water. Then add 50 mL glycerol and mix (*see* **Note 4**).

2.3 Renaturing Buffer

To prepare 500 mL, measure out 3.94 g Tris base (65 mM) and add to 300 mL of deionized water with a stir bar on a stir plate. After it has dissolved, fill to 400 mL with deionized water. Then, while still stirring, add 100 mL glycerol [20% (v/v)] to the beaker. Then adjust pH to 7.4 and store in a labeled bottle with a screw top. Solution can be stored at room temperature.

2.4 Assay Buffers

(a) Assay buffer pH 4.0 is a sodium acetate buffer, 0.1 M with 1 mM EDTA, and freshly added 2 mM DTT. Two solutions will need to be prepared first as stock solutions, then combined at the appropriate ratios to produce a stable buffer.

Solution A: 5.755 mL glacial acetic acid, then fill to 500 mL (0.2 M) in water.

Solution B: Measure 13.6 g of sodium acetate ($NaC_2H_3O_2 \cdot 3H_2O$) and dissolve in 300 mL of water, then add water to fill to 500 mL to produce a 0.2 M solution (or make appropriate weight measure calculations if using the anhydrous form).

For a pH 4.0 solution, mix 205 mL of Solution A with 45 mL of Solution B. Then dilute with water to a total of 500 mL. Afterward, add 2 mL of 0.25 M EDTA. This solution will be stable. The DTT must be added fresh before incubating with the zymogram. Test the pH to be sure that it is 4.0 either with pH paper or a pH meter.

(b) Assay buffer pH 6.0 is a sodium phosphate buffer, 0.1 M, with 1 mM EDTA, and freshly added 2 mM DTT.

Solution C: Measure out 13.8 g $NaH_2PO_4 \cdot H_2O$, then dissolve in 300 mL of water (or make appropriate weight measure calculations if using the anhydrous form). Then fill to 500 mL to make a 0.2 M final solution.

Solution D: Measure out 26.825 g $Na_2HPO_4 \cdot 7H_2O$, then dissolve in 300 mL of water (or make appropriate weight measure calculations if using the anhydrous form). Then fill to 500 mL to make a 0.2 M final solution.

For a pH 6.0 solution, mix 219.25 mL of Solution C with 30.75 mL of Solution D. Then dilute with water to a total of 500 mL. Then add 2 mL of 0.25 M EDTA to make it 1 mM EDTA. This solution will be stable. The DTT must be added fresh before incubating with the zymogram. Test the pH to be sure that it is 6.0 either with pH paper or a pH meter.

2.5 Staining and Destaining

1. *Coomassie Stain solution*: To make a 1 L volume of this solution, measure 650 mL of water, and add 100 mL acetic acid, 250 mL 2-propanol, and 0.45 g Coomassie Blue R250.
2. *Destain solution*: To make a 1 L volume of this solution, measure 800 mL water, and add 100 mL 2-propanol and 100 mL acetic acid.

3 Methods

Carry out all procedures at room temperature unless otherwise specified.

3.1 Cell Extraction and Sample Preparation

1. Add 60 μL of zymo lysis buffer to each well of a 12-well dish and scale accordingly for larger size dishes. Goal is to get a final protein concentration greater than 1 μg/μL. If homogenizing tissue, then use appropriate amount of zymo lysis buffer based on tissue size and expected mass, still with the goal of getting final solubilized protein concentration of greater than 1 μg/mL.
2. Perform cell extraction on ice. Use a cell scraper to ensure collecting all cell extracts and collect in a microcentrifuge tube.
3. Sonicate the cell extract briefly using a probe sonicator to ensure disruption of cell membranes.
4. Centrifuge sonicated cell extracts at high speed (>13,000 × *g*) to pellet cellular debris and any insoluble fractions.
5. Remove soluble supernatant and place in a separate, new microcentrifuge tube.
6. Perform a protein assay to determine total protein concentration (Pierce Micro-BCA kit is recommended, but other versions will do).

7. Prepare normalized protein extract samples and add appropriate volumes of 5× nonreducing sample buffer such that a final protein concentration of at least 1 μg/μL is obtained (*see* **Note 5**).

3.2 Substrate Gel Electrophoresis

1. Prepare assembly and plates for gel electrophoresis, with a 0.75 mm spacer plate and a short plate. Fill with water to test for leakage while preparing the gel solution.
2. Gather the appropriate combs to prepare wells for the 0.75 mm thick polyacrylamide gel with cross-linked gelatin.
3. Prepare two substrate polyacrylamide standard size gels [approximately 92 × 68 × 0.75 mm^3 (width × height × thickness) which are the standard Bio-Rad dimensions], mix the following solutions in this order: 860 μL deionized water, 3.34 mL Protogel (acrylamide/bis-acrylamide), 1.6 mL 5× separating buffer, 1.72 mL of 5 mg/mL gelatin solution, 400 μL 1.5% APS, and 80 μL 10% SDS.
4. At this point, be sure to check the plate assembly for any leaks, and if no leaks, then pour out the water and use a tissue to remove the remaining water in the assembly by capillary action.
5. Then add 7.5 μL TEMED to the polyacrylamide gel solution to initiate the polymerization reaction and quickly inject into the plate assembly up to a level that still allows for the combs to be inserted and for the stacking gel to be added.
6. Inject about 150 μL of butanol onto the separating gel to make an airtight seal, remove any air bubbles in the solution, and to create a flat line for integration with the stacking gel.
7. While the separating gel polymerizes, prepare the gel solution for the stacking gel by mixing in this order: 1.9 mL water, 1 mL Protogel, 1 mL 5× stacking buffer, 1 mL 1.5% APS, and 50 μL 10% SDS.
8. Be sure the separating gel has polymerized by tipping the assembly slightly to watch only the butanol layer flow and not the polymerized gel (it will take at least 10–15 min for complete polymerization).
9. At this point, be sure to remove the butanol from the gel plate assembly first by decanting, then by adding deionized water, and swirling it side to side to be sure that the butanol is displaced above the water layer. Then decant and repeat until it has been assured that all butanol has been removed.
10. Then prepare the comb, and add 4 μL of TEMED to the stacking gel solution, swirl briefly, then using the pipette, inject into the plate assembly over the separating gel. Fill the plates and then insert the combs immediately prior to polymerization of the gel.

11. After polymerization of the stacking gel, carefully remove the comb from the gels being sure not to disturb the wells, by pulling them straight out.
12. Assemble the gels onto the electrode assembly with the short plates on the inside (if using the Bio-Rad Mini-Tetran apparatus).
13. Place into the electrophoresis chamber and fill with 1× SDS-PAGE running buffer to the appropriate volume according to the apparatus being used.
14. Carefully load equal protein amounts of samples into each lane of the gel. Also, load a protein ladder and a positive control (*see* **Note 6**).
15. Add a matching volume of 1× buffer to remaining wells to ensure an even running front during the electrophoresis.
16. Attach the electrodes to the voltage box and apply 200 V for about 1 h to separate the proteins in the sample, watching the gel until the bromophenol blue line front reaches the bottom of the gel (*see* **Note 7**).

3.3 Renaturing of Enzymes and Cleavage of Substrate Proteins

1. After electrophoresis, carefully remove polyacrylamide gels from the apparatus being sure not to rip the thin gels.
2. Place them into containers with renaturing buffer, and place on a shaker rotating at slow to medium speed.
3. After 10 min, decant renaturing buffer, add fresh renaturing buffer, and place back on the shaker. Repeat two more times for a total of three times (*see* **Note 8**).
4. During this renaturation, thaw one frozen aliquot of 0.5 M DTT in preparation of the assay buffer.
5. During the third rinse in renaturing buffer, prepare the assay buffer (pH 6.0 for cathepsins K, L, S, and V, or pH 4.0 to distinguish cathepsin L uniquely) by adding 200 μL of 0.5 M DTT to 50 mL of assay buffer to get to a final concentration of 2 mM DTT (*see* **Note 9**).
6. After the third rinse, decant the renaturing buffer and add 25 mL of the assay buffer to the container to equilibrate the gel for 30 min (*see* **Note 10**).
7. Prepare plastic containers to incubate gels overnight with fresh assay buffers at 37 °C. This can be done in old pipette tip boxes with the inside racks removed, or the preferable way is to use airtight plastic bags sealed with the plastic sealer. If using plastic bags, then the bags can be incubated in a 37 °C water bath overnight. If using a container, then it can be placed inside of an incubator overnight, but agitation is recommended for best results (*see* **Note 11**).

3.4 Coomassie Staining

1. After 24 h incubation, stain with Coomassie blue for 1 h (*see* **Note 12**).
2. Rinse with deionized water.
3. Add destain buffer to observe cleared white bands on a dark blue background indicative of cathepsin activity.

3.5 Imaging and Densitometry

1. Imaging of the gel can be done with a lightbox and a digital camera, a scanner, or any number of types of gel imaging apparatus such as the ImageQuant.
2. Densitometry of the signals can be performed using NIH ImageJ, or other software that might come with your imaging equipment, being sure to invert the image to convert the cleared white bands on a dark band to dark bands on a white background (*see* **Notes 13–16** for data interpretation).

4 Notes

1. To warm the water, either use a hot plate, or for faster preparations, add 50 mL of deionized water to a small beaker and microwave for 30 s. This will sufficiently warm the water for quick dissolution of the gelatin. Gelatin can be used for 48 h before it is recommended to make a fresh batch.
2. Be careful when measuring SDS because it can aerosolize and cause coughing fits.
3. APS is an initiator for the polymerization reaction and may need to be made freshly as this solution can go bad. Therefore, it is recommended to make only small volume solutions that will be depleted over shorter periods of times (within the month), such as 100 mL.
4. SDS can precipitate while mixing the 5× nonreducing loading buffer solution as it will become quite thick viscous. It is easier to mix this in a 50 mL conical tube to ensure proper mixing, then to aliquot into 500–1000 μL tubes and store at −20 °C until ready to use. Prior to use, warm each aliquot to 37 °C to be sure to redissolve any SDS that might precipitate out of solution while being frozen.
5. Normalizing protein concentrations is important to ensure loading of equal amounts of protein into each individual well of the polyacrylamide gel. Then, when the signal is developed at the end of the protocol, it is fair to compare the intensity of the cleared white bands among the different conditions being tested by that experiment. If one gel is overloaded with a larger amount of protein and there is a greater signal, then it could be interpreted not that the treatment induced greater amounts of active cathepsins, but that more cathepsins were

loaded into the gel. When testing for amount of active cathepsins in secreted/conditioned media, equal volumes are usually a more fair way to normalize the signals since under different treatment conditions, cells may secrete more or less protein which would nullify loading gels based on total protein concentrations.

6. Protein ladders are important for proper determination of cathepsin identities at the end of the assay. To ensure its visibility after the Coomassie staining, we recommend loading at least 10–15 μL of protein ladder. Positive controls—samples loaded that ensure cathepsin activity, visible in the completed zymogram—are important. If there is no signal in the samples that were loaded, it could just be that buffers were not properly made, not necessarily that there were no active cathepsins in the sample. Recombinant cathepsins can be purchased and loaded into the positive control lane, but this can become costly over time. A much more inexpensive alternative is to use extract from RAW264.7 cells, a murine macrophage cell line that produces an abundant amount of active cathepsins. Then only 2–4 μg of RAW macrophage cell extract should be loaded to provide a sufficiently detectable signal that will not oversaturate the zymogram. Also, with careful loading, be sure the positive control does not overflow into neighboring wells as it will produce a signal in one of the experimental samples that may not be from that sample. For recombinant enzymes, we recommend loading about 10–20 ng of mature cathepsins to provide sufficient signal. Cathepsin L requires at least 50 ng for sufficient signal.

7. By 200 V, it will take about 1 h to run when using the dimensions described for the gel. However, there can be a loss of resolution if trying to distinguish intracellular cathepsin K from cathepsin V. Cathepsin K separates to about 37 kDa by molecular ladder and cathepsin V is 35 kDa, so there is an important distinction that can be better visualized by running the electrophoresis at 100 V for 2 h. When cathepsin K is secreted into the conditioned media, it actually runs at much higher molecular size, most likely due to differential processing/tagging that targets it for secretion. This is shown in Fig. 2 where A is a zymogram from conditioned medium from either macrophages or osteoclasts and cathepsin K is running much higher (around 75 kDa) compared to its location at 37 kDa, just above the 35 kDa cathepsin V signal inside of endothelial cells stimulated with tumor necrosis factor alpha (TNFα), which we have shown to turn on cathepsin K [16, 17].

8. Three washes with renaturing buffer after 10 min incubations for each is the standard protocol, but there is a faster protocol that uses only 5 min incubations, but still three washes, and

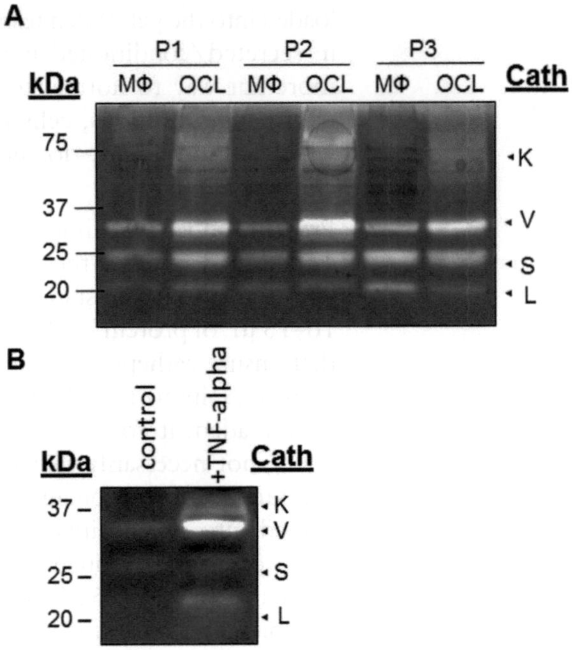

Fig. 2 Multiplex cathepsin zymography signals different cells. (**a**) Conditioned media from differentiated macrophages (MF) and osteoclasts (OCL) from peripheral blood mononuclear cells were collected and tested by zymography. Cathepsin K, V, S, and L signals are shown with cathepsin K at the high molecular weight of 75 kDa from conditioned media. (**b**) Human aortic endothelial cells were cultured in the absence (control) or presence of 10 ng/mL TNFα for 24 h prior to lysis and loading for multiplex cathepsin zymography. The appearance of the cathepsin K band at 37 kDa just above the 35 kDa cathepsin V band is shown by that M-shaped band just above the other. (P-person; P1, P2, P3 are from samples from three different people)

will provide a sufficient signal [15]. This should be tested first under your own specific experimental cells or tissue conditions before becoming routine.

9. DTT must be added freshly at this step. It is the reducing agent and cathepsins require a reducing environment for the cysteine at the active site to participate in hydrolysis. By thawing an aliquot once the renaturing stage begins, it should be ready to add by the time of incubation in the assay buffer.
10. For the assay buffer incubation step, 25 mL is sufficient based on the size of the container in which the gel is being incubated. This volume may need to be increased or decreased depending on the size of the container. Most important is to use a volume that will cover the gels.
11. We prefer to incubate them in airtight plastic-sealed bags filled with the assay buffer, then submerged in 37 °C water overnight. However, if that is not available, it also works with placing gels in a container that is covered, and not airtight, that is put on

an overnight shaker at 37 °C (could be a bacterial incubator or a shaker in a warm room). It is not necessary to use a shaker, but that does provide the best results. For better results, keep gels incubating in separate containers for the renaturing, assay, and staining steps. There can be artifacts that affect the final image when the gels are touching or if diffusion is limited.

12. Incubation can be as few as 4 h, depending on the concentration of enzymes in the sample, but this needs to be determined in each case. In most cases, this is an overnight incubation that will provide sufficient signal. Coomassie staining may not take the full 1 h to begin to see the signal, but it is recommended to allow for 1 h of staining at least the first time, and then to continue to monitor for the appearance of cleared white bands in the positive control before removing the Coomassie stain and adding destain solution. Observing the stained gel over a light box is one way to view the staining progress of the zymogram.
13. Images can be captured by digital cameras on lightboxes, scanners, or imaging darkboxes. To protect the electronic equipment, a transparent plastic sheet protector can be used to protect the staining/destaining solutions from the gels from damaging or staining the electronics. Then images can be adjusted in Photoshop or other imaging software, and some imaging softwares have their own image optimization algorithms. It is important to invert the image prior to densitometry since the zymogram is a white band on a dark background, and most densitometry programs are best with dark bands on a white background. NIH ImageJ is free software available to quantify the intensity of the white bands to determine the amount of active cathepsins by one treatment over another. It is available at http://imagej.nih.gov/ij/download.html.
14. Important information about data interpretation: Interpretation of pH 6.0 zymogram bands: cathepsin K 37 kDa, cathepsin V 35 kDa, cathepsin S 25 kDa, and cathepsin L at 20 kDa. These are from the intracellular/cell extract signals. From secreted/conditioned media, these positions may change: catK 37–50 kDa. For pH 4.0 zymogram bands, cathepsin L generates three signal bands 27, 25, and 20 kDa, with a cathepsin V band appearing higher at 35 kDa, and a slight cathepsin S overlap signal at 26 kDa, just above the main cathepsin L band (Fig. 3). Recombinant cathepsins being used provide different ranges of signals between secreted and isolated forms vs. intracellular isolated forms. Electrophoretic migration distance is based on enzyme size, not necessarily weight. It is important to remember that each cathepsin contains between 5 to 7 disulfide bonds, but by running a nonreducing SDS-PAGE, these bonds are not broken and therefore

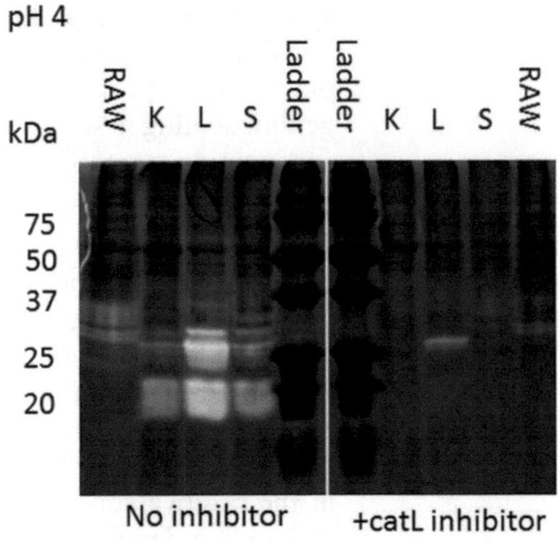

Fig. 3 Cathepsin inhibitors can be used to selectively identify active cathepsin bands. Recombinant cathepsins K, L, and S were loaded for parallel zymograms at pH 4 which select for cathepsins L and V, where, after electrophoresis, one of the gels was incubated with the cathepsin L inhibitor Z-FY-DMK, and the other was incubated only with vehicle controls. Incubation with Z-FYDMK, a selective inhibitor of cathepsin L, inhibited many of the cathepsin active bands, but did not block all active cathepsin bands. This suggested that the residually active band was not cathepsin L or susceptible to inhibition by 5 μM Z-FY-DMK

there is no complete linearization of the proteins of interest. Glycosylation of the enzymes can affect electrophoretic migration of cathepsins [3]. This will matter for recombinant enzymes isolated from *E. coli* or from eukaryotic cells; also whether the cathepsins are secreted or intracellular also will affect the electrophoretic migration distance. Fair comparisons of cathepsin activity of different samples loaded in the same gel can be made, but absolute standards can also be loaded to fit the quantified densitometry signal and calculate an actual value to compare across different gels [12, 23].

15. Other notes on optimization: 0.75 mm gel thickness provides the best signals, but 1 mm gel thickness can work. It is important to remember that diffusion of the buffers through the thickness of the gels helps control renaturation of the proteases and better equilibration of both pH and ion concentrations. The fresher the samples are, the better the results. Freeze–thaw cycles cause loss of cathepsin activity.

16. To confirm identity of cathepsins in your samples under specific conditions, nonreducing Western blots should be run with antibody confirmation. The cysteine cathepsins can run at different distances when under reduced vs. nonreduced conditions. It is also possible to incubate small molecule inhibitors in the assay buffer during the overnight incubation stage of

the zymography to selectively block the signal of one cathepsin over another (cathepsin L inhibitor: Z-FY-DMK), cathepsin K inhibitor (Z-L-NHNHCONHNH-LF-Boc); serine protease inhibitor (PMSF); MMP inhibitor (GM6001); broad cathepsin inhibitor E-64. Example of the inhibition of the cathepsin L signal by 5 μM Z-FY-DMK is shown in Fig. 3 with only one band remaining after pH 4 incubation.

References

1. Chapman HA, Riese RJ, Shi GP (1997) Emerging roles for cysteine proteases in human biology. Annu Rev Physiol 59:63–88
2. Reiser J, Adair B, Reinheckel T (2010) Specialized roles for cysteine cathepsins in health and disease. J Clin Invest 120:3421–3431
3. Bromme D, Li Z, Barnes M, Mehler E (1999) Human cathepsin V functional expression, tissue distribution, electrostatic surface potential, enzymatic characterization, and chromosomal localization. Biochemistry 38:2377–2385
4. Lafarge JC, Naour N, Clement K, Guerre-Millo M (2010) Cathepsins and cystatin C in atherosclerosis and obesity. Biochimie 92:1580–1586
5. Sukhova GK, Zhang Y, Pan JH, Wada Y, Yamamoto T, Naito M, Kodama T, Tsimikas S, Witztum JL, Lu ML, Sakara Y, Chin MT, Libby P, Shi GP (2003) Deficiency of cathepsin S reduces atherosclerosis in LDL receptor-deficient mice. J Clin Invest 111:897–906
6. Lecaille F, Bromme D, Lalmanach G (2008) Biochemical properties and regulation of cathepsin K activity. Biochimie 90:208–226
7. Lutgens E, Lutgens SP, Faber BC, Heeneman S, Gijbels MM, de Winther MP, Frederik P, van der Made I, Daugherty A, Sijbers AM, Fisher A, Long CJ, Saftig P, Black D, Daemen MJ, Cleutjens KB (2006) Disruption of the cathepsin K gene reduces atherosclerosis progression and induces plaque fibrosis but accelerates macrophage foam cell formation. Circulation 113:98–107
8. Platt MO, Ankeny RF, Shi GP, Weiss D, Vega JD, Taylor WR, Jo H (2007) Expression of cathepsin K is regulated by shear stress in cultured endothelial cells and is increased in endothelium in human atherosclerosis. Am J Physiol Heart Circ Physiol 292:H1479–H1486
9. Littlewood-Evans AJ, Bilbe G, Bowler WB, Farley D, Wlodarski B, Kokubo T, Inaoka T, Sloane J, Evans DB, Gallagher JA (1997) The osteoclast-associated protease cathepsin K is expressed in human breast carcinoma. Cancer Res 57:5386–5390
10. Snoek-van Beurden PA, Von den Hoff JW (2005) Zymographic techniques for the analysis of matrix metalloproteinases and their inhibitors. Biotechniques 38:73–83
11. Wilder CL, Park K-Y, Keegan PM, Platt MO (2011) Manipulating substrate and pH in zymography protocols selectively distinguishes cathepsins K, L, S, and V activity in cells and tissues. Arch Biochem Biophys 516:52–57
12. Li WA, Barry ZT, Cohen JD, Wilder CL, Deeds RJ, Keegan PM, Platt MO (2010) Detection of femtomole quantities of mature cathepsin K with zymography. Anal Biochem 401:91–98
13. McGrath ME (1999) The lysosomal cysteine proteases. Annu Rev Biophys Biomol Struct 28:181–204
14. Barry ZT, Platt MO (2012) Cathepsin S cannibalism of cathepsin K as a mechanism to reduce type I collagen degradation. J Biol Chem 287:27723–27730
15. Dumas JE, Platt MO (2013) Systematic optimization of multiplex zymography protocol to detect active cathepsins K, L, S, and V in healthy and diseased tissue: compromise among limits of detection, reduced time, and resources. Mol Biotechnol 54:1038–1047
16. Keegan PM, Surapaneni S, Platt MO (2012) Sickle cell disease activates peripheral blood mononuclear cells to induce cathepsins K and V activity in endothelial cells. Anemia 2012:201781
17. Keegan PM, Wilder CL, Platt MO (2012) Tumor necrosis factor alpha stimulates cathepsin K and V activity via juxtacrine monocyte-endothelial cell signaling and JNK activation. Mol Cell Biochem 367:65–72
18. Parker IK, Roberts LM, Hansen L, Gleason RL Jr, Sutliff RL, Platt MO (2014) Pro-atherogenic shear stress and HIV proteins synergistically upregulate cathepsin K in endothelial cells. Ann Biomed Eng 42:1185–1194
19. Platt MO, Ankeny RF, Jo H (2006) Laminar shear stress inhibits cathepsin L activity in endothelial cells. Arterioscler Thromb Vasc Biol 26:1784–1790

20. Park KY, Li G, Platt MO (2015) Monocyte-derived macrophage assisted breast cancer cell invasion as a personalized, predictive metric to score metastatic risk. Sci Rep 5:13855
21. Park KY, Li WA, Platt MO (2012) Patient specific proteolytic activity of monocyte-derived macrophages and osteoclasts predicted with temporal kinase activation states during differentiation. Integr Biol (Camb) 4:1459–1469
22. Burton LJ, Smith BA, Smith BN, Loyd Q, Nagappan P, McKeithen D, Wilder CL, Platt MO, Hudson T, Odero-Marah VA (2015) Muscadine grape skin extract can antagonize snail-cathepsin L-mediated invasion, migration and osteoclastogenesis in prostate and breast cancer cells. Carcinogenesis 36:1019–1027
23. Chen B, Platt MO (2011) Multiplex zymography captures stage-specific activity profiles of cathepsins K, L, and S in human breast, lung, and cervical cancer. J Transl Med 9:109
24. Porter KM, Wieser FA, Wilder CL, Sidell N, Platt MO (2016) Cathepsin protease inhibition reduces endometriosis lesion establishment. Reprod Sci 23:623–629
25. Hansen L, Parker I, Roberts LM, Sutliff RL, Platt MO, Gleason RL Jr (2013) Azidothymidine (AZT) leads to arterial stiffening and intima-media thickening in mice. J Biomech 46: 1540–1547
26. Hansen L, Parker I, Sutliff RL, Platt MO, Gleason RL Jr (2013) Endothelial dysfunction, arterial stiffening, and intima-media thickening in large arteries from HIV-1 transgenic mice. Ann Biomed Eng 41:682–693
27. Caulk AW, Soler J, Platt MO, Gleason RL Jr (2015) Efavirenz treatment causes arterial stiffening in apolipoprotein E-null mice. J Biomech 48:2176–2180
28. Platt MO, Xing Y, Jo H, Yoganathan AP (2006) Cyclic pressure and shear stress regulate matrix metalloproteinases and cathepsin activity in porcine aortic valves. J Heart Valve Dis 15:622–629
29. Balachandran K, Sucosky P, Jo H, Yoganathan AP (2009) Elevated cyclic stretch alters matrix remodeling in aortic valve cusps: implications for degenerative aortic valve disease. Am J Physiol Heart Circ Physiol 296:H756–H764

Chapter 24

Transfer Zymography

Daniel Pan, Karl A. Wilson, and Anna Tan-Wilson

Abstract

The technique described here, transfer zymography, was developed to overcome two limitations of conventional zymography. When proteolytic enzymes are resolved by nonreducing SDS-PAGE into a polyacrylamide gel with copolymerized protein substrate, the presence of the protein substrate can result in anomalous, often slower, migration of the protease and an estimated mass higher than its actual mass. A further drawback is that the presence of a high background of substrate protein interferes with proteomic analysis of the protease band by excision, tryptic digestion, and LC-MS/MS analysis. In transfer zymography, the proteolytic enzymes are resolved by conventional nonreducing SDS-PAGE, without protein substrate in the gel. The proteins in the resolving gel are then electrophoretically transferred to a receiving gel that contains the protein substrate, by a process similar to western blotting. The receiving gel is then processed in a manner similar to conventional zymography. SDS is removed by Triton X-100 and incubated in conditions suitable for the proteolytic activity. After protein staining, followed by destaining, bands representing regions with active protease are visualized as clear bands in a darkly stained background. For proteomic analysis, electrophoresis is carried out simultaneously on a second resolving gel, and the bands corresponding to the clear regions in the receiving gel after zymogram development are excised for proteomic analysis.

Key words Zymography, Electrophoretic transfer, Proteolytic enzyme, Protease, Proteomics, Nonreducing SDS-PAGE

1 Introduction

Protein zymography, in which samples containing proteolytic enzymes are separated on nonreducing SDS-PAGE gels copolymerized with the protein substrate, has been very useful for the detection and characterization of proteolytic enzymes with respect to pH optimum and mechanistic class [1]. Following removal of the SDS with Triton X-100 [2] and incubation in buffers that favor protease activity, protein staining reveals clear bands where the protein substrate had been digested by active proteolytic enzymes. In this conventional method, however, the molecular mass of the protease, determined from the migration distance of the band in the zymogram, is often inaccurate, an anomaly that has been attributed

Jeff Wilkesman and Liliana Kurz (eds.), *Zymography: Methods and Protocols*, Methods in Molecular Biology, vol. 1626,
DOI 10.1007/978-1-4939-7111-4_24, © Springer Science+Business Media LLC 2017

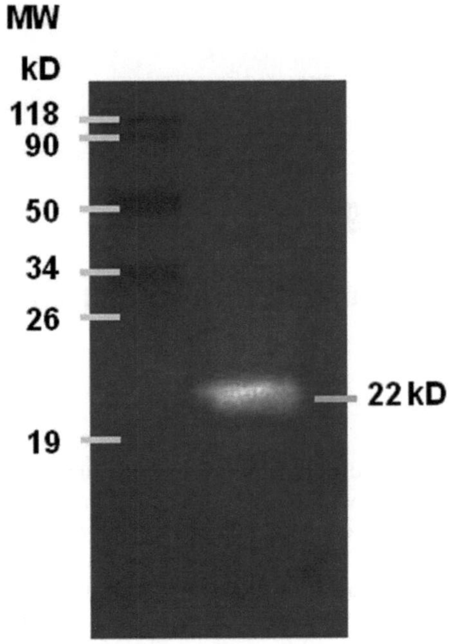

Fig. 1 Developed electrophoretic transfer zymogram. Bovine chymotrypsin (80 pg) on a resolving gel (15% T, 2.6% C) was transferred to a receiving gel (15% T, 2.6% C) containing 0.025% (w/v) gelatin for protein substrate. The zymogram was developed in 0.1 M Tris–HCl pH 8, at 37 °C with gentle agitation for 24 h. The *dark background* represents undigested protein substrate stained with Coomassie blue. The clear band marks the region where the protease from the resolving gel had transferred to the receiving gel, then digested the protein substrate

to the presence of the protein substrate in the gel [3]. Transfer zymography overcomes this limitation (reviewed in [4]). In this method, enzymes are first resolved by nonreducing SDS-PAGE in a gel without protein substrate. This is followed by the electrophoretic transfer of the proteins on the resolving gel to a receiving gel, the latter incorporating protein substrate; in a process that is very similar to protein blotting [5]. This was first applied toward detection of amylase [6], then adapted and improved for the study of proteolytic enzymes [7]. After the electrophoretic transfer, the receiving gel is incubated with buffer suitable for protease activity followed by protein staining. The presence of active proteolytic enzymes is manifested as clear bands in a dark background (Fig. 1). The positions of the bands correspond to the migration of the proteases in nonreducing SDS-PAGE, without the anomalies introduced by the incorporation of protein substrate in the resolving gel. This also allows for identification of the protease [7]. Aliquots of the same samples are resolved simultaneously on two SDS-PAGE gels. One gel is used to locate the active proteolytic enzymes by transfer zymography. Bands corresponding to positions of active protease are excised from the second gel for proteomic analysis.

This chapter includes detailed protocols for preparing the receiving and resolving gels, the electrophoretic transfer, and development of the zymogram, including steps required if proteomic analysis is to be carried out.

2 Materials

All solutions described here should be prepared with reagent-grade chemicals and purified water (resistance at least 13–16 MΩ cm). Powder-free gloves should be used.

2.1 Nonreducing SDS-PAGE Components

The electrophoretic system used is that of Laemmli [8].

1. Acrylamide solution: 40% Acrylamide/Bis solution 37.5:1 (Bio-Rad Laboratories, Hercules, CA). Alternatively, dissolve 38.96 g acrylamide and 1.04 g bis-acrylamide in water to a final volume of 100 mL. Store at 4 °C protected from light. Avoid inhalation of dust and contact with skin.
2. Running gel buffer: 1.5 M Tris–HCl, pH 8.8. Store at 4 °C.
3. Stacking gel buffer: 0.5 M Tris–HCl, pH 6.8. Store at 4 °C.
4. SDS solution: 10% (w/v) sodium dodecylsulfate (Bio-Rad Laboratories). Alternatively, dissolve 100.0 g SDS in water to a final volume of 1.0 L (avoid foaming). Store at room temperature. Avoid inhalation of the dust by using care in handling and by wearing a filter mask.
5. TEMED: (*N*,*N*,*N*′,*N*′-tetramethylethylenediamine). Store at room temperature.
6. Ammonium persulfate solution: 10% (w/v) ammonium persulfate. Dissolve 0.5 g ammonium persulfate in water to a final volume of 5.0 mL. Prepare just before using.
7. Electrophoresis tank buffer: 0.025 M Tris base, 0.192 M glycine, 0.1% (w/v) SDS, pH 8.3. Dissolve 12.1 g Tris base and 57.6 g glycine in water. Add 40.0 mL 10% (w/v) SDS solution, and adjust with water to a final volume of 4.0 L. Store at 4 °C. Do not reuse.
8. 2× Laemmli sample buffer (Bio-Rad Laboratories). Do *not* add 2-mercaptoethanol. Alternatively, mix together 2.5 mL stacking gel buffer, 4.0 mL SDS solution, 2.0 mL glycerol, and water to a final volume of 10.0 mL. Add 1.0 mg bromophenol blue and mix until dissolved. Store at room temperature.
9. Commercial prestained protein standards suitable for the mass range of the proteases of interest. Store at −20 °C.

2.2 Components for Electrophoretic Transfer

1. Transfer buffer: 0.025 M Tris base, 0.192 M glycine, pH 8.3. Dissolve 6.0 g Tris base and 28.8 g glycine in water to a final volume of 2.0 L. Store at 4 °C. Do not reuse.

2.3 Zymography Components

1. Protein substrate stock solution (*see* **Note 1**): In the procedure given here, we use 1.0% (w/v) EIA-grade gelatin (Bio-Rad Laboratories). Dissolve 0.25 g of the protein substrate in water to a final volume of 25.0 mL. Heat to dissolve, but do not boil. Cool to room temperature before using.
2. Triton X-100 solution: 2.5% (v/v) Triton X-100. Dissolve 25 mL Triton X-100 detergent in water to a final volume of 1.0 L. Store at room temperature.
3. Zymogram development buffer (*see* **Note 2**): This buffer would set the conditions for optimum activity of the protease(s) of interest.
4. Protein stain: 0.5% (w/v) Coomassie Brilliant Blue R-250 in 40% (v/v) methanol, 10% (v/v) acetic acid. Dissolve 5.0 g Coomassie Brilliant Blue R-250 in 0.4 L methanol, 0.1 L glacial acetic acid, and water to a final volume of 1.0 L. Store at room temperature in a tightly closed container.
5. Destain solution: 40% (v/v) methanol, 10% (v/v) acetic acid. Mix 1.6 L methanol, 0.4 L glacial acetic acid, and water to a final volume of 4.0 L. Store at room temperature in a tightly closed container.
6. Zymogram storage solution: 5% (v/v) methanol, 7% (v/v) acetic acid. Mix 5.0 mL methanol, 7.0 mL glacial acetic acid, and water to a final volume of 100.0 mL. Store at room temperature in a tightly closed container.

2.4 Electrophoresis and Electrophoretic Transfer Equipment

1. Vertical gel electrophoresis apparatus and power supply: This protocol was developed using the Mini-Protean Electrophoresis System (Bio-Rad Laboratories, Hercules, CA).
2. Equipment for casting and loading electrophoresis gels: gel casting stands, clamps, glass plates, spacers (or glass plates with integrated spacers), and gel combs to cast 1.5 mm thick gels in the mini-gel format (8×7.3 cm^2), 125 mL vacuum flask with stopper, water aspirator or small diaphragm vacuum pump, felt-tip pen with water-insoluble ink, 50 mL beaker of water and glass syringe with needle or other delivery device for layering water on top of poured gel, 50–100 μL glass micro-syringe or pipettor with long narrow tips for sample loading, paper towels, plastic wrap.
3. Tank transfer apparatus for performing western blots: This protocol was developed using the Mini Trans-Blot Electrophoretic Transfer Cell (Bio-Rad Laboratories).
4. Equipment for the transfer assembly. Transfer cassette compatible with the transfer cell, shallow pan to accommodate the open cassette, filter paper and fiber pads cut to the size of the gels.

2.5 Equipment for Development of Zymograms

1. Gel cutter, scalpel.
2. Plastic boxes with lids that seal. The size should allow for incubation of a mini-gel with shaking.
3. Shaking water bath set at the temperature for zymogram development and modified to secure the plastic boxes.

2.6 Solution for Proteomic Analysis

1. Dilute protein stain: 0.1% (w/v) Coomassie Brilliant Blue R-250 in 40% (v/v) methanol, 10% (v/v) acetic acid. Dissolve 0.1 g Coomassie Brilliant Blue R-250 in 40 mL methanol, 10 mL glacial acetic acid, and water to a final volume of 100 mL. Store at room temperature in a tightly closed container.

3 Methods

3.1 Selecting Conditions for Transfer Zymography

1. Select (1) the formulation of the running gel layer of the resolving gel, (2) the formulation of the receiving gel, and (3) the duration of the electrophoretic transfer step (*see* **Note 3**).
2. Select the protein substrate to be copolymerized with the acrylamide in the receiving gel (*see* **Note 1**).
3. Select the buffer to be used for development of the zymogram (*see* **Note 2**).
4. Plan the duration and temperature for zymogram development in accordance with the amounts of samples to be loaded in the gel lanes (*see* **Note 4**).

3.2 Casting the Receiving gel

1. To ensure complete copolymerization of the protein substrate with the acrylamide, cast the receiving gel at least 24 h before use. The materials listed in Subheading 2.1, **items 1**, **2**, **4–6**; Subheading 2.3, **item 1**, and Subheading 2.4, **items 1** and **2** will be needed. It is advisable to cast at least two gels (*see* **Note 5**).
2. Using thoroughly clean, dry, and dust-free glass plates, assemble the gel mold on the casting stand on a paper towel set on a level surface. Using a felt-tip laboratory pen with water-insoluble ink, place a mark on the outside surface of the glass plate 2 mm from the top edge of the glass plate. The gel solution will be poured to this mark.
3. Perform the following steps quickly (*see* **Note 6**). Work at room temperature. Wear powder-free gloves. Measure the appropriate volumes (Table 1) of the acrylamide solution, running gel buffer, SDS solution, water, and protein substrate stock solution. Mix with gentle swirling in the vacuum flask. Stopper the flask, and degas for 30 s to 1 min using a water aspirator or small diaphragm vacuum pump. While degassing, swirl contents gently, avoiding foaming. Disconnect the flask

Table 1
Formulation of the receiving gel solution with protein substrate[a]

	Total acrylamide monomer concentration[b]		
	10%(w/v)	12.5%(w/v)	15%(w/v)
Components	Volume (mL)		
Acrylamide solution	5.00	6.25	7.50
Resolving gel buffer	5.00	5.00	5.00
SDS solution	0.20	0.20	0.20
Water	7.60	6.35	5.10
Protein substrate stock solution	2.00	2.00	2.00
TEMED	0.01	0.01	0.01
Ammonium persulfate solution	0.20	0.20	0.20

[a]Makes 20 mL of gel solution, sufficient for two mini-gels
[b]All formulations are with 2.6% cross-linker

and add the TEMED and ammonium persulfate solution. Swirl gently to mix and immediately pour into the gel casting mold to the level of the felt-tip pen mark. Carefully lower a gel comb into the gel solution at a slant, slowly leveling out to avoid trapping of air bubbles (*see* **Note 7**). Some overflow of the gel solution may occur. Avoid skin contact, especially with the solutions containing acrylamide and bis-acrylamide.

4. Leave the gels undisturbed until the gel is well-polymerized, as indicated by a very distinct interface between the gel and the comb. Wrap with damp paper towels and seal tightly with plastic wrap; store at 4 °C.

3.3 Casting the Resolving Gel (See Note 8)

1. This gel will be poured in two layers—the running gel, then the stacking gel layer. It is advisable to cast two resolving gels (*see* **Note 5**). Gather together all the materials listed in Subheading 2.1, **items 1–6** and Subheading 2.4, **items 1** and **2**.
2. Use a gel mold of the same size as the one used to cast the receiving gels. Select the gel comb according to the number of wells required. After assembling the mold, position the comb between the glass plates; then using the felt-tip pen, mark the outer surface of the glass plate 1 cm below the teeth of the comb. The running gel layer will be poured to this level. Place another mark 2 mm below the top edge of the glass plate. The stacking gel layer will be poured to this level. Remove the comb.

3. While casting the resolving gel, wear powder-free gloves. Mix the components for the running gel layer (*see* **Note 9**), following the procedure in Subheading 3.2, **step 3**, but use the solutions and volumes in Table 2 and pour upto the lower mark. Immediately use the syringe and needle to carefully layer water several mm deep on top of the gel solution. Avoid mixing the water with the underlying gel. Leave undisturbed until the gel has polymerized, as indicated by a distinct line between the water layer and the gel (~1 h).
4. Remove the water overlay by tilting the gel assembly over a beaker and using a small piece of filter paper to soak up the water that clings to the glass plates. Use the solutions and volumes in Table 3 (*see* **Note 10**) to prepare the stacking gel solution. Mix and degas as before (Subheading 3.2, **step 3**)

Table 2
Formulation of the running gel portion of the resolving gel[a]

	Total acrylamide monomer concentration[b]		
	10%(w/v)	12.5%(w/v)	15%(w/v)
Components	Volumen (mL)		
Acrylamide solution	5.00	6.25	7.50
Resolving gel buffer	5.00	5.00	5.00
SDS Solution	0.20	0.20	0.20
Water	9.60	8.35	7.10
TEMED	0.01	0.01	0.01
Ammonium persulfate solution	0.20	0.20	0.20

[a]Yields 20 mL of gel solution, sufficient for two mini-gels
[b]All formulations are with 2.6% cross-linker

Table 3
Formulation of the stacking gel solution[a]

Components	Volumen (mL)
Acrylamide solution	1.00
Stacking gel buffer	2.50
SDS solution	0.10
Water	6.40
TEMED	0.005
Ammonium persulfate solution	0.050

[a]4% (w/v) total acrylamide monomer, 2.6% cross-linker

and pour upto the higher mark. Immediately insert the comb carefully into the stacking gel solution first on a slant and straightening out later. If there are trapped air bubbles between the comb and the gel solution, maneuver the comb to a slanting position and lower again (*see* **Note 11**).

5. Leave undisturbed until polymerization is indicated by a distinct interface between the comb and the stacking gel layer (~1 h). Wrap the gels (preferably still on the gel casting stand) with damp paper towels and plastic wrap. Store at 4 °C.

3.4 Preparation and Loading of Samples on the Resolving Gel

1. Mix each sample with an equal volume of 2× Laemmli sample buffer without 2-mercaptoethanol or other reducing agent. These samples should not be heat-treated. Samples should be kept cold (*see* **Note 12**).
2. Prepare prestained molecular mass standards for at least one lane of each gel according to the manufacturer's instructions, using the highest suggested amount. Do not add reducing agent. Do not heat. If possible, load the standards in wells in an asymmetric pattern from left to right or mark the glass plate at the side closest to the leftmost well (*see* **Note 13**).

3.5 Electrophoresis of the Resolving Gel

1. This procedure must be carried out at 4 °C.
2. Set the cold resolving gel into the electrophoresis apparatus. Carefully remove the comb. Add cold electrophoresis tank buffer to the top and bottom reservoirs of the apparatus. Use a Pasteur pipette to gently flush the wells with tank buffer (*see* **Note 14**). Use a micropipette or micro-syringe (*see* **Note 15**) to deliver samples under the buffer in the wells. Add a similar volume of 1X Laemmli sample buffer to wells without sample.
3. Connect the electrophoresis apparatus to the power supply, and set the power supply to 87 V. Run at constant voltage until the bromophenol blue approaches the bottom of the gel (~4 h) (*see* **Notes 16** and **17**). Turn off, then disconnect the power supply, and retrieve the resolving gel. Follow the equipment manufacturer's instructions and heed all precautions.
4. While waiting, prepare for zymogram development by bringing the shaking water bath to the desired temperature and equilibrate the zymogram development buffer to this temperature.

3.6 Electrophoretic Transfer

1. The assembly of the sandwich and the electrophoretic transfer is done at 4 °C (*see* **Note 12**).
2. Wear powder-free gloves throughout. Use a gel cutter or spatula to separate and discard the stacking gel layer. Carefully transfer the running gel layer to cold transfer buffer in a plastic container. Agitate the submerged gel gently on an orbital shaker for 10 min at 4 °C. Repeat with fresh cold transfer buffer for another 10 min.

3. Unwrap the receiving gel. In a separate container, gently agitate the receiving gel in cold transfer buffer (2 × 10 min) at 4 °C.
4. After equilibration, leave both gels submerged in cold transfer buffer. Also soak, in cold transfer buffer, the fiber pads (*see* **Note 18**) and filter paper sheets that have been cut to the size of the sandwich. Continue working at 4 °C. Lay the open cassette in a large shallow pan and pour cold transfer buffer to about 1 cm above the cassette. The components for the sandwich will be assembled in a specific order (*see* Fig. 2). On top of the part of the cassette that will be nearest to the anode, lay one wet fiber pad, then one sheet of wet filter paper. Next position the receiving gel, followed by the resolving gel, aligning the lower edges of the two gels (*see* **Notes 19** and **20**). Lay one sheet of wet filter paper on top of the gels, followed by another fiber pad. After addition of each component, readjust if needed to release trapped air bubbles (*see* **Note 21**), then add just enough cold transfer buffer to keep the buffer 1 cm above the assembly. Working carefully so as not to shift the stacked components, close, and latch the cassette.
5. Continue to work at 4 °C. Position the sandwich in the electrophoretic transfer apparatus. Check to make sure that the receiving gel is closer to the anode and the resolving gel is closer to the cathode. Use a stirring motor and magnetic stirring bar in the tank to keep the transfer buffer circulating. Set the apparatus to run at constant voltage of 15 V per cm

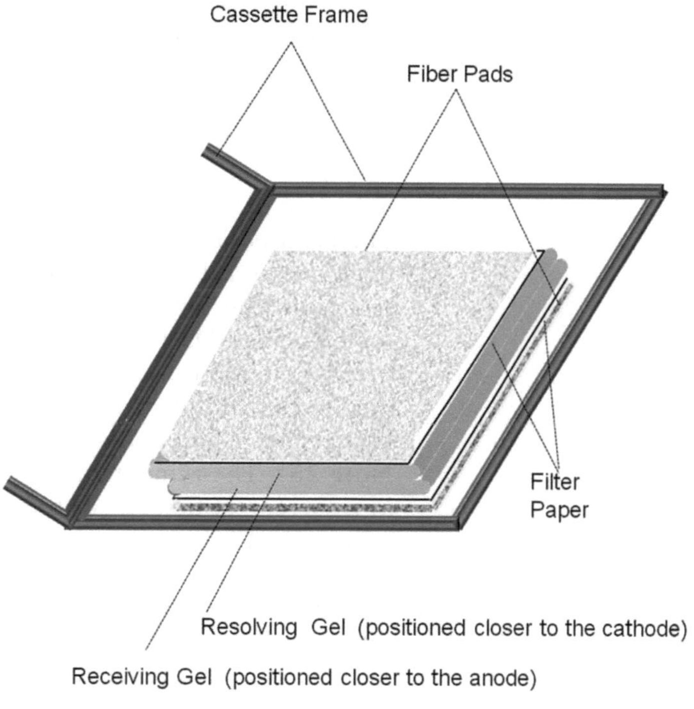

Fig. 2 Assembly of the electrophoretic transfer sandwich

between electrodes (60 V for the Bio-Rad mini trans-blot electrophoretic transfer cell) for the predetermined length of time. The same voltage setting is used whether one or two cassettes are set up. When transfer time is reached, turn off the electric current, disconnect the power supply, and remove the cassettes. Follow the equipment manufacturer's instructions and heed all precautions.

3.7 Zymogram Development

1. At the end of the planned transfer time, turn off and disconnect the power supply. Open the electrophoretic transfer cassette. The prestained protein molecular mass standards should be visible on the receiving gel and not on the resolving gel (*see* **Note 22**). Notch the edge of the receiving gel with a scalpel to mark the positions of the molecular mass standards (*see* **Note 23**).
2. Immerse the receiving gel in Triton X-100 solution in plastic container with a tight-fitting lid, using one container for each gel. With an orbital shaker, gently agitate the containers for 15 min at room temperature. Change to fresh Triton X-100 solution and repeat for another 15 min.
3. Remove the Triton X-100 solution and immerse the gel for 20 min in the temperature-equilibrated zymogram development buffer, shaking in the water bath for 20 min. Repeat with fresh buffer for another 20 min. Replace with fresh buffer and incubate in the water bath for the predesignated temperature and time period.
4. Immerse the receiving gel in protein stain. Agitate gently overnight at room temperature. Follow with destaining solution, changing periodically at regular intervals until clear bands appear in the dark background. To be consistent, measure the volumes of the solutions and maintain the same level of agitation. Scan the zymogram or capture the image with a camera. The band positions on the receiving gel will be at the same distances traveled by the proteases in the resolving gel; therefore, molecular mass can be determined to the accuracy possible on nonreducing SDS-PAGE (*see* **Note 24**). For quantification of band intensities, set parameters in the image analysis program such that the intensities in the region of the clear bands have higher values than the intensity of the background (*see* **Note 25**). The zymograms can be kept for future reference in zippered plastic bags with 3–4 mL of zymogram storage solution at 4 °C.

3.8 Protocol for Proteomic Analysis

1. Cast and run two identical resolving gels at the same time in the same electrophoresis apparatus.
2. After electrophoresis of the samples (*see* **Note 26**), immerse one of the gels (gel P) in dilute protein stain for proteomic analysis.

Stain overnight. Follow with destaining until dark bands are visible against a clear background.

3. At the same time, prepare the other resolving gel (gel T) for electrophoretic transfer of proteins to a receiving gel (gel Z) as described (Subheading 3.6). Develop the zymogram as described (Subheading 3.7) to visualize the clear bands of active protease against a dark background.
4. Lay gel P and the developed gel Z side by side. On gel P, locate and excise the bands that correspond to the active protease bands on gel Z, then carry out in-gel trypsin digestion and LC-MS/MS analysis.

4 Notes

1. The ideal protein substrate in conventional zymography is one that copolymerizes well with the acrylamide and does not migrate during the electrophoretic run. If the protein substrate migrates, the developed zymogram will have zones of light background stain indicating where the gel has been depleted of protein substrate and dark background stain where there has been no net loss of protein substrate [9]. In transfer zymography, protein substrate that migrates in the electric field will do so uniformly out of the gel; therefore, zones with different protein substrate concentrations do not form. This should allow for experimentation beyond the use of a heterologous substrate such as gelatin, to include proteins that serve as actual physiologically relevant substrates of the proteolytic enzymes studied. The concentration of the protein substrate is another parameter that can be varied. A concentration of 0.1% (w/v) is commonly used; slight reduction in concentration can expand the lower limits of detection of protease activity.
2. For the zymogram development buffer, select one that promotes activity of the proteolytic enzyme(s) of interest. Manipulating pH allows for the detection of different proteases in the same sample [10]. An initial study done with traditional enzyme assays or conventional zymography is recommended to determine the buffer composition that is optimum for proteolytic activity.
3. Select the formulation of the running gel layer of the resolving gel so as to achieve resolution of the proteolytic enzymes of interest, while also allowing for efficient electrophoretic transfer of proteins out of the resolving gel. For a given percentage of cross-linker [(g bis-acrylamide/(g acrylamide + g bis-acrylamide)) × 100], e.g., 2.6% C as in the protocol given, proteins of a given size transfer out more readily with lower

total acrylamide monomer concentration (% T) [7], when gel pore size is larger [11]. Select the formulation of the receiving gel and the duration for electrophoretic transfer with a view toward efficient entry of proteins into the receiving gel, a process that is also more rapid with lower % T. This has to be counter-balanced with the understanding that proteins can pass through, then out of the receiving gel, a process that is minimized in gels with higher total acrylamide monomer concentration. These conditions can be determined by analyzing the efficiency by which prestained protein mass standards are transferred. This is illustrated for transfer out of a resolving gel with the formulation 12.5% T, 2.6% C into receiving gels that were either of the same formulation or that were 15% T, 2.6% C (Fig. 3). Band intensities of transferred proteins of different masses were plotted as a function of transfer time. To illustrate use of such data, for analysis of proteolytic enzymes expected to have masses of 25–50 kDa, we would prepare receiving gels with the formulation 15% T, 2.6% C and set a transfer time of 60 min. If proteolytic enzymes of interest range in size from 20 to 200 kDa, we would use two sets of transfer conditions, one for enzymes of lower mass and another for enzymes of higher mass.

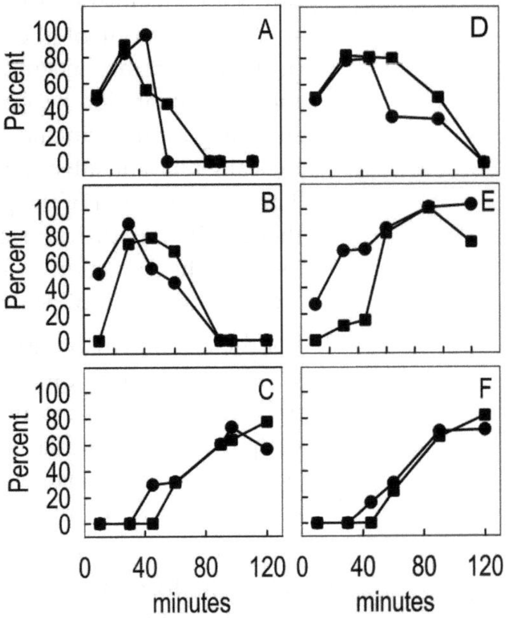

Fig. 3 Relative amounts of proteins of different molecular mass transferred as a function of protein mass, gel formulation, and transfer time. The resolving gel was 12.5% T, 2.6% C. The receiving gel was (**a**)–(**c**): 12.5% T, 2.6% C; (**d**)–(**f**): 15% T, 2.6% C. Molecular masses: (**a**) and (**d**): 19 kDa (*circle*), 26 kDa (*square*); (**b**) and (**e**): 34 kDa (*circle*), 50 kDa (*square*); (**c**) and (**f**): 90 kDa (*circle*), 118 kDa (*square*). Adapted from [7] with permission from Elsevier

4. The reaction occurs more rapidly with higher temperature so long as the protease is not denatured. A longer time period for zymogram development increases sensitivity so that less abundant proteases can be detected [12]; however, the temperature and time should be selected such that the band intensities are linear with respect to the amount of protease present in the sample under the specific conditions for zymogram development [13]. The upper limits can be determined by loading increasing amounts of the same sample in different wells of the resolving gel, then quantifying the intensities of the band after transfer and zymogram development (Fig. 4). The lower limit of detection for transfer zymography compared to conventional zymography depends upon two factors [7]. The sensitivity is decreased if the proteolytic enzyme passes through the receiving gel. On the other hand, the absence of protein substrate in the resolving gel reduces the loss of protease due to protein substrate binding, usually manifested as a streak above the band in conventional zymography. Schedules should also be considered

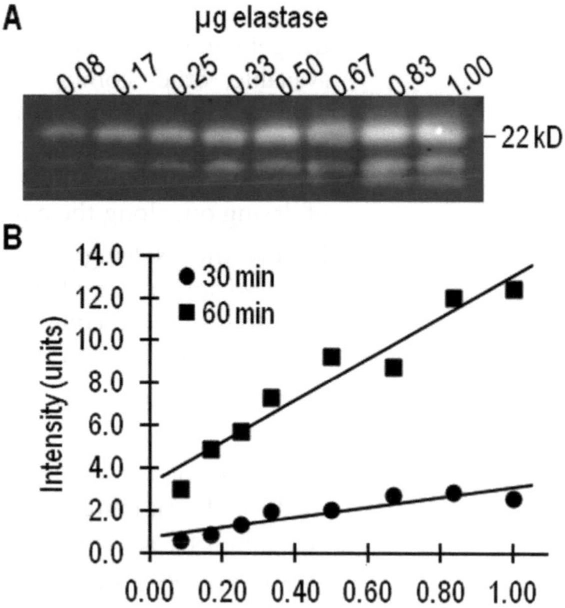

Fig. 4 Linear relationship between band intensity and amount of protease. (**a**) Transfer zymogram showing porcine elastase after 60 min transfer from the resolving gel (15% T, 2.6% C) to a receiving gel (15% T, 2.6% C) with 0.1% (w/v) casein. Different amounts of elastase solution (µg) with specific activity of 4 µmol *N*-succinyl-Ala-Ala-*p*-nitroanilide hydrolyzed per min per mg were loaded in each lane. The receiving gel was incubated for 24 h at 37 °C in 0.1 M Tris–HCl pH 8.0. (**b**) Band intensities of the major 22 kDa band in the transfer zymogram after transfer for 30 and 60 min are plotted vs. amount of elastase loaded on the lanes. Adapted from [7] with permission from Elsevier

in planning the conditions for zymography development. The procedure from loading samples onto the resolving gel to the start of zymogram development takes 6–7 h; therefore, it would be practical to set conditions so that zymogram development occurs overnight.

5. We routinely run two resolving gels and prepare two receiving gels, in case one or the other gel tears during the assembly of the transfer sandwich. In the event that two pairs of gels are set up successfully in the transfer cassette, the second set can serve as a duplicate or used to vary parameters such as time for transfer or a condition for zymogram development.
6. To avoid polymerization before the gel solution has been poured into the mold, gel casting steps must be performed quickly and without pausing between steps until the gel comb is in place. This is especially important once TEMED and the ammonium persulfate solution have been added.
7. The comb minimizes contact of the gel solution with oxygen, which tends to slow down polymerization; and along with the damp paper towels and plastic wrap, to prevent desiccation of the gel before use. A one-tooth comb is preferable; but if not available, other combs or a water overlay as described in Subheading 3.3, **step 3** can be used. In the latter situation, the receiving gels must be stored in an upright position. The gels can be stored for about 2–4 days, so long as there are no signs of drying out along the edges.
8. Because the resolving gel must be cold before use, and because the receiving and resolving gels are made using many of the same solutions and equipment, we routinely cast the receiving and resolving gels at the same time.
9. Protein substrate stock solution will not be needed for the resolving gel.
10. Regardless of the total acrylamide monomer concentration used for the running gel layer, the stacking gel is always 4% (w/v) in total acrylamide monomer concentration.
11. In addition to preventing slowing down of polymerization by oxygen, the elimination of air bubbles will ensure that the bottom of the wells are flat so that the bands will have straight edges.
12. The samples to be analyzed should be in buffers that are conducive to keeping the proteolytic enzymes active. If it is necessary to freeze the samples, it is advisable to determine the conditions for optimal retention of activity and freeze in aliquots so that samples undergo only one freeze–thaw cycle. Once thawed, the solutions should be kept cold to prevent autolysis or digestion by other proteases in the sample. Furthermore, because the proteolytic enzymes need to be active for zymogram

development, all steps from loading the samples into a cold resolving gel, through electrophoresis, and electrophoretic transfer to a cold receiving gel are carried out at 4 °C.

13. Keeping track of the left → right orientation of the resolving and receiving gels can be challenging. One way to do this is to always load the prestained molecular mass standards in the leftmost well. If it is important to have molecular mass standards at the two extreme ends, the following steps will make it easy to keep track of the orientation: Mark the glass plate close to the leftmost sample with a felt-tip pen. Cut off the lower left hand corner of the resolving gel after electrophoresis. After the transfer process, before separating the gels, cut off the corner of the receiving gel adjacent to the cut corner of the receiving gel.
14. This removes traces of partially polymerized gel and ensures that the bottom of the wells will be flat. Otherwise, the bands may have a wavy perimeter.
15. It is easier to control sample loading with a micro-syringe; but the syringe has to be flushed at least five times with water between samples.
16. It takes longer to complete the electrophoresis at 4 °C than at room temperature.
17. Electrophoretic transfer must be carried out immediately after electrophoresis of the resolving gel. If the operator's schedule does not allow that, the voltage could be set lower (30 V) so that electrophoresis proceeds overnight (~10 h), postponing electrophoretic transfer and zymography development to the next day.
18. Transfer cassettes are designed to be used with one gel and a thin membrane. To accommodate the thickness of the second gel, we use fiber pads that have been compressed through repeated previous use.
19. Gels tend to tear when they stick to gloves. To prevent this, wet the fingertips of the gloves with buffer before handling the gels.
20. If the tendency of two wet gels to slide past each other makes assembly of the transfer sandwich too difficult, equilibrate only the resolving gel in transfer buffer. After setting up the cassette with fiber pad and filter paper, position the resolving and receiving gels against each other outside of the buffer in the pan, then place them simultaneously on top of the filter paper, making sure that the resolving and receiving gels will end up closer to the cathode and anode, respectively. Continue adding filter paper and fiber pad, removing bubbles, and closing up the cassette. After setting up in the transfer apparatus, initially set the power supply to run at 110 V. As transfer buffer passes through the gel that was not equilibrated to transfer

buffer (~5–10 min), reset the voltage to 60 V and run at constant voltage.

21. Rolling a test tube or gently pressing the side of a gloved finger along the surface can help to release trapped air bubbles. Take care not to shift the components in the stack.
22. Protein staining of the resolving gel after electrophoretic transfer can verify that the proteins have been transferred out.
23. We mark the mass standards with notches on the receiving gel after transfer because the bands can be difficult to detect against the dark background of the developed zymogram after staining.
24. Although the molecular mass determined by transfer zymography is more accurate than the mass determined by conventional protein zymography, it should be noted that in order to preserve protease activity, disulfide bonds on the proteins are not reduced and the samples are not heat-treated. Under such conditions, the proteolytic enzymes might not be fully denatured, and thus might not bind SDS to attain a uniform charge density, resulting in slight deviations from the mass that would be determined by reducing SDS-PAGE with heat treatment.
25. Interpretation of changes in intensities of bands should be confined to bands on the same zymogram and bands that exhibit the same mobility.
26. These samples would contain the protease(s) to be identified and, ideally, would constitute a partially purified sample with a minimum amount of other proteins.

References

1. Heussen C, Dowdle EB (1980) Electrophoretic analysis of plasminogen activators in polyacrylamide gels containing sodium dodecyl sulfate and copolymerized substrates. Anal Biochem 102:196–202
2. Lee LT, Deas JE, Howe C (1978) Removal of unbound sodium dodecyl sulfate (SDS) from proteins in solution by electrophoresis through Triton X-100-agarose. J Immunol Methods 19:69–75
3. Hummel KM, Penhelter AR, Gathman AC, Lilly WW (1996) Anomalous estimation of protease molecular weights using gelatin-containing SDS-PAGE. Anal Biochem 233:140–142
4. Wilkesman J, Kurz L (2012) Advances in zymography techniques and patents regarding protease analysis. Recent Pat Biotechnol 6:106–114
5. Towbin H, Staehlin T, Gordon J (1979) Electrophoretic transfer of proteins from polyacrylamide gels to nitrocellulose sheets: procedure and some applications. Proc Natl Acad Sci U S A 76:4350–4354
6. Steup M, Gerbling KP (1983) Multiple forms of amylase in leaf extracts: electrophoretic transfer of the enzyme forms into amylose-containing polyacrylamide gels. Anal Biochem 184:96–100
7. Pan D, Hill AP, Kashou A, Wilson KA, Tan-Wilson A (2011) Electrophoretic transfer protein zymography. Anal Biochem 411: 277–283
8. Laemmli UK (1970) Cleavage of structural proteins during the assembly of the head of bacteriophage T4. Nature 227:680–685
9. Fernandez-Resa P, Mira E, Quesada A (1995) Enhanced detection of casein zymography of matrix metalloproteinases. Anal Biochem 224: 434–435

10. Wilder CL, Park KY, Keegan PM, Platt MO (2011) Manipulating substrate and pH in zymography protocols selectively distinguishes cathepsins K, L, S, and V activity in cells and tissues. Arch Biochem Biophys 516:52–57
11. Chrambach A, Rodbard D (1971) Polyacrylamide gel electrophoresis. Science 172:440–451
12. Lantz MS, Ciborowski P (1994) Zymographic techniques for detection and characterization of microbial proteases. Methods Enzymol 235:563–594
13. Kleiner DE, Stetler-Stevenson WG (1994) Quantitative zymography: detection of picogram quantities of gelatinases. Anal Biochem 218:325–329

Chapter 25

Sequential Detection of Thermophilic Lipase and Protease by Zymography

Liliana Kurz, Zully Hernández, Lellys M. Contreras, and Jeff Wilkesman

Abstract

Lipase and protease present in cell-free fractions of thermophilic *Bacillus* sp. cultures were analyzed by polyacrylamide gel (PAG) electrophoresis. After run, the gel is electrotransferred to another PAG copolymerized with glycerol tributyrate, olive oil, and gelatin. This multi-substrate gel was incubated first for lipase detection, until bands appeared, and then stained with Coomassie for protease detection. Advantages of this sequential procedure are the detection of two different enzyme activities on a single PAG, beside time and resource saving.

Key words Lipase, Protease, Sequential zymography, Thermophiles

1 Introduction

When performing zymography, normally, only one substrate is copolymerized in the gel-matrix for the electrophoresis. Because of the different steps involved in sample treatment when submitted to electrophoresis, the enzyme must be reactivated afterwards, generally by incubating the gel in a specific buffer. The enzyme will then degrade the copolymerized substrate, and depending on the catalytic reaction executed by the enzyme, a specific staining method must be applied to visualize the process. Here, we focus on two types of hydrolysis. In one hand, we have lipases/esterases (EC 3.1.1.x), which when assayed under proper incubation conditions, bands are seen directly on the gel [1, 2]. On the other hand, for proteases (EC 3.4.x.x), translucent bands are visualized under a deep blue background when staining with Coomassie blue [3]. The detection of these enzyme activities has been commonly performed separately. The method described here allows to detect both enzyme activities in only one electrophoresis gel [4]. Therefore, the enzymatic activities to be tested are submitted to polyacrylamide gel electrophoresis (PAGE) and then the gel is electrotransferred to a different polyacrylamide gel, containing the substrates.

Jeff Wilkesman and Liliana Kurz (eds.), *Zymography: Methods and Protocols*, Methods in Molecular Biology, vol. 1626,
DOI 10.1007/978-1-4939-7111-4_25,

As substrate, a mixture of glycerol tributyrate, olive oil, and gelatin is used for the sequential detection of lipase and protease. Following electrotransference, the gel is incubated in an activation buffer containing Tris–HCl and NaCl, until bands are visualized as clear bands under a grayish background, corresponding to lipolytic activity. At this stage, the image must be scanned or photographed, as the next staining step will cover this result. Once Coomassie blue is applied over the zymogram, proteolytic activity is seen as clear bands with a deep blue background.

Electrotransference has been used before in PAGE under nondenaturing conditions for lipase analysis [5]. Enzymes were transferred to a nitrocellulose membrane and then detected after reaction with 1-naphtolpalmitate. The use of electrotransference for zymographical purposes has also been published elsewhere [6].

2 Materials

2.1 Special Reagents and Equipment

1. Enzyme controls: commercial trypsin (e.g., from porcine pancreas) and lipase.
2. Centrifuge, electrophoresis equipment, electrotransference unit.

2.2 Sample Preparation

1. Sample resuspension buffer: 20 mM Tris–HCl pH 8.

2.3 Electrophoresis

1. Resolving gels without substrate: 0.375 M Tris–HCl, pH 8.8, 12% acrylamide/bis-acrylamide (29:1), 0.1% SDS, 0.05% ammonium persulfate (APS), 0.005% TEMED.
2. Resolving gels with substrate: 0.375 M Tris–HCl, pH 8.8, 12% acrylamide/bis-acrylamide (29:1), 0.1% SDS, tributyrate 2%, olive oil 0.67%, 0.05% ammonium persulfate (APS), 0.005% TEMED.
3. Stacking gels: 0.125 M Tris–HCl pH 6.8, 4% acrylamide/bis-acrylamide (29:1), 0.1% SDS, 0.05% APS, 0.001% TEMED.
4. Sample buffer under denaturing conditions (1×): 24 mM Tris–HCl (pH 6.8), 0.8% (w/v) SDS, 10% (v/v) glycerol, and 0.06% (w/v) bromophenol blue.
5. Running buffer without SDS: 25 mM Tris, 192 mM glycine pH 8.8.

2.4 Silver-Coomassie Staining

1. Fixing solution: 50% (v/v) methanol–10% (v/v) acetic acid.
2. Coomassie-TCA staining solution: 0.25% (w/v) Coomassie Blue G-250, 50% (v/v) methanol, 12.5% (v/v) trichloroacetic acid.
3. Decoloring solution: 50% (v/v) methanol.

4. Incubation solution A: 5% (v/v) methanol, 10% (v/v) acetic acid.
5. Incubation solution B: 30% (v/v) ethanol.
6. Silver solution: 0.1% (w/v) silver nitrate (fresh prepared).
7. Developing solution: 3.75% (w/v) sodium carbonate, 0.03% (v/v) formaldehyde (fresh prepared).
8. Stop solution: 1% (v/v) acetic acid.

2.5 Electro-transference

1. Electrotransference buffer: same running buffer without SDS: 25 mM Tris, 192 mM glycine, final pH 8.8 (without adjustment).

2.6 Zymography

1. Activation buffer: Tris–HCl 50 mM, NaCl 25 mM, pH 8.8.
2. Coomassie staining solution: 0.1% (w/v) Coomassie G-250, 45% (v/v) methanol, 10% (v/v) acetic acid.
3. Destaining solution: 45% (v/v) methanol, 10% (v/v) acetic acid.

3 Methods

3.1 Biological Material

1. Prepare a protein extract containing the lipase and protease activity (*see* **Note 1**).
2. In the case of bacterial thermophilic cultures, incubate at 55 °C with constant agitation.
3. After maximal protein expression is achieved (*see* **Note 2**), samples are centrifuged at 9500 × *g* for 30 min at 4 °C.
4. Supernatant is decanted and saved, while pellet is resuspended in 2 mL 20 mM Tris–HCl pH 8, then sonicated and centrifuged at 8160 × *g* for 30 min.
5. Supernatants are ten times heat-concentrated at 60 °C. All fractions are kept at 4 °C until assay.
6. As controls, commercial trypsin from porcine pancreas and lipase can be used.

3.2 Electrophoresis

1. PAGE was performed according to Laemmli [7]. Table 1 shows the amounts used for the resolving and stacking gels enough to prepare one gel.
2. Commercial lipase (~5 μg) and protease (~3 μg) are dissolved separately in sample buffer under denaturing conditions.
3. Micrograms amounts of cell-free extracts from the thermophilic cultures were assayed under same conditions.
4. Two 12% gels are run at 100 V for 90 min at 4 °C (*see* **Note 3**).

Table 1
Volumes (mL) for preparation of resolving and stacking gel

Solution	12% Resolving gel	4.5% Stacking gel
Distilled water		
1.5 M Tris–HCl pH 8.8	1.00	–
0.5 M Tris–HCl pH 6.8	–	0.50
Acrylamide-bisacrylamide solution	1.60	0.30
APS solution	0.04	0.015
TEMED	0.01	0.005

5. After the run, one gel is stained by the silver-Coomassie protocol (*see* **Note 4**). The other gel is submitted to electrotransference.

3.3 Silver-Coomassie Staining

1. The Silver-Coomassie staining method according to De Moreno et al. [8] was used (*see* **Note 4**).
2. Place gel in fixing solution for at least 30 min. Discard solution appropriately.
3. Now incubate in Coomassie-TCA staining solution for at least 1 h with gentle shaking. Discard solution appropriately.
4. Wash three times with decoloring solution, 15 min each. Discard solution appropriately.
5. Add incubation solution A and shake gently for 15 min. Discard solution appropriately.
6. Add incubation solution B and shake gently for 15 min. Discard solution appropriately.
7. Wash four times with ddH_2O, 5 min each.
8. Add silver solution, place in the dark and leave still for 45 min. Remove solution and discard appropriately.
9. Rinse briefly with water (~15 s).
10. Add developing solution and shake gently until bands appear (generally 5–10 min).
11. Stop reaction by adding stop solution.
12. Gel may be kept in stop solution until its final destination (scanning, drying, or discarding).

3.4 Electro-transference

1. Prepare the substrate gel. The first ingredients of the mixture—glycerol, olive oil, water, and acrylamide solution—are first sonicated (5 pulses, 18% amplitude) and then gelatin, APS, and TEMED are added (Table 2).

Table 2
Preparation of the substrate mixture gel for zymography (12%)

Component	Volume (mL)
Water (dd)	2.7
Acrylamide/bis-acrylamide solution	2.4
Glycerol tributyrate	0.12
Olive oil	0.04
Sonication	
2% Gelatin	0.700
10% APS	0.060
TEMED	0.003

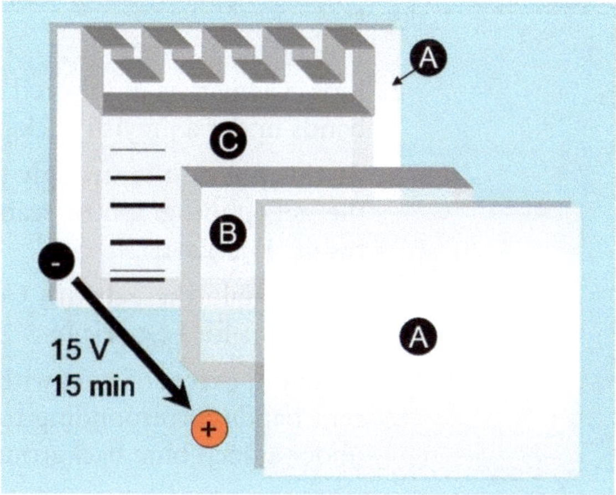

Fig. 1 General scheme for the electrotransference setup. (**A**) Cellulose paper sheet. (**B**) Mixed substrate PAG. (**C**) PAG containing the enzyme activities to be tested, previously run. Gel (**C**) is not stained. Protein bands seen is only in the case a prestained standard was used. Note direction of the current from the negative electrode towards the positive electrode. Optimal electrotransference times oscilate between 15 and 30 min

2. Gel is cast correspondingly and assembled into the sandwich (Fig. 1, *see* **Note 5**).
3. The proteins contained in the electrophoretic gels are ready to be electrotransferred to the substrate gel (containing the substrate mixture). Set equipment at 15 V to electrotransfere for 25 min at 4 °C, using the same running buffer (*see* **Note 5**).

3.5 Zymography

1. After transference, incubate the substrate gel for 24 h at 55 °C in activation buffer (Tris–HCl 50 mM, NaCl 25 mM, pH 8.8) (*see* **Note 6**).

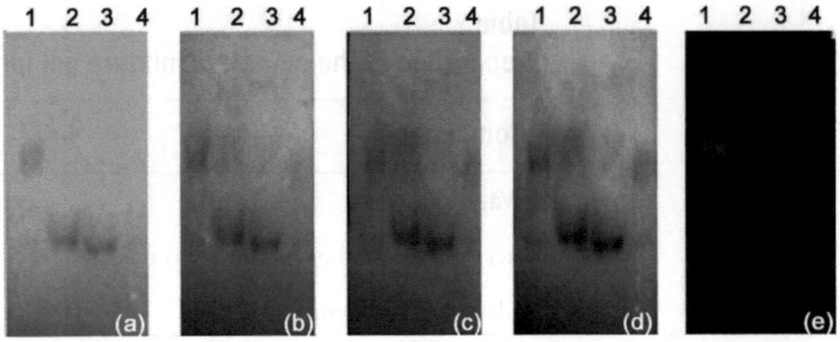

Fig. 2 Zymogram (12%) by electrotransference using glycerol tributyrate, olive oil, and gelatin as substrates. Thermophilic strains used as sample were culture under different media, and intra- and extracelullar fractions were tested. Lanes: (*1*) intracellular fraction [medium 1] (106 μg); (*2*) extracellular fraction [medium 1] (35 μg); (*3*) extracellular fraction [medium 2] (35 μg); (*4*) extracellular fraction [medium 3] (5 μg). Sequential detection of the enzymatic activities was performed First (**a**–**d**), lipase activity was visualized after: (**a**) 2 h, (**b**) 3 h; (**c**) 21 h; and (**d**) 24 h incubation in activation buffer. (**e**) Finally, the gel was stained with Coomassie for protease detection

2. Lipase activity is monitored for 24 h until optimal translucent bands under a grayish background are seen (*see* **Note 7**).
3. At this point, it is strongly suggested to capture an image of the gel (photograph or scan), as the next step will cover this result (Fig. 2a–d).
4. Finally, stain the gel with Coomassie staining solution. Leave overnight for best results.
5. Wash out excess of stain with destaining solution until translucent bands corresponding to proteolytic activity are visualized under a deep blue background (Fig. 2e).

4 Notes

1. In our case, strains from thermophilic *Bacillus* sp. were used. It is strongly recommended to test for enzyme activity with an alternative method (photometry, fluorescence spectroscopy, etc.) before performing zymography. SDS is reported to be a lipase inhibitor [9]. Its presence must be separately tested over lipase activity.
2. Culture time must be previously determined, in order to determine when maximal growth and/or maximal protein expression occurs.
3. Voltage and times used vary among the literature. The best is to run control experiments in order to determine the best parameters. We have found that by fixing voltage at 100 V and placing the electrophoresis chamber inside the fridge, the run took 90 min until the blue front reached less than 1 cm before

the end of the gel. Please notice that electrophoresis is done under nonreducing conditions, i.e., 2-mercaptoethanol is absent, and samples are not boiled. However, denaturing conditions prevail, as SDS is present in the sample and running buffer.

4. Many other staining techniques are published [10]. The one reported here will not be used anymore in our laboratory due to ecological reasons. Methanol and formaldehyde are toxic for the environment. Its use must be avoided when possible.
5. Consult manufacturer's indication according to the model available in your lab. It is recommended to use 15–20 V, for 15–30 min. Optimal conditions must be tested separately.
6. Incubation time may vary. Bands can be seen already after 2 h incubation (Fig. 2).
7. If band visualization is not possible under the naked eye, use a UV lamp as illumination source.

Acknowledgments

This work was supported by the *Consejo de Desarrollo Científico y Humanístico* of the University of Carabobo (CDCH-UAC 373-2010) and was benefited by the Venezuelan Law of Science and Technology (LOCTI) and the Researcher Stimulation Program (PEII).

References

1. Kwon MA, Kim HS, Hahm DH, Song JK (2011) Synthesis activity-based zymography for detection of lipases and esterases. Biotechnol Lett 33:741–746
2. Singh R, Gupta N, Goswami VK, Gupta R (2006) A simple activity staining protocol for lipases and esterases. Appl Microbiol Biotechnol 70:679–682
3. Wilkesman J, Kurz L (2009) Protease analysis by zymography: a review on techniques and patents. Recent Pat Biotechnol 3:175–184
4. Wilkesman J, Hernández Z, Fernández M, Contreras LM, Kurz L (2014) Enhancement of sequential zymography technique for the detection of thermophilic lipases and proteases. Amino Acids 46(5):1409–1413
5. Brahimi-Horn MC, Guglielmino ML, Gaal AM, Sparrow LG (1991) Nondenaturing protein electrotransfer of the esterase activity of lipolytic preparations. Anal Biochem 196:76–79
6. Pan D, Hill AP, Kashou A, Wilson KA, Tan-Wilson A (2011) Electrophoretic transfer protein Zymography. Anal Biochem 411: 277–283
7. Laemmli UK (1970) Cleavage of structural proteins during the assembly of the head of bacteriophage T4. Nature 227:680–685
8. De Moreno MR, Smith JF, Smith RV (1985) Silver staining of proteins in polyacrylamide gels: increased sensitivity through a combined Coomassie blue-silver stain procedure. Anal Biochem 151:466–470
9. Kim MH, Kim HK, Lee JK, Park SY, Oh TK (2000) Thermostable lipase of *Bacillus stearothermophilus*: high-level production, purification, and calcium-dependent thermostability. Biosci Biotechnol Biochem 64:280–286
10. Manchenko GP (2003) Handbook of detection of enzymes on electrophorectic gels, 2nd edn. CRC Press LLC, Boca Raton, FL

Chapter 26

Calpain Zymography: General Methodology and Protocol

Kevin K.W. Wang

Abstract

Casein zymography has become one of the gold standard assays for monitoring mammalian calcium-activated proteases (calpains) in purified enzyme, cell, or tissue samples. This calpain zymography method takes advantages of (1) casein is an excellent substrate for major isoforms of calpains (Calpain-1, 2 and 3), (2) the embedded casein is digested into small peptides where the calpain bands are located, thus creating a clear zone upon Commassie blue gel staining, and (3) the calpain isoforms have different gel mobility under native gel conditions. Casein zymography is also useful in studying reversibility of inhibitor binding to calpains.

Key words Calpain, Caseinolysis, Protease substrates

1 Introduction

Gelatin zymography was first developed as a sensitive assay for matrix metalloproteases (MMP) such as MMP-2, and MMP-9, often in a biological sample [1–3]. Samples are run on denaturing SDS-polyacrylamide gel and the MMPs are renatured by removal of SDS from the gel, followed by incubation of a MMP activation buffer. Calpains are emerging as prominent calcium-activated cytosolic proteases in muscle, brain, and other tissues since 1980. There are at least 14 isoforms of calpain identified, with calpain-1 and calpain-2 being the ubiquitously expressed isoforms; however, at the time, similar zymography assays were not available for studying calpains initially. This is mainly because (1) gelatin is not a good substrate for calpain, and (2) the two major calpains (calpain-1 and calpain-2) exist as functional enzyme only as a heterodimer by associating with its small subunit (capns1 or capns2), thus denaturing SDS-gel did not adequately preserve calpain activity.

We thus developed a calpain zymography method by substituting gelatin with major milk protein casein as the latter is a preferred substrate for calpains [4]. We also adopted native (non-denaturing) gel electrophoresis conditions to keep the calpain heterodimer

Jeff Wilkesman and Liliana Kurz (eds.), *Zymography: Methods and Protocols*, Methods in Molecular Biology, vol. 1626,
DOI 10.1007/978-1-4939-7111-4_26,

together and to avoid denaturing calpains. The gels are incubated with buffer containing reducing agent (e.g., DTT) and calcium chloride. Finally, the gels are developed by staining with Coomassie Brilliant blue G-250. The lightly stained or nonstained gel bands (clear zones) over a blue-stained background indicate active calpain bands in the samples.

2 Materials

2.1 Stacking Gel Solution

To prepare 5 mL for four gels (*For casting your own gels only)*: 4% (w/v) acrylarmide solution, 0.10% (w/v) *N,N'*-methylene-bisacrylamide, 330 mM Tris–HCl (pH 6.8).

2.2 Separating Gel Solution

Make 25 mL for four gels (*For casting your own gels only)*: 12% (w/v) acrylamide, 0.32% (w/v) *N,N'*-methylene-bisacrylamide, 375 mM Tris–HCI (pH 8.8), 0.2% (w/v) casein, 3.5% glycerol.

2.3 Non-denaturing 4× Sample Buffer

20% (v/v) glycerol, 2 mM 2-mercaptoethanol, 0.004% (w/v) bromophenol blue, 200 mM Tris–HCI (pH 7.0).

2.4 Non-denaturing Gel Running Buffer (pH 8.3)

25 mM Tris-base, 192 mM glycine, 1 mM EGTA, 1 mM DTT.

2.5 Calpain Reactivation Buffer C

20 mM Tris–HCl (pH 7.4), 10 mM DTT, 4 mM $CaCl_2$. Keep at room temperature.

2.6 Fixing/Destaining Solution

Methanol/water/acetic acid (5:4:1).

2.7 Staining Solution

0.25% (w/v) Coomassie blue R-250 in Fixing/Destaining solution.

2.8 Special Reagents and Equipments

The following items must be available to perform the experiments:

1. Bovine casein (sodium salt).
2. Calpain 1, human.
3. Calpain 2, native bovine (CAPN2-350B).
4. E64c (irreversible calpain inhibitor).
5. SJA6017 (reversible, calpain inhibitor) [Calpain Inhibitor VI (CAS 190274-53-4) (sc-293,979)].
6. PD150606 (reversible, calpain inhibitor).
7. Calcium chloride ($CaCl_2$).
8. Bolt® Empty Mini Gel Cassettes (NW2010).
9. Bolt® Empty Mini Gel Cassette Combs, 10-well (NW3010).

10. Precast Novex™ 12% Zymogram (Casein) Protein Gels, 1.0 mm, 12-well (EC64052BOX).
11. Xcell SureLock Mini-Cell electrophoresis unit (EI0002).
12. PowerEase® 300 W Power Supply (115 VAC) (PS0300).

3 Methods

3.1 Zymogram Preparation

1. To initiate polymerization, ammonium persulfate (0.04%, w/v) and TEMED (0.028%, v/v) are added to the separating casein gel solution.
2. The separating casein gel solution mixture is immediately poured into the empty gel cassettes and allowed to polymerize for 30–60 min.
3. After the well-forming combs are inserted into the top of the cassettes, ammonium persulfate (0.08%, w/v) and TEMED (0.028%, v/v) are added to the stacking gel solution.
4. The top part of the cassettes is filled with the stacking gel solution mixture. The gel is allowed to polymerize for 15–30 min. *Alternative:* Commercially available precast casein polyacrylamide gels can be used (*see* **Note 1**).

3.2 Calpain-Containing Sample Preparation

1. Calpain-containing samples in 50 mM Tris–HCl pH 7.4, 1 mM DTT, l mM EGTA is either untreated or incubated in 50 mM Tris–HCl (pH 7.4), 3 mM DTT, 2 mM $CaCl_2$ in the presence of a calpain inhibitor (30 μM, if desired) for 10 min on ice (final volume 32 μL) (*see* **Note 2**).
2. Examples of calpain inhibitors (if used) are E64c (irreversible calpain inhibitor), SJA6017 (reversible, calpain inhibitor), and PD150606 reversible, calpain inhibitor).
3. Three microliters of 100 mM EGTA is added to chelate the calcium ion (total volume 35 μL).
4. Five microliters of the non-SDS-sample buffer is added (total volume 40 μL).

3.3 Running of Non-Denaturing Casein Gel

1. The casein gel is prerun with the non-SDS running buffer for 15 min in an ice water bath. This removes residual polymerization reagents from the gel that might inactivate calpain.
2. Protease-containing samples are then loaded into the wells. Samples are purified calpain protein (0.5–2 μg), cell lysate (30 μg protein), or tissue lysate (30 μg) [4, 5] (*see* **Note 3**).
3. Rainbow molecular weight markers are also run alongside as indicators for the progress of electrophoresis.

4. Electrophoresis is run at constant voltage (125 V) for about 2 h in an ice-water bath covering at least 50% (in height) of the gel running unit.
5. The gel is then removed and incubated in Reactivation Buffer at ambient temperature (21–24 °C) with two buffer changes of 30 min each.
6. The gel is then further incubated overnight (20–24 h) at ambient temperature.

3.4 Staining and Destaining of Zymogram and Calpain Activity Visualization

1. At the end of the proteolysis reaction, incubate the gels in water for 1 h with two changes (15 min each), followed by 30 min in the fixing solution.
2. The gel is stained with the staining solution for 30 min, followed by the destaining solution with several changes in 2–5 h.
3. Calpain-1 and/or -2 heterodimer will migrate to a distinct band in the casein gel and upon addition of reactivation buffer, where it will digest the embedded casein. Thus, a clear zone will be detected over a uniform blue-stained casein background (Fig. 1).
4. If both calpain-1 and calpain-2 are present in the same samples, they can be distinguished by their different mobility in the casein gel (Fig. 1). Calpain-3 can also be detected with this method (Fig. 1). It is possible to detect other isoforms of calpain (*see* **Note 4**).
5. To achieve optimal results, some assay optimization is recommended. In addition, some more sensitive zymogram alternatives are available (*see* **Note 5–8**).

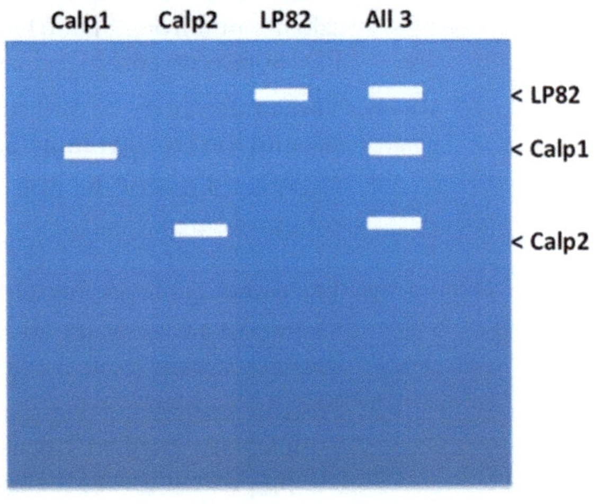

Fig. 1 Calpain zymography using casein as embedded substrate. Human calpain 1, human calpain-2, or rat lens calpain-3 isoform LP82 were loaded (2 μg) onto casein gel, zymogram was developed according to the described protocol. Position of calpain-1 vs calpain-2 enzymes is shown

6. The gel is stored in 2.5% (v/v) acetic acid until imaging and quantification using densitometry method (e.g., NIH ImageJ software).

4 Notes

1. Commercially available precast casein polyacrylamide gels can be used [Precast Novex™ 12% Zymogram (Casein) Protein Gels, 1.0 mm, 12-well (Cat. # EC64052BOX, ThermoFisher)].
2. Casein zymogram can be used to distinguish reversible versus irreversible inhibitor. In this case, inhibitor at optimal concentration (e.g., 30 μM) was preincubated with calpain samples with 1 mM $CaCl_2$ and 10 mM DTT for 10 min [4]. 5 mM EDTA is then added to the solution to chelate the calcium. Samples are then mixed with sample buffer and run onto the casein gel. Irreversible inhibitor (e.g., E64c) will block calpain activity in casein gel, while the inhibition of reversible inhibitor (e.g., SJA6017, PD150606) will be lifted with the long incubation in the reactivation buffer (Fig. 2).
3. Cell or tissue lysate can be prepared by subjecting cell pellet or pulverized frozen tissue samples to lysis buffer (1% Triton-X100, 20 mM Tris–HCl (pH 7.4), 5 mM EGTA, 1 mM DTT). Recommended protein concentration is 1–2 mg/mL for the lysate [4, 5]. The addition of protease inhibitor cocktail should be avoided. Samples should be freshly prepared before running for zymogram. If sample storage is needed, 50% by volume of

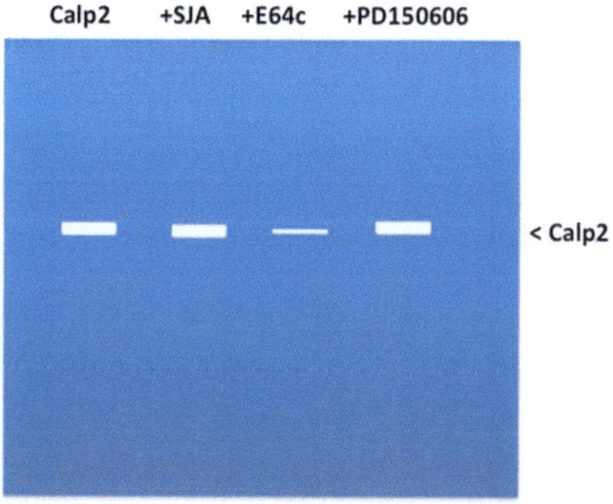

Fig. 2 Studying reversibility of calpain inhibitors. Reversible calpain inhibitor SJA6017, PD150606, and irreversible calpain inhibitor E64c (30 μM each) were preincubated with calpain-2 (2 μg) in the presence of 2 mM calcium chloride and DTT for 10 min at room temperature before mixing with native gel sample buffer and loading onto casein gel

glycerol should be added and samples stored at −20 °C freezer for up to 1–3 months.

4. There are at least 14 different isoforms of calpains [6, 7]. Many of them can potentially proteolyze casein. Thus, casein zymography can be used. For example, calpain-3 isoform LP82 and rat retina [Rt88 (homodimer)] can be detected in the tissue sample [8, 9]. Calpain-10 is also detectable by casein zymography from lens tissue lysate from a rat cataractogenesis model.
5. For better sensitivity, FITC- or Fluorescein-labeled casein can be used, and fluorescence images can be obtained as endpoints [8].
6. Different pH (7.2–7.8) of the reactivation buffer can be used to enhance the detection of calpain-1 versus calpain 2 in brain samples after experimental brain injury [5].
7. This assay can also be used to study calpain activation and subcellular translocation. For example, Zhao et al. [10] showed that in a rat traumatic brain injury (TBI) study, there were marked increases in calpain-1 and calpain-2 activity in cytosolic fraction in the ipsilateral cortex as early as 15 min and became maximal at 6 h postinjury. In addition, there is also a shift in calpain-1 activity from the cytosolic fraction to the membrane fraction occurred at 3 h after injury and became maximal at 24 h. In contrast, calpain-2 was only detected in cytosolic fraction.
8. In addition, a variant calpain zymography has been developed by Dr. Schnellmann's group using a calpain-preferred fluorogenic peptide substrate Succinyl-Leu-Leu-Val-Tyr-7-amino-4-methylcoumarin (SLLVY-AMC) instead of casein. With this SLLVY-zymogram method, they were able to find that rat kidney mitochondria contained several calpain-10 splice variants 75 kDa (calpain 10a), 56 kDa (calpain 10c or 10d), and 50 kDa (calpain 10e) [11].

Acknowledgment

This study is supported in part by NIH grant R21 NS085455-01 (K.K.W.).

References

1. Romanic AM, White RF, Arleth AJ et al (1998) Matrix metalloproteinase expression increases after cerebral focal ischemia in rats: inhibition of matrix metalloproteinase-9 reduces infarct size. Stroke 29:1020–1030
2. Wang KKW (2002) Assaying proteases in cellular environments. Curr Protoc Protein Sci Chapter 21:Unit 21.12. doi:10.1002/0471140864.ps2112s27
3. Wang X, Mori T, Jung JC et al (2002) Secretion of matrix metalloproteinase-2 and -9 after mechanical trauma injury in rat cortical cultures and involvement of MAP kinase. J Neurotrauma 19:615–625. doi:10.1089/089771502753754082
4. Raser KJ, Posner A, Wang KKW (1995) Casein zymography: a method to study mu-calpain,

m-calpain, and their inhibitory agents. Arch Biochem Biophys 319:211–216

5. Zhao X, Newcomb JK, Posmantur RM et al (1998) pH dependency of mu-calpain and m-calpain activity assayed by casein zymography following traumatic brain injury in the rat. Neurosci Lett 247:53–57
6. Huang Y, Wang KKW (2001) The calpain family and human disease. Trends Mol Med 7:355–362
7. Sorimachi H, Hata S, Ono Y (2011) Calpain chronicle—an enzyme family under multidisciplinary characterization. Proc Jpn Acad Ser B Phys Biol Sci 87:287–327
8. Ma H, Shih M, Hata I et al (1998) Protein for Lp82 calpain is expressed and enzymatically active in young rat lens. Exp Eye Res 67:221–229. doi:10.1006/exer.1998.0515
9. Azuma M, Fukiage C, Higashine M et al (2000) Identification and characterization of a retina-specific calpain (Rt88) from rat. Curr Eye Res 21:710–720
10. Zhao X, Posmantur R, Kampfl A et al (1998) Subcellular localization and duration of μ-calpain and m-calpain activity after traumatic brain injury in the rat: a casein zymography study. J Cereb Blood Flow Metab 18:161–167. doi:10.1097/00004647-199802000-00006
11. Giguere CJ, Covington MD, Schnellmann RG (2008) Mitochondrial calpain 10 activity and expression in the kidney of multiple species. Biochem Biophys Res Commun 366:258–262. doi:10.1016/j.bbrc.2007.11.133

Chapter 27

CTAB Zymography for the Analysis of Aspartic Proteases from Marine Sponges

Oscar González and Jeff Wilkesman

Abstract

Electrophoresis under denaturing conditions in the presence of SDS is a standard method for the protein and enzyme scientist. Nevertheless, there are special situations where this method may originate nonoptimal results. SDS may cause protein aggregation or precipitation. Beyond this, depending on the type of protein, some just do not resolve well or migrate abnormally in SDS gels. SDS, an anionic detergent, may be however substituted by a cationic detergent, like CTAB (cetyltrimethylammonium bromide), in order to solubilize and electrophorize proteins. CTAB electrophoresis allows the separation of proteins based on molecular weight and can be carried out at neutral or acidic pH. Here, we describe the development of a CTAB zymography method to analyze aspartic proteases from marine sponges, which present an abnormal high R_f value when run in SDS-PAGE. The special feature of using CTAB is that it binds proteins, making them positively charged and thus migrating in the opposite direction compared to SDS-PAGE.

Key words Aspartic proteases, CTAB, Cationic zymography, Electrophoresis

1 Introduction

Aspartic protease analysis is relevant in medicine and biological science. For example, it is known that HIV presents aspartic protease activity [1], thus the implementation and enhancement of methods of higher resolution and effective separation will allow a more detailed analysis of these enzymes and its inhibition.

Electrophoresis is an analytical method that allows the separation of biomolecules according to its mass and electric charge employing a polymeric matrix. There are, however, studies where atypical molecular weight values of proteases are reported when using SDS-PAGE [2]. It has become relevant to verify if these values are valid, and if not, to check the source that originated this shift in M_r.

Though SDS-PAGE has been a very common choice to perform electrophoresis, there is an alternative to employ a cationic detergent—instead of an anionic one like SDS—in an acidic buffer

Jeff Wilkesman and Liliana Kurz (eds.), *Zymography: Methods and Protocols*, Methods in Molecular Biology, vol. 1626,
DOI 10.1007/978-1-4939-7111-4_27, © Springer Science+Business Media LLC 2017

system, named cationic electrophoresis [3]. This technique is employed preferentially in the analysis of very acidic proteins that are not able to interact appropriately with SDS. It is also used in the analysis of very basic nucleoproteins that behave abnormally in SDS-PAGE [4]. In this case, the direction of the electric field must be inverted.

At neutral pH, alkaline proteases as well as very acidic proteases do not migrate as expected under the electric field when applying electrophoresis, resulting in the accumulation of proteins with similar characteristics in the upper part of the resolving gel. This migration pattern is a hindrance for the protein identification after the separation. Glycoproteins and lipoproteins also migrate abnormally in SDS-PAGE, as their non-protein moieties do not uniformly bind the detergent. Proteins with an unusual amino acid sequence, especially those containing high amounts of Lys, Pro, acidic, or basic residues, are prone to behave erratically under standard SDS-PAGE conditions [3, 5].

Cationic electrophoresis may result as a more complicated method compared to SDS-PAGE [4], as polymerization of acidic gels is somehow more difficult than regular basic gels. As cationic detergent, CTAB has been commonly used [3, 6, 7]. However, the use of 16-BAC is increasing [8–10] (Fig. 1).

A new multiphase system has been described for 16-BAC-PAGE, in which the phosphate buffer is substituted by acetic acid [11]. This feature plus the optimization of the 16-BAC concentration and the use of a new tracking dye allows a highly efficient performance of the electrophoresis [12]. This leads to a system that is as easy to manage and is as good—in terms of resolution quality—as a standard SDS-PAGE.

Here, we have enhanced the cationic electrophoresis by copolymerizing gelatin in the gel matrix in order to develop a cationic zymography method. Zymography is a technique that enables one to detect enzymatic activity, under nonreducing conditions. Enzyme

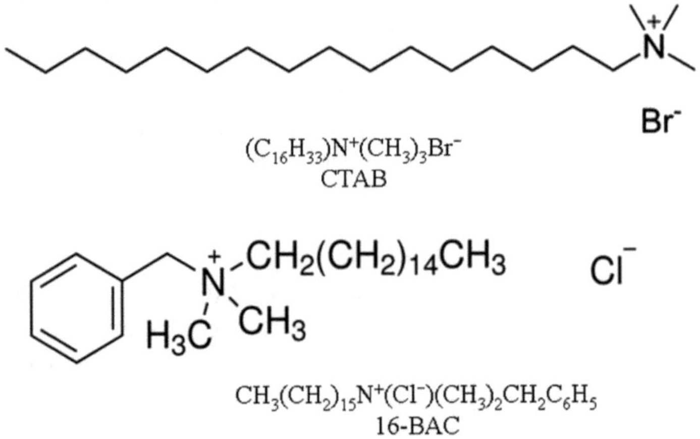

Fig. 1 Common detergents used in cationic electrophoresis

activity may be inhibited by the presence of the detergent. In this regard, after the run, the gel is washed with Triton X-100, in order to remove the detergent and allowing renaturalization of the enzyme. Afterwards, that gel is placed in an activation buffer, allowing the enzyme to catalyze hydrolysis reaction of the substrate. The end stage consists of staining the gel with Coomassie Brilliant Blue (CBB) in order to visualize the enzyme activity as translucid bands under deep blue background [13].

2 Materials

2.1 Sample Homogenization

1. Homogenization buffer: 0.1 M Tris–HCl pH 7.4, 0.05 M NaCl.

2.2 Protein Quantitation

1. 1 mg/mL BSA solution, Bradford protein determination kit.

2.3 CTAB Electrophoresis

1. Resolution gel: 10% acrylamide-bisacrylamide, 60 mM KOH; 0.376 M [~2.15% (v/v)] acetic acid buffer pH 4.3.
2. Stacking gel: 5% acrylamide-bisacrylamide, 60 mM KOH, 0.36% (v/v) acetic acid buffer pH 6.7.
3. Sample buffer 2×: 120 mM KOH, 127 mM [~0.725% (v/v)] acetic acid buffer pH 6.7, 4% (m/v) CTAB, 20% (v/v) glycerol, 0.02% (m/v) methyl green (*see* **Note 1**).
4. Running buffer: 350 mM β-alanin, 140 mM acetic acid final pH 4.5.
5. Activation buffer: 0.1 M acetic acid–sodium acetate pH 4.0.
6. CBB staining solution: 0.25% (m/v) Coomassie brilliant blue R-250, 40% (v/v) methanol, 10% (v/v) acetic acid.
7. Destaining solution: 40% (v/v) methanol, 10% (v/v) acetic acid.
8. Triton X-100 stock solution 10% (v/v): dilute 10 mL Triton X-100 until a final volume of 100 mL.
9. Triton X-100 washing solution (1%): dilute 5 mL of Triton X-100 stock solution (10%) with water until a final volume of 50 mL.

2.4 General Lab Equipment

Electrophoresis chamber, power supply, ultracentrifuge, photometer, ultraturrax.

3 Methods

3.1 Marine Sponge Homogenate

1. Marine sponge specimens of *Amphimedon erina* were collected from the Caribbean sea at the Isla Larga, San Esteban National Park, Carabobo State (10°29′18″N 67°56′40″O).

2. Weigh a 20 g sample of sponge. Cut in small pieces and homogenize in a mortar with liquid nitrogen and 25 mL homogenization buffer.
3. Once a soft paste is obtained, homogenize further with an ultraturrax.
4. Centrifuge the homogenate at 20,000 × *g* for 30 min at 4 °C [2].

3.2 Quantification of Total Proteins

1. Perform total protein determination according to the Bradford protocol [14].
2. Prepare a calibration curve employing 1 mg/mL BSA as standard protein.
3. Read absorbance at 595 nm.
4. Make duplicate of the calibration curve.

3.3 CTAB-PAGE

1. Set up the glasses in order to polymerize the gels. We have adapted the CTAB-PAGE protocol according to Shi and Jackowski [15] and to Díaz-López et al. [16].
2. Polymerize the gels according to Table 1.
3. Dissolve sample in sample buffer 2×.
4. Fill the chamber with running buffer.
5. Apply ~0.5 μg of sample per well.
6. Place the lid of the chamber as usual. When connecting the cables to the power supply, invert the positive and negative plugs. That is, place the (+) red cable in the black (−) hole of the power supply, and similarly, place the (−) black cable in the red (+) hole of the power supply.
7. Begin the run at 60 V for 30 min (at room temperature ~ 28 °C; *see* **Note 2**).
8. Increase the voltage to 120 V and let it run for further 1.5–2 h, or until methyl green indicator reaches the bottom of the gel (*see* **Note 3**).
9. After run, stain the gel in CBB staining solution. Destain the excess of dye until bands are visualized.

3.4 CTAB Zymography

1. Follow the same steps described in Subheading 3.3.
2. For gel polymerization, prepare 10% gels containing 0.1% gelatine (Table 1).
3. Once the run is finished, place the gel in activation buffer and incubate overnight (*see* **Note 4**).
4. Stain the gel as in **step 9** from Subheading 3.3 (Fig. 2).

Table 1
Preparation of resolving and stacking gels

Gel component	Resolving gel (10%)	Resolving gel (10%) for zymography	Stacking gel (4%)
Water	1.735 mL	0.745 mL	1.82 mL
2.15% (v/v) acetic acid—60 mM KOH pH 4.3	2.580 mL	2.58 mL	–
0.36% (v/v) acetic acid—60 mM KOH pH 6.7	–	–	1.44 mL
1% (m/v) Gelatin	–	0.6 mL	–
Acrylamide–Bisacrylamide solution (29:1)%	2 mL	2 mL	0.66 mL
10APS	60 μL	60 μL	60 μL
TEMED	15 μL	15 μL	15 μL

Fig. 2 Preliminary result of CTAB electrophoresis

4 Notes

1. Methyl green is included as a track dye, in order to monitor the run of the gel. Glycerol is included to minimize diffusion during sample application in the wells.
2. CTAB may precipitate under 15 °C [15]. Thermolabile enzymes may be an issue. Check first how thermo-resistant your enzyme is. However, the system may still be placed in a special refrigerated compartment between 15 and 20 °C.
3. Other positively charged tracking dyes may be used as methylene blue or pyronin Y [16].
4. The exact incubation time must be determined. Sometimes we obtained bands after 36 h incubation.

Acknowledgments

We acknowledge partial funding from the Consejo de Desarrollo Científico y Humanístico de la Universidad de Carabobo (CDCH-2014). We kindly thank the donation of CTAB from Prof. Dr. C. Cabrele, University of Salzburg, Austria. Sponge samples were kindly provided by Prof. J.G. Rodríguez, Biology Department, FACYT, University of Carabobo.

References

1. Brik A, Wong CH (2003) HIV-1 protease: mechanism and drug discovery. Org Biomol Chem 1:5–14
2. Wilkesman J, Schröder HC (2002) Heat-stable protease from the marine sponge *Geodia cydonium*. Cell Mol Biol (Noisy-le-Grand) 48:379–383
3. Buxbaum E (2012) Cationic electrophoresis. Methods Mol Biol 869:55–63
4. Janson JC (2011) Protein purification: principles, high resolution methods, and applications. John Wiley & Sons, New York, p 375
5. Garfin DE (2003) Gel electrophoresis of proteins. In: Davey J, Lord M (eds) Essential cell biology, vol 1: Cell structure, a practical approach. Oxford University Press, Oxford, UK, pp 197–268
6. Eley MH, Burns P, Kannapell CC, Campbell P (1979) Cetyltrimethylammonium bromide polycrylamide gel electrophoresis: estimation of protein subunit molecular weights using cationic detegents. Anal Biochem 92:411–419
7. Akin DT, Shapira R, Kinkade JM Jr (1985) The determination of molecular weights of biologically active proteins by cetyltrimethylammonium bromide-polyacrylamide gel electrophoresis. Anal Biochem 145:170–176
8. Hartinger J, Stenius K, Högemann D, Jahn R (1996) 16-BAC/SDS-PAGE: a two-dimensional gel electrophoresis system suitable for the separation of integral membrane proteins. Anal Biochem 240:126–133
9. Macfarlane DE (1989) Two dimensional benzyldimethyl-n-hexadecylammonium chloride-sodium dodecyl sulfate preparative polyacrylamide gel electrophoresis: a high capacity high resolution technique for the purification of proteins from complex mixtures. Anal Biochem 176:457–463
10. Nothwang HG, Schindler J (2009) Two-dimensional separation of membrane proteins by 16-BAC-SDS-PAGE. Methods Mol Biol 528:269–277
11. Kramer ML (2006) A new multiphasic buffer system for benzyldimethyl-n-hexadecylammonium chloride polyacrylamide gel electrophoresis of proteins providing efficient stacking. Electrophoresis 27:347–356

12. Braun R, Kinkl N, Beer M, Ueffing M (2007) Two-dimensional electrophoresis of membrane proteins. Anal Bioanal Chem 389:1033–1045
13. Snoek-van Beurden PA, Von den Hoff JW (2005) Zymographic techniques for the analysis of matrix metalloproteinases and their inhibitors. Biotechniques 38:73–83
14. Bradford MB (1976) A rapid and sensitive method for the quantitation of micrograms quantities of protein utilizing the principle of protein-dye binding. Anal Biochem 72:248–254
15. Shi Q, Jackowski G (1998) One-dimensional polyacrylamide gel electrophoresis. In: Hames BD (ed) Gel electrophoresis of proteins: a practical approach, 3rd edn. Oxford University Press, New York, pp 1–52
16. Díaz-López M, Moyano-López FJ, Alarcón-López FJ, García-Carreño FL, Navarrete del Toro MA (1998) Characterization of fish acid proteases by substrate-gel electrophoresis. Comp Biochem Physiol B Biochem Mol Biol 121:369–377

Chapter 28

Zymography Detection of a Bacterial Extracellular Thermoalkaline Esterase/Lipase Activity

María Tapizquent, Marleny Fernández, Georgina Barreto, Zully Hernández, Lellys M. Contreras, Liliana Kurz, and Jeff Wilkesman

Abstract

Lipases are esterases that occur widely in nature, yet those with commercial relevance are exclusively from microbial origin. Glycerol and long-chain fatty acids are the products after hydrolysis of esters bonds in saponifiable lipids catalyzed by lipases. In this work, we describe lipase/esterase activity contained in cell-free fractions from thermophilic bacteria, cultured in medium containing olive oil. Analysis of the cell-free fractions by electrotransference zymography, using tributyrin as substrate, revealed bands corresponding to lipase activity. The method is simple, fast, and inexpensive.

Key words Electrophoresis, Electrotransference, Esterase, Lipases, Thermophiles, Zymography

1 Introduction

Lipases are a versatile type of enzymes able to catalyze the hydrolysis and—in some cases—the synthesis of triglycerides at the water/oil surface [1–3]. These multitalented enzymes carry out esterifications, transesterifications, and a wide range of other reactions [4]. Generally, lipases do not necessarily obey the Michaelis–Menten model, as catalysis is performed in heterogeneous media [5, 6]. Temperature, organic solvents, and other chemical substances like detergents (SDS) affect the enzyme structure and its activity. Still, some lipases are able to conserve their structure and activity after exposition towards harsh conditions. Using zymography, cstcrasc/lipase activity can be detected and an average molecular mass determined. Though some methods have been described, herein we focus on the electrotransference method for the zymography accomplishment, as it allows the enzyme to be embedded with substrate in a separated gel, without altering the *Rf* value.

Jeff Wilkesman and Liliana Kurz (eds.), *Zymography: Methods and Protocols*, Methods in Molecular Biology, vol. 1626, DOI 10.1007/978-1-4939-7111-4_28, © Springer Science+Business Media LLC 2017

2 Materials

2.1 General Equipment

Electrophoresis chamber, electrotransfer chamber.

2.2 Growth Medium

Culture medium: 8 g/L nutrient broth (commercial brand) supplemented with 1% (v/v) olive oil (commercial brand).

2.3 Protein Quantification Kit

Commercial kits are available. Standard protein solution 1 g/L BSA.

2.4 Electrophoresis and Zymography Solutions

1. 12% resolving gels: 0.375 M Tris–HCl, pH 8.8, 12% acrylamide/bis-acrylamide (29:1), 0.05% ammonium persulfate (APS), 0.005% TEMED.
2. 5% stacking gels: 0.125 M Tris–HCl pH 6.8, 5% acrylamide/bis-acrylamide (29:1), 0.05% APS, 0.001% TEMED.
3. Running buffer: 25 mM Tris–HCl, 192 mM glycine, final pH 8.0, without SDS.
4. Sample buffer (4×): 125 mM Tris–HCl pH 6.8, 4% (w/v) SDS without 2-mercaptoethanol.
5. Substrate gel: 12% acrylamide, 2% tributyrin, 0.1% PSA, and 0.05% TEMED.
6. Transference buffer: 25 mM Tris–HCl, 192 mM glycine, final pH 8.0.
7. Activation buffer: 30 mM Tris–HCl pH 7.4, 200 mM NaCl.

3 Methods

3.1 Bacterial Strains and Culture Conditions

1. Choose the biological source. *Geobacillus* sp. isolated from Las Trincheras (Venezuela) hot springs is a Gram positive thermophilic strain used in these assays.
2. Culture bacteria in nutrient broth supplemented with 1% (v/v) olive oil, at 55 °C, and with constant agitation for 192 h.
3. Centrifuge culture at 9500 × *g* for 30 min.
4. Collect the resulting supernatant (cell-free fraction) and keep for further analysis.
5. Concentrate the cell-free fraction sample tenfold using a commercial concentrator (Eppendorf) at 60 °C.
6. After concentration, keep samples on ice until further processing for the gels.

3.2 Protein Quantification

1. Perform a protein quantification following the Bradford method [7] (*see* **Note 1**).

Table 1
Volumes (mL) for preparation of resolving and stacking gel

Gel component	12% Resolving gel	5% Stacking gel
Distilled water	1.37	1.10
1.5 M Tris-HCl pH 8.8	1.00	–
0.5 M Tris-HCl pH 6.8	–	0.50
30% Acrylamide-bisacrylamide solution	1.60	0.33
10% APS	0.04	0.015
TEMED	0.01	0.005

2. Build calibration curve with BSA standard protein solution.
3. Measure absorbance at 590 nm.

3.3 Electrophoresis and Zymography

1. Prepare two gels (12% resolving gel, 5% stacking gel) for the PAGE (Table 1). Take into account that it will be run under nonreducing (sample buffer without 2-mercaptoethanol and no heating) and non-denaturing conditions (absence of SDS in running buffer) (*see* **Note 2**). One of the gels will be stained and the other will be electrotransferred.
2. Prepare the samples by dissolving them in sample buffer (4×).
3. Apply the samples in the wells and run the gel at 100 V for ~90 min on a vertical camera.
4. After the run, take one of the gels and stain it using the silver staining method [8] (*see* **Note 3**).
5. For the zymography process, the remaining gel will be electrotransferred [9].
6. Rinse the gel with distilled water for 10 min and place it into the electrotransference set.
7. Previously, a 12% substrate gel must have been casted (Table 2) (*see* **Note 4**).
8. Set the transference sandwich according to the specifications of the manufacturer (Fig. 1).
9. Carry out the transference in the electrotransference equipment, at 15 V for 15 min, with transference buffer.
10. After transference, incubate the substrate gel in activation buffer for 1 h at 55 °C (*see* **Note 5**).
11. Activity is visualized as a transparent band in an opaque background.

Table 2
Preparation of the substrate gel for transfer zymography (12%)

Component	Volume (mL)
Water (dd)	1.7
Acrylamide/bis-acrylamide solution	1.2
Glycerol tributyrate	0.06
Olive oil	0.02
Sonication	
10%APS	0.030
TEMED	0.003

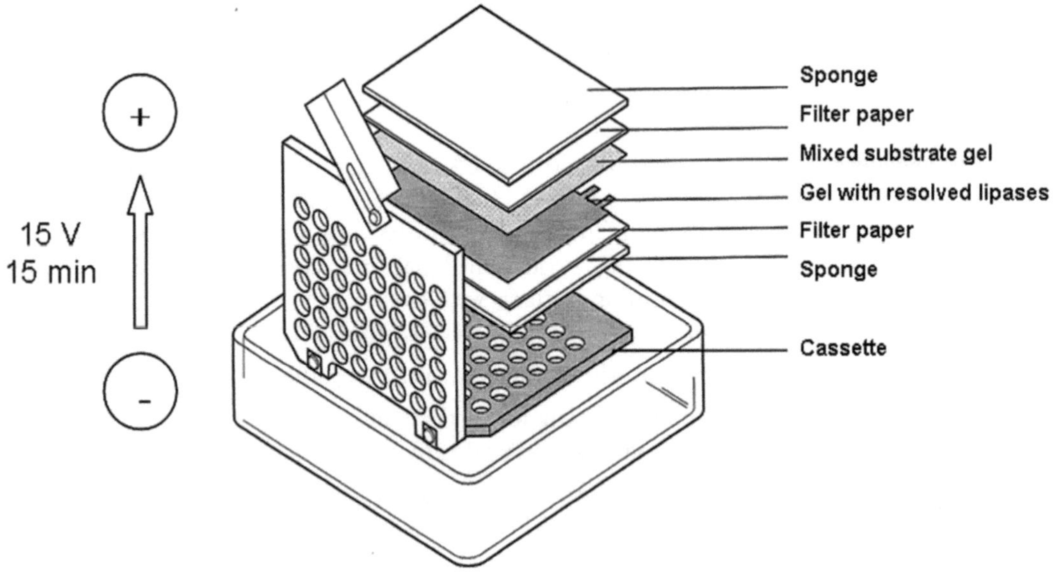

Fig. 1 Diagram for the electrotransference setup

12. Register the image with a transilluminator or scanner. UV (280 nm) irradiation may be applied to obtain better gel images, with a grayish background and dark bands.
13. SDS-PAGE and zymography examples are shown in Fig. 2 (*see* **Note 6**).

4 Notes

1. Regard that total protein is being measured (so other proteins present will be measured as well). In parallel, it is recommended to determine enzyme activity, employing, e.g.,

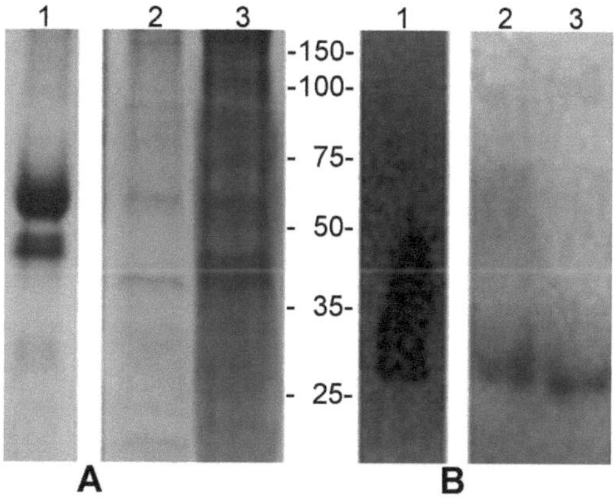

Fig. 2 Electropherograms showing lipase/esterase activity present in cell-free fractions. (**a**) SDS-PAGE (12%) under nonreducing conditions stained with CBB. (**b**) Zymogram (12%) with tributyrin as substrate, accomplished by electrotransference. Lanes: (*1*) Lipase control (9 μg). (*2* and *3*) Cell-free fractions (~35 μg)

p-nitrophenyl palmitate (*p*-NPP) as substrate and measuring absorbance at 420 nm [10–12]. Porcine lipase type II can be used as positive control.

2. These running conditions assure lipase activity at the end stage. However, some lipases have been shown to be SDS-tolerant. In our case, we allow SDS to present in the sample buffer, but not in the running buffer. This creates a semi non-denaturing condition in the run that has not been described in the literature so far.
3. Many staining methods are described [8]. However, you may choose also the Coomassie staining procedure. The staining method chosen will depend upon the amount of lipase contained in your sample applied per lane.
4. The electrotransference receptor gel holds the copolymerized substrate. For our thermophilic lipase, substrates tried were tributyrin and olive oil. The substrate's lipophilic nature was not a disadvantage because copolymerization process allowed homogeneous distribution in the gel.
5. Timing must be optimized. In our case, bands were possible to be seen after 1 h incubation. Darker bands were obtained after 24 h incubation [9]. Sometimes activity is seen at room temperature (~25 °C), depending on lipase activity and substrate affinity. Transparent bands can spread quickly in all the gel. So, it is recommended to run different enzyme concentrations to optimize bands revealing.
6. Olive oil may also be used as substrate in the gel. However, we have seen a preference for tributyrin over olive as substrate. This reveals that our enzyme is most likely an esterase, according

to reports that define true lipases as those acting only over long-chain acylglycerol substrates [13]. Nonetheless, other descriptions of lipases are less restrictive and define them as proteins that convert triacylglycerol at the interface between aqueous and nonaqueous phases [11, 13, 14].

Acknowledgments

This research was partially funded by the *Consejo de Desarrollo Científico y Humanístico* (CDCH-UAC-373-2010), University of Carabobo and by LOCTI. The authors thank the kind assistance of M.Sc. V. Storaci and D. Quintero.

References

1. Houde A, Kademi A, Leblanc D (2004) Lipases and their industrial applications: an overview. Appl Biochem Biotechnol 118:155–170
2. Pandey A, Benjamin S, Soccol C, Nigam P, Krieger N, Soccol V (1999) The realm of microbial lipases in biotechnology. Biotechnol Appl Biochem 29:119–131
3. Jaeger K, Eggert T (2002) Lipases for biotechnology. Curr Opin Biotechnol 13:390–397
4. Gupta R, Gupta N, Rathi N (2004) Bacterial lipases: an overview of production, purification and biochemical properties. Appl Microbiol Biotechnol 64:763–781
5. Jaeger K, Ransac S, Dijkstra B, Colson C, van Heuvel M, Misset O (1994) Bacterial lipases. FEMS Microbiol Rev 15:29–63
6. Zaks A, Klibanov A (1985) Enzyme-catalyzed processes in organic solvents. Proc Natl Acad Sci U S A 82:3192–3196
7. Bradford MM (1976) A rapid and sensitive method for the quantitation of microgram quantities of protein utilizing the principle of protein-dye binding. Anal Biochem 72: 248–254
8. Chevallet M, Luche S, Rabilloud T (2006) Silver staining of proteins in polyacrylamide gels. Nat Protoc 1(4):1852–1858. doi:10.1038/nprot.2006.288
9. Wilkesman J, Hernández Z, Fernández M, Contreras LM, Kurz L (2014) Enhancement of sequential zymography technique for the detection of thermophilic lipases and proteases. Amino Acids. doi:10.1007/s00726-014-1707-1
10. Sharma R, Chisti Y, Benerjee UC (2001) Production, purification, characterization, and applications of lipases. Biotechnol Adv 19:627–662
11. Nawani N, Kaur J (2000) Purification, characterization and thermostability of lipase from a thermophilic Bacillus sp. J33. Mol Cell Biochem 206:91–96
12. Salameh M, Wiegel J (2007) Purification and characterization of two highly thermophilic alkaline lipases from Thermosyntropha lipolytica. Appl Environ Microbiol 73:7725–7731
13. Jaeger K, Dijkstra B, Reetz M (1999) Bacterial biocatalysts: molecular biology, three-dimensional structures, and biotechnological applications of lipases. Annu Rev Microbiol 53:315–351
14. Mahadevan GD, Neelagund SE (2013) Thermostable lipase from *Geobacillus* sp. Iso5: bioseparation, characterization and native structural studies. J Basic Microbiol. doi:10.1002/jobm.201200656

Chapter 29

Amylase Zymography

Adarelys Andrades and Lellys M. Contreras

Abstract

Amylase zymography was carried out for the detection of amylases produced by a *Geobacillus stearothermophilus* strain isolated from the Thermal Center "Las Trincheras" in Venezuela. Zymography is an electrophoretic technique used to study hydrolases by means of thin gels containing copolymerized-specific substrates, under nonreducing conditions. In this study, 0.1% starch was incorporated into the gel as substrate. The formation of clear zones against a dark background in the gel stained with iodine indicated the presence of amylolytic activity. The thermophilic bacteria released several extracellular amylases to a selective growth medium supplemented with 1% soluble starch at 55 °C after 40 h incubation. The amylolytic enzymes showed an optimum temperature of 60 °C and an optimum pH at 6.0. The amylases were partially purified by cold acetone precipitation followed by two chromatographic techniques. These purified amylases showed different molecular masses which were determined by sodium dodecyl sulfate gel electrophoresis and confirmed by zymography.

Key words Zymography, Gel electrophoresis, Hydrolases, Amylases, Thermophiles, *Geobacillus stearothermophilus*

1 Introduction

α-Amylases (EC 3.2.1.1) act as starch-degrading enzymes by catalyzing the hydrolysis of internal α-1,4-*O*-glycosidic bonds in polysaccharides with the retention of α-anomeric configuration in the products. They are mostly metalloenzymes and require calcium ions for activity, structural integrity, and stability [1]. α-Amylases can be divided into four basic categories: (1) Endoamylases (cleave internal α-1,4 bonds resulting in a α-anomeric products), (2) Exoamylases (cleave α- 1,4 or α-1,6 bonds of the external glucose residues resulting in α- or β-anomeric products, (3) Debranching Enzymes (hydrolyze α-1,6 bonds exclusively leaving long linear polysaccharides), and (4) Transferases (cleave α-1,4 glycosidic bond of the donor molecule and transfer part of the donor to a glycosidic acceptor forming a new glycosidic bond) [2–4]. The potential application of this enzyme in industries such as paper, textile, and food has increased during the last two decades [2].

Jeff Wilkesman and Liliana Kurz (eds.), *Zymography: Methods and Protocols*, Methods in Molecular Biology, vol. 1626,
DOI 10.1007/978-1-4939-7111-4_29, © Springer Science+Business Media LLC 2017

As enzymatic liquefaction and saccharification of starch are performed at high temperatures, thermostable amylolytic enzymes are required and have been currently investigated to improve industrial processes of starch degradation. For example, α-amylases produced from *Bacillus coagulans*, *B. stearothermophilus*, *B. caldolyticus*, *B. brevis*, *B. acidocaldarius*, and *B. thermoamyloliquefaciens* are used extensively for conversion of starch into sugar, syrups and dextrins, which forms the major part of the starch processing industry [5–9].

In an attempt to identify new amylases with potential application to industry, we have partially purified and characterized amylases obtained from thermophilic bacteria *Geobacillus stearothermophilus*, isolated from the Thermal Center "Las Trincheras" located in Venezuela.

In this chapter, we describe the characterization of several thermostable amylases by means of a zymographic technique, on the basis of starch degradation, which is copolymerized in a polyacrylamide gel electrophoresis. The bacterial strain was grown in minimal medium supplemented with 1% starch (*see* Subheading 2.3), exhibiting a high amylolytic activity after 40–48 h of growth at 55 °C and pH 7 [10–13]. Cell-free supernatant obtained by centrifugation was used for estimation of total amylolytic activity, using the dinitrosalicylic acid (DNS) method [14]. Some properties related to the effect of temperature and pH on enzyme production was also determined. Then, we have partially purified these amylases employing three conventional methods: (a) cold acetone precipitation, (b) ion-exchange chromatography, and (c) gel filtration chromatography. Finally, we have detected the presence of several amylolytic enzymes on zymograms. Four molecular entities were identified, with molecular masses corresponding to 64.6, 61.6, 60.3, and 23.4 kDa. Possibly, the association of these in nonspecific complexes was detected in zones of high molecular weight.

2 Materials

All chemicals were of analytical grade.

2.1 Microorganisms

Bacterial *Geobacillus stearothermophilus* strains were isolated from the Thermal Center "Las Trincheras" (hot springs 55 and 87 °C), located in central-north Venezuela. The strain with amylolytic activity was identified as *Geobacillus stearothermophilus* (Spanish Type Culture Collection, CECT, Valencia, Spain).

2.2 Stock Solutions and Reagents

Amylase zymography for the determination of amylase activity was performed according to the method described by Laemmli, with modifications [15]. Zymogram gels are based on 10% polyacrylamide gel electrophoresis which is embedded with 0.1% soluble starch—a specific substrate for amylases.

1. *1.5 M Tris–HCl, pH 8.8*: Weigh 27.23 g Tris base and dissolve with 80 mL deionized water. Adjust to pH 8.8 with 6 M HCl. Bring total volume to 150 mL with deionized water. Store at +4 °C.
2. *0.5 M Tris–HCl, pH 6.8*: Weigh 6 g Tris base and dissolve with 60 mL deionized water. Adjust to pH 6.8 with 6 N HCl. Bring total volume to 100 mL with deionized water. Store at +4 °C.
3. *30% Acrylamide-Bis-acrylamide*: Weigh 87.6 g acrylamide, 2.4 g *N′ N′*-bis-methylene-acrylamide. Mix and make up to volume 300 mL with deionized water. Filter and store at +4 °C in the dark (*see* **Note 1**).
4. *10% SDS*: Dissolve 10 g SDS in 90 mL water with gentle stirring and bring to 100 mL with deionized water.
5. *0.1% Soluble starch*: Weigh 1 g soluble starch; dissolve in 10 mL deionized water. Prepare it fresh.
6. *10% APS*: Weigh 100 mg ammonium persulfate, dissolve in 1 mL of deionized water. Store the stock solutions in the freezer at −20 °C.
7. *N,N,N′,N′ - tetramethylethylenediamine (TEMED)*.
8. *Nonreducing loading sample buffer 4×* (without β-mercaptoethanol) (*see* **Note 2**): Mix 1.25 mL 0.5 M Tris pH 6.8 with 2.0 mL 10% SDS, 2.5 mL glycerol, and 0.2 mL 0.5% bromophenol blue. Add 3.55 mL deionized water to a total volume 9.5 mL.
9. *10× Running buffer, pH 8.3*: Weigh 30.3 g Tris base, 144.0 g glycine, and 10.0 g SDS, dissolve and bring total volume up to 1000 mL with deionized water. Store at +4 °C. To use: dilute to a final concentration of 1× running buffer before use (*see* **Note 3**). Generally, we prepare 1 L by mixing 100 mL 10× running buffer and 900 mL deionized water.
10. *Renaturing buffer*: 1% (v/v) Triton X-100 in deionized water (*see* **Note 4**).
11. *Zymogram development buffer*: 50 mM Tris–HCl, pH 7.0, 1 mM $CaCl_2$.
12. *KI/I2 Staining solution*: Weigh 2 g KI; dissolve in 50 mL deionized water. Add 1 g iodine and shake it thoroughly to dissolve. Make up to 100 mL total volume. To use: dissolve 1:10 with deionized water (*see* **Note 5**).
13. *α-Amylase from barley malt 5.0 mg/mL* (commercially available, used as control).

14. *Molecular weight standards, Broad Range* (commercially available): Myosin 200,000 Da, β-galactosidase 116,250 Da, phosphorylase b 97,400 Da, serum albumin 66,200 Da, ovalbumin 45,000 Da, carbonic anhydrase 31,000 Da, trypsin inhibitor 21,500 Da, lysosyme 14,400 Da, aprotinin 6,500 Da.

2.3 Medium Composition for the Production of Amylases from Thermophiles

Geobacillus stearothermophilus strain was cultured in 500 mL Erlenmeyer flask containing 150 mL of a selective medium containing 0.2 M phosphate buffered saline (PBS) pH 7.0, 0.00001% $CuSO_4$, 0.00010% $FeSO_4$, 0.00002% $MgSO_4$•$7H_2O$, 0.00005% $ZnSO_4$, 0.00005% NaCl, 0.00010% NH_4SO_4, 0.00005% $CaCl_2$•$2H_2O$, 0.3% yeast extract, 0.3% tryptone, and 1%soluble starch (all % by mass).

2.4 Equipments

Following devices are necessary:

1. Shaker incubator with temperature control.
2. Centrifuge.
3. Concentrator.
4. FPLC—Chromatography system.
5. Electrophoresis system.
6. Imaging System.

3 Methods

3.1 Extraction and Partial Purification of Extracellular Amylases

Geobacillus stearothermophilus strain was grown in selective medium supplemented with 1% soluble starch (described in Subheading 2.3).

1. Incubate 150 mL cultures medium on a shaker incubator (70 rpm) at 55 °C for 40 h. Perform colorimetric assays for the determination of total amylase activity (*see* **Note 6**).
2. Harvest 1 L culture with high amylolytic activity and centrifuge at 10,000 × *g* for 15 min at 4 °C. Carefully collect the cell-free supernatant.
3. Concentrate this supernatant tenfold at 55 °C in glass Petri dishes (*see* **Note** 7).
4. Transfer the tenfold concentrated supernatant to Eppendorf vials in 1 mL aliquots. Then, add gradually 1 mL of cold acetone (50% final concentration), let it stay for 2 h at 4 °C.
5. Centrifuge at 10,000 × *g* for 10 min, carefully decant the supernatant, and keep the pellet (containing amylase activity).
6. Allow the acetone to evaporate at room temperature (~25 °C) about 15–30 min.
7. Resuspend the pellet in 40 mL 50 mM Tris–HCl pH 7.0 to obtain the *enzyme solution*, which is used subsequently for the purification of the enzyme.

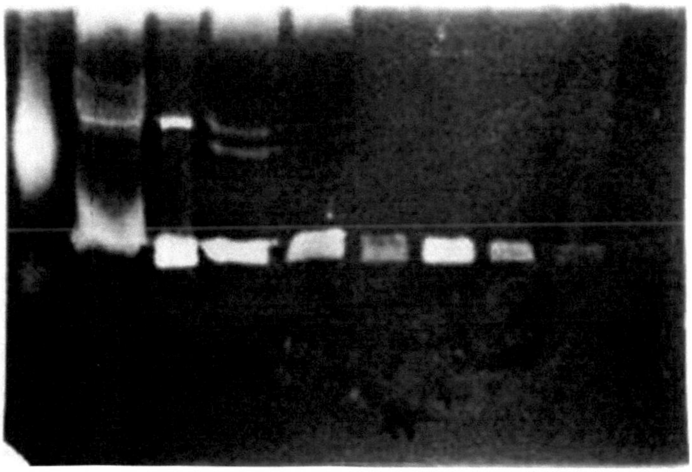

Fig. 1 Amylase zymogram using 0.1% starch as substrate. Active fractions from an ion-exchange chromatography were analyzed. First lane (*left*) shows the α-amylase standard as enzymatic control

8. Perform 10% zymograms to detect amylase activity of the *enzyme solution*. A representative zymogram is shown in Fig. 1.

3.2 Ion-Exchange Chromatography as First Purification Step

1. 5 mL of *enzyme solution* are diluted with 50 mM Tris–HCl, pH 7.0 to make up 10 mL of total volume.
2. These 10 mL are loaded onto a DEAE column (1.5 × 20 cm), previously equilibrated with 50 mM Tris–HCl pH 7.5.
3. The column is eluted with the same buffer, employing a linear NaCl gradient from 0 to 1.0 M.
4. The eluate is collected in 1 mL fractions (flow rate 0.2 mL/min).
5. The fractions are concentrated 2.5-fold using a concentrator and then the amylase activity is detected by zymography (Fig. 1).

3.3 Gel Filtration Chromatography as Second Purification Step

1. The active fractions obtained from ion-exchange chromatography are pooled, concentrated 2.5-fold, and then subjected to gel filtration on a Sephacryl 100 column (0.7 × 30 cm), previously equilibrated with 50 mM Tris–HCl, pH 7.5.
2. The eluate was collected in 0.5 mL fractions (flow rate 0.1 mL/min) using the same chromatographic system as mentioned previously.
3. The amylolytic activity of the fractions is also detected by zymography (Fig. 2).

3.4 Amylases by Zymography

1. Load the fractions from the different purification steps onto 10% polyacrylamide gels supplemented with 0.1% starch solution.

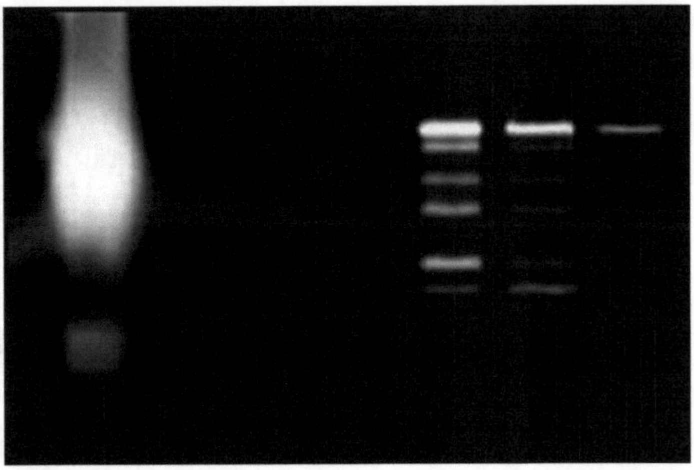

Fig. 2 Amylase zymogram using 0.1% starch as substrate. Active fractions from gel filtration chromatography. First lane (*left*) shows the α-amylase standard as enzymatic control

2. Use α-amylase from barley malt (5.0 mg/mL) as a positive control.
3. Prepare the resolving gel (10% PAGE copolymerized with 0.1% soluble starch) by mixing 1 mL 1.5 mM Tris–HCl pH 8.8 with 0.400 mL 10% SDS, 1.330 mL 30% Acrylamide/Bisacrylamide, 0.845 mL deionized water and add 0.400 mL soluble starch (1% stock solution), swirl gently with a vortex. Carefully add 0.020 mL 10% APS followed for 0.005 mL TEMED. Mix thoroughly the gel solution and carefully pour in between the electrophoretic glasses. Add a thin layer of deionized water on top of the gel solution to allow polymerization (*see* **Note 8**). Allow 1,5 h for complete polymerization.
4. Prepare stacking gel (4.5%) by mixing 0.500 mL 0.5 M Tris–HCl pH 6.8 with 0.200 mL 10% SDS, 0.300 mL 30% Acrylamide/Bisacrylamide, and 0.900 mL deionized water.
5. Now, remove the water layer on the resolving gel by using filter paper or tissue.
6. Mix gently the polymerizing reagents of the stacking gel with 0.010 mL 10% APS and 0.005 mL TEMED. Carefully pour the stacking gel on top of the polymerized resolving gel.
7. Insert the comb, trying not to get bubbles stuck underneath and allow 1 h for complete polymerization.
8. After polymerization, introduce the gel plate into the electrophoresis system containing 1× running buffer. Remove carefully the combs.
9. Prepare the samples by mixing 20 μL sample with 5 μL nonreducing loading sample buffer 4×. Mix by briefly vortexing.

10. If thermostability is going to be additionally tested, submit the samples to heat (in our case 60 °C for 8 min) (*see* **Note 9**).
11. Prepare 5 mg/mL α-amylase standard in nonreducing loading buffer 4× and include in each zymogram well as positive control for enzymatic activity.
12. Load 25 μL of the sample per lane and run the gel at constant voltage (100–150 V), until the bromophenol blue tracking dye reaches the bottom of the gel (approx. 60–90 min).
13. Remove the zymogram gels from the chamber and place them in a chamber or Petri plate with 1% Triton X-100 solution for 10 min, with gentle agitation.
14. Briefly, rinse the gels at least five times with deionized water.
15. Incubate zymograms in development buffer for 3 h at room temperature (~25 °C) on an orbital shaker.
16. Finally, decant the development buffer and stain the zymogram gels with KI/I_2 staining solution until appearance of clear bands on a uniformly dark background, indicating amylases activity.
17. Analysis of the amylolytic activity can be performed by an automated ChemiDoc Imaging System (or similar) in order to visualize the amylase bands on the zymogram gels (Figs. 1 and 2).
18. Molecular mass of the active bands is determined in parallel by polyacrylamide gel electrophoresis (PAGE) under nonreducing conditions (*see* **Note 10**).

4 Notes

1. Wear gloves and use all safety precautions routinely used when handling acrylamide solutions.
2. Removing β-mercaptoethanol from the buffer will guarantee enzyme activity afterwards. Nonreducing loading sample buffer 4× should be prepared fresh just prior to use.
3. 10× Running buffer should not be long-term stored. Therefore, 1× running buffer should be prepared fresh und filtered just prior to use.
4. Triton X-100 is used to remove the SDS, allowing enzyme renaturation. Dissolve 1 mL Triton X-100 in 80 mL water with gentle stirring and bring to 100 mL with deionized water.
5. The KI/I_2 staining solution may stain skin, clothing, or surfaces if spilled. The stain is not harmful, but may require several days to wear off. Wear protective gloves, protective clothing, and eye protection. KI/I_2 solutions must be protected from light. Prepare in amber bottles or cover the bottle with aluminium foil.

6. We have used the classic dinitrosalicylic acid (DNS) method for the estimation of total amylase activity [14].
7. 1 L supernatant from several simultaneously incubated cultures, was heat-concentrated at 55 °C to a final volume of 100 mL.
8. Adding water on top of the resolving gel will prevent the contact of the polymerizing gel with the air and keep the gel straight without any air bubbles.
9. Do not heat the samples above 65 °C, to preserve enzyme activity.
10. It is suggested to always run in parallel a SDS-PAGE with molecular weight standards, in order to correlate R_f and calculate molecular mass of enzymes.

Acknowledgments

This work was supported by Project Pem 2001002268 from the Venezuelan Fondo Nacional de Ciencia, Tecnología e Innovación.

References

1. Srivastava RAK, Baruah JN (1986) Culture conditions for production of thermostable amylase by *Bacillus stearothermophilus*. Appl Environ Microbiol 52:179–184
2. Aiyer PV (2005) Amylases and their applications. Afr J Biotechnol 4:1525–1529
3. Zaferanloo B, Bhattacharjee S, Ghorbani MM, Mahon PJ, Palombo EA (2014) Amylase production by *Preussia minima*, a fungus of endophytic origin: optimization of fermentation conditions and analysis of fungal secretome by LC-MS. BMC Microbiol 14:55
4. Shaw JF, Lin FP, Chen SC, Chen HC (1995) Purification and properties of an extracellular α-amylase from *Thermus* sp. Bot Bull Acad Sin 36:195–200
5. Campbell LL (1955) Purification and properties of an α-amylase from facultative thermophilic bacteria. Arch Biochem Biophys 54:154–161
6. Malhotra R, Noorwez SM, Satyanarayana T (2000) Production and partial characterization of thermostable and calcium-independent α-amylase of an extreme thermophile *Bacillus thermooleovorans* NP54. Lett Appl Microbiol 31:378–384
7. Egelseer E, Shocher I, Sára M, Sleytr UB (1995) The S-layer from *Bacillus stearothermophilus* DSM 2358 functions as an adhesion site for a high-molecular-weight amylase. J Bacteriol 177: 1444–1451
8. Mielenz JR (1983) *Bacillus stearothermophilus* contains a plasmid-borne gene for α-amylase. Proc Natl Acad Sci U S A 80:5975–5979
9. Hatman P, Wellerson R, Tetrault PA (1955) *Bacillus stearothermophilus*. I. Thermal and pH stability of the amylase. Appl Microbiol 3:7–10
10. Freer SN (1993) Purification and characterization of the extracellular α-amylase from *Streptococcus bovis* JB1. Appl Environ Microbiol 59:1398–1402
11. Ohdan K, Kurik T, Kaneko H, Shimada J, Takada T, Fujimoto Z, Mizuno H, Okada S (1999) Characteristics of two forms of α-amylases and structural implication. Appl Environ Microbiol 65:4652–4658
12. Najafi MF, Deobagkar D, Deobagkar D (2005) Purification and characterization of an extracellular α-amylase from *Bacillus subtilis* AX20. Protein Expr Purif 41:349–354
13. Pfueller SL, Elliot WH (1969) The extracellular α-amylase of *Bacillus stearothermophilus*. J Biol Chem 244:48–54
14. Miller GL (1959) Use of dinitrosalicylic acid reagent for determination of reducing sugar. Anal Chem 31:426–428
15. Laemmli UK (1970) Cleavage of structural proteins during the assembly of the head of bacteriophage T4. Nature 227:680–685

Index

Jeff Wilkesman and Liliana Kurz (eds.), *Zymography: Methods and Protocols*, Methods in Molecular Biology, vol. 1626,
DOI 10.1007/978-1-4939-7111-4, © Springer Science+Business Media LLC 2017

MIX
Papier aus verantwortungsvollen Quellen
Paper from responsible sources
FSC® C105338

If you have any concerns about our products,
you can contact us on
ProductSafety@springernature.com

In case Publisher is established outside the EU,
the EU authorized representative is:
Springer Nature Customer Service Center GmbH
Europaplatz 3, 69115 Heidelberg, Germany

Printed by Libri Plureos GmbH
in Hamburg, Germany